BTEC First Engineering
Mandatory and Selected
Optional Units for BTEC Firsts
in Engineering

BTEC First Engineering

Mandatory and Selected Optional Units for BTEC Firsts in Engineering

Mike Tooley
Formerly Vice Principal
Brooklands College of Further and Higher Education

LRC Stoke Park
GUILDFORD COLLEGE

AMSTERDAM • BOSTON • HEIDELBERG • LONDON • NEW YORK • OXFORD
PARIS • SAN DIEGO • SAN FRANCISCO • SINGAPORE • SYDNEY • TOKYO

Newnes is an imprint of Elsevier

Newnes is an imprint of Elsevier
Linacre House, Jordan Hill, Oxford OX2 8DP, UK
30 Corporate Drive, Suite 400, Burlington MA 01803, USA

First published 2006
Second edition 2010

British Library Cataloguing in Publication Data
A catalogue record for this book is available from the British Library

Library of Congress Cataloging-in-Publication Data
A catalog record for this book is available from the Library of Congress

ISBN: 978-1-85617-685-9

For information on all Newnes publications
visit our website at www.books.elsevier.com

Typeset by MPS Limited, a Macmillan Company, Chennai, India.
www.macmillansolutions.com

Printed and bound in Great Britain

Working together to grow
libraries in developing countries

www.elsevier.com | www.bookaid.org | www.sabre.org

ELSEVIER   BOOK AID
International   Sabre Foundation

CONTENTS

Welcome to the challenging and exciting world of engineering! This book is designed to help you succeed on a course leading to a BTEC First award in Engineering. It contains all of the essential underpinning knowledge required of a student who may never have studied engineering before and who wishes to explore the subject for the first time.

ABOUT YOU

Have you got what it takes to be an engineer? The BTEC First course in Engineering will help you find out and still keep your options open. Successful completion of the course will provide you with a route into studying engineering at Level 3, for example, a BTEC Level 3 Certificate or Diploma in Engineering, or an NVQ award studied as part of an Engineering apprenticeship.

Engineering is an immensely diverse field but, to put it simply, engineering, in whatever area that you choose, is about thinking *and* doing. The 'thinking' that an engineer does is both logical and systematic. The 'doing' that an engineer does can be anything from building a bridge to testing a space vehicle. In either case, the essential engineering skills are the same. You do not need to have studied engineering before starting a BTEC First award. All that is required to successfully complete the course is an enquiring mind, an interest in engineering, and the ability to explore new ideas in a systematic way. You also need to be able to express your ideas and communicate these in a clear and logical way to other people.

As you study your BTEC First course in Engineering you will be learning in a practical environment as well as in a classroom. This will help you to put into practice the things that you learn in a formal class situation. You will also discover that engineering is fun—it's not just about learning a whole lot of meaningless facts and figures!

ABOUT THE BTEC FIRST AWARDS IN ENGINEERING

The BTEC First awards in Engineering will help you to build and apply knowledge in a wide variety of engineering contexts. It will provide you with the understanding and skills necessary to prepare you for employment or to provide you with developing your career if you are already working. On successful completion of a BTEC First qualification, you may progress into employment or progress your career within employment by continuing to study in your chosen vocational area.

BTEC Firsts are now part of the new Qualifications and Credit Framework (QCF) at Level 2. Within the Level 2 framework there are three different sizes of qualification; Awards (1 to 12 credits), Certificates (13 to 36 credits), and Diplomas (37 credits and above). The 15-credit BTEC Level 2 Certificate offers a specialist qualification that focuses on particular aspects of employment within the chosen vocational sector and is broadly equivalent to one GCSE. The 30-credit BTEC Level 2 Extended Certificate offers more flexibility and allows a choice of emphasis through optional units. The Level 2 Extended Certificate is broadly equivalent to two GCSEs. The 60-credit BTEC Level 2 Diploma further extends the specialist work-related focus and is broadly equivalent to four GCSEs.

HOW TO USE THIS BOOK

This book provides full coverage of all of the Mandatory units of the BTEC First Awards in Engineering. One chapter is devoted to each unit and each of these chapters contain text, illustrations, examples, Test your knowledge questions, activities and a set of review questions.

The Mandatory units that make up the Edexcel BTEC First Awards in Engineering are as follows:

BTEC Level 2 Certificate in Engineering:

- *Unit 1* Working Safely and Effectively in Engineering (5 credits)

BTEC Level 2 Extended Certificate in Engineering:

- *Unit 1* Working Safely and Effectively in Engineering (5 credits)
- *Unit 2* Interpreting and Using Engineering Information (5 credits)

BTEC Level 2 Diploma in Engineering:

- *Unit 1* Working Safely and Effectively in Engineering (5 credits)
- *Unit 2* Interpreting and Using Engineering Information (5 credits)
- *Unit 3* Mathematics for Engineering Technicians (5 credits).

In addition to the mandatory units for the Level 2 Certificate, Level 2 Extended Certificate and Level 2 Diploma, the book also provides coverage of five popular optional units. These are Unit 4 (*Applied Electrical and Mechanical Science*), Unit 7 (*Electronic Devices and Communication Applications*), Unit 8 (*Selecting Engineering Materials*), Unit 10 (*Computer Aided Drawing Techniques*) and Unit 19 (*Electronic Circuit Construction*).

Test your knowledge questions are interspersed with the text throughout the book. These questions allow you to check your understanding of the preceding text. They also provide you with an opportunity to reflect on what you have learned and consolidate this in manageable chunks.

Most Test your knowledge questions can be answered in only a few minutes and the necessary information can be gleaned from the surrounding text. Activities, on the other hand, require a significantly greater amount of time to complete. Furthermore, they often require additional library or resource area research coupled with access to computing and other information technology resources.

As you work through this book, you will undertake a programme of activities as directed by your teacher or lecturer. Don't expect to complete *all* of the activities in this book—your teacher or lecturer will ensure that those activities that you do undertake relate to the resources available to you and that they can be completed within the timescale of the course. Activities also make excellent vehicles for improving your skills and for gathering the evidence that can be used to demonstrate that you are competent in core skills.

The Review questions presented at the end of each chapter are designed to provide you with an opportunity to test your understanding of each unit. These questions can be used for revision or as a means of generating a checklist of topics with which you need to be familiar. Here again, your tutor may suggest that you answer specific questions that relate to the context in which you are studying the course.

The book ends with some useful information and data presented in the form of seven appendices. These include abbreviations for common terms used in engineering, information on how to use a scientific calculator, data on selected engineering materials, useful web addresses and answers to the numerical Test your knowledge questions in Units 3 and 4.

Finally, here are a few general points worth keeping in mind:

- Allow regular time for reading—get into the habit of setting aside an hour, or two, at the weekend to take a second look at the topics that you have covered during the week.
- Make notes and file these away neatly for future reference—lists of facts, definitions and formulae are particularly useful for revision!
- Look out for the inter-relationship between subjects and units—you will find many ideas and a number of themes that crop up in different places and in different units. These can often help to reinforce your understanding.
- Don't be afraid to put your new ideas into practice. Remember that engineering is about thinking *and* doing—so get out there and *do* it!
- Lastly, I hope that you will find some useful support material at the book's companion website, http://www.key2study.com. Tutors will find answers to the numerical Review Questions at http://textbooks.elsevier.com.

Good luck with your BTEC First Engineering studies!

Mike Tooley

Working Safely and Effectively in Engineering

SUMMARY

The unit provides you with an essential tool kit that will help you to embark on a career in engineering and the skills and knowledge that you gain will be put to good use in your everyday working life. When you complete this unit you should be able to:

1. Understand statutory regulations and organisational safety requirements
2. Work efficiently and effectively in engineering.

The ability to work safely in an engineering environment is essential for your own well-being as well as the well-being of others around you. This unit will introduce you to the essential working practices of engineering and it will also help you to appreciate some of the potential hazards that exist in the workplace.

The unit starts by considering how materials and equipment should be handled as well as the most appropriate personal protective equipment (e.g. eye or hand protection) to use when undertaking particular engineering activities. You will examine some of the hazards and risks associated with an engineering activity including the environment in which engineering activities are performed (e.g. working at height), the use of tools and equipment, as well as working with materials and substances that may cause harm.

The unit will also take you through some of the typical incidents that you may have to deal with at some point in your career (e.g. summoning help, contacting the first aider, sounding alarms, stopping machinery). Because most work in engineering requires the co-operation of others, the unit also looks at ways in which good working relationships can be maintained with colleagues and other people who provide support, assistance and advice.

1.1 REGULATIONS AND SAFETY

Before you get started on developing your engineering skills it is essential to have an understanding of the appropriate statutory regulations as well as the safety rules that apply in your school or college (and also the company in which you might be working). We begin by taking a look at the Health and Safety at Work Act and the duties that it imposes on both the employer and the employee. Later you will put this knowledge to good use as you begin to practice your skills and get to experience some real engineering activities.

BTEC First Engineering: Mandatory and Selected Optional Units for BTEC Firsts in Engineering. DOI: 10.1016/B978-1-85617-685-9.00001-9

Materials and Equipment Handling

WHAT DO YOU ALREADY KNOW?

You are probably already well aware that many engineering processes, such as welding, casting, forging and grinding, can be potentially dangerous. You will probably also be aware that some of the materials used in engineering can be dangerous. These materials include fuels and fluids such as isopropyl alcohol and ferric chloride.

What you might not be aware of is that many processes and materials that are usually thought of as being 'safe' can become dangerous as a result of misuse or mishandling. For example, soldering is generally considered to be a safe process, however the fumes produced from molten solder can be highly toxic. Here, the combination of the process (soldering) with the material (flux) can result in a hazardous condition (the generation of toxic fumes).

HEALTH AND SAFETY AT WORK ACT

All work activities are covered by the Health and Safety at Work Act 1974 (HASAWA). The Act seeks to promote greater personal involvement coupled with the emphasis on individual responsibility and accountability.

You need to be aware that the Health and Safety at Work Act applies to *people*, not to premises. The Act covers all employees in all employment situations. The precise nature of the work is not relevant, neither is its location. The Act also requires employers to take account of the fact that other persons, not just those that are directly employed, may be affected by work activities. The Act also places certain obligations on those who manufacture, design, import or supply articles or materials for use at work to ensure that these can be used safely and do not constitute a risk to health.

DUTY OF THE EMPLOYER

Under the Act, It is the *duty of the employer* to ensure, so far as is reasonably practicable, the health, safety and welfare at work of all the employees. The employer also needs to ensure that all plant and systems are maintained in a manner so that they are safe and without risk to health. The employer is also responsible for:

- the absence of risks in the handling, storage and transport of articles and substances
- instruction, training and supervision to ensure health and safety at work
- the maintenance of the workplace and its environment to be safe and without risk to health
- to provide where appropriate a statement of general policy with respect to health and safety and to provide where appropriate, arrangements for safety representatives and safety committees
- conduct his or her undertakings in such a way so as to ensure that members of the public (i.e. those not in his or her employment) are not affected or exposed to risks to their health or safety
- give information about such aspects of the way in which he conducts his undertakings to persons who are not his employees as might affect their health and safety
- in addition to having responsibilities for employees, the employer also has responsibilities to persons such as the general public (including clients, customers, visitors to engineering facilities and passers-by).

> **Test your knowledge 1.1**
>
> List THREE duties of an employer under the Health and Safety at Work Act.

DUTY OF THE EMPLOYEE

Under the Act, it is the *duty of every employee* whilst at work, to take all reasonable care for the health and safety of himself and other persons who may be affected by his acts and omissions. Employees are required to:

- co-operate with the employer to enable the duties placed on him (the employer) to be performed

- have regard of any duty or requirement imposed upon his employer or any other person under any of the statutory provisions
- not interfere with or misuse any thing provided in the interests of health, safety or welfare in the pursuance of any of the relevant statutory provisions.

OTHER LEGISLATION

The Health and Safety at Work Act is augmented by a number of other pieces of legislation including the Control of Substances Hazardous to Health Regulations 2002 (COSHH), the Reporting of Injuries, Diseases and Dangerous Occurrences Regulations 1995 (RIDDOR), the Manual Handling Operations Regulations 1992 and the Workplace (Health, Safety and Welfare) Regulations 1992. All of these important regulations have major implications for both employers and employees.

ACTIVITY 1.1

Visit the COSHH section of the UK Government's Health and Safety Executive website (you will find this at www.hse.gov.uk/coshh). View or download a copy of 'COSHH: a brief guide to the regulations' and use it to identify the EIGHT steps that help to ensure that a company complies with the COSHH legislation. Present your findings in the form of an A4 'fact sheet'.

ACTIVITY 1.2

Obtain a copy of 'RIDDOR Explained' (available on-line from the Health and Safety Executive website at www.hse.gov.uk) and use it to answer the following questions:

1. What is RIDDOR?
2. What does RIDDOR apply to?
3. In what circumstances should a report be made and who should make it?
4. Give FOUR examples of major injuries that are reportable under RIDDOR.
5. Give FOUR examples of dangerous occurrences that are reportable under RIDDOR.

PRODUCTION AND MANUFACTURING PROCESSES

Many engineering processes are potentially hazardous and these include activities such as casting, cutting, soldering, welding, etc. In addition, some processes involve the use of hazardous materials and chemicals. Furthermore, even the most basic and straightforward activities can potentially be dangerous if carried out using inappropriate tools, materials, and methods.

In all cases, the correct tools and protective equipment should be used and proper training should be provided. In addition, safety warnings and notices should be prominently placed in the workplace and access to areas where hazardous processes take place should be restricted and carefully controlled so that only appropriately trained personnel can be present. In addition, the storage of hazardous materials (chemicals, radioactive substances, etc.) requires special consideration and effective access control.

Processes that are particularly hazardous include:

- casting, forging and grinding
- welding and brazing
- chemical etching
- heat treatment
- use of compressed air.

Test your knowledge 1.2
List THREE duties of an employee under the Health and Safety at Work Act.

Test your knowledge 1.3
List THREE examples of hazardous engineering processes.

3

OPERATION AND MAINTENANCE

Hazards are associated with the operation and maintenance of many engineering systems. These include those that use single-phase or three-phase AC mains supplies, compressed air, fluids, gas, or petrochemical fuels as energy sources. They also include those systems that are not in themselves particularly hazardous but which are used in hazardous environments (for example in mining or in the oil and gas industries). Operations such as refuelling a vehicle can be potentially dangerous in the presence of naked flames or if electrical equipment is used nearby. A static discharge, for example, can be sufficient in the presence of petroleum vapour, to cause an explosion. Similar considerations apply to the use of mobile phones or transmitting apparatus in the vicinity of flammable liquids.

ACTIVITY 1.3

Medium density fibreboard (MDF) is a widely used material in wood working and furniture construction that may appear to be perfectly safe from a simple external inspection. Use the Internet to find out more about this material and answer the following questions:

1. What processes involving MDF are considered to be hazardous?
2. What constituents of MDF cause it to be a hazardous material?
3. What are the symptoms of exposure to MDF dust and airborne particles?
4. What protection should be used when processing MDF?

LIFTING AND MANUAL HANDLING

Safe working practices need to be adopted for manual operations (such as lifting, stacking and loading). Personnel need to be trained to perform these tasks and managers should insist that the correct procedures and protective clothing are used at all times. The use of hard hats should be made obligatory wherever overhead work is being carried out and when personnel are working underground or on building sites.

ELECTRICAL SAFETY

It should go without saying that, with the exception of portable low-voltage equipment (such as hand-held electronic test equipment) all electrical equipment should be considered potentially dangerous. This applies to *all* mains operated equipment. There are three basic hazards associated with electricity:

- electric shock
- explosion due to the presence of flammable vapours, and
- fire due to overheating of cables and appliances.

HOUSEKEEPING AND TIDINESS

A sign of a good worker is a clean and tidy working area. Only the minimum of tools for the job should be laid out at any one time. These tools should be organised in a tidy and logical manner so that they are immediately accessible. Tools not immediately required should be cleaned and properly stored away (see Fig. 1.1). All hand tools should be regularly checked and kept in good condition. Spillages, either on the workbench or on the floor, should always be cleaned up immediately.

HUMAN CARELESSNESS

Most accidents are caused by human carelessness or negligence. This can range from 'couldn't care less' and 'macho attitudes', to the deliberate disregard of safety regulations and codes of practice. Carelessness can also result from fatigue and ill-health and these, in turn, can result from a poor working environment.

FIGURE 1.1
Example of a tidy
workshop layout with
specialist tooling clearly
displayed

PERSONAL HABITS

Personal habits such as alcohol and drug abuse can render workers a hazard not only to themselves but also to other workers. Fatigue due to a second job (or 'moonlighting') can also be a considerable hazard, particularly when operating machines. Smoking in prohibited areas where flammable substances are used and stored can cause fatal accidents involving explosions and fire.

SUPERVISION AND TRAINING

Another cause of accidents is lack of training as well as inadequate or poor quality training. Lack of supervision can also lead to accidents if it leads to safety procedures being disregarded.

ENVIRONMENT

Unguarded and badly maintained plant and equipment are obvious causes of injury. However the most common causes of accidents are falls on slippery floors, poorly maintained stairways, scaffolding and obstructed passageways in overcrowded workplaces. Noise, bad lighting, and inadequate ventilation can lead to fatigue, ill-health and carelessness. Dirty surroundings and inadequate toilet and washing facilities can lead to a lowering of personal hygiene standards.

ELIMINATION OF HAZARDS

The workplace should be tidy with clearly defined passageways. It should be well lit and ventilated. It should have a well maintained non-slip flooring. Noise should be kept down to acceptable levels. Hazardous processes should be replaced with less dangerous and more environmentally acceptable alternatives. For example, asbestos clutch and brake linings should be replaced with safer materials.

GUARDS

Rotating machinery, drive belts and rotating cutters must be securely fenced to prevent accidental contact. Some machines have interlocked guards. These are guards coupled to the machine drive in such a way that the machine cannot be operated when the guard is open for loading and unloading the work. All guards must be set, checked and maintained by appropriately qualified and certificated staff. They must not be removed or tampered with by operators. Some examples of guards are shown in Figure 1.2.

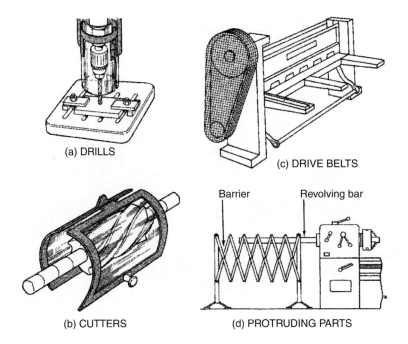

FIGURE 1.2
Different types of guard used with machine tools

(a) DRILLS

(c) DRIVE BELTS

(b) CUTTERS

(d) PROTRUDING PARTS

Barrier   Revolving bar

FIGURE 1.3
A grinder with appropriate warning and prohibition signs. Notice also that a copy of the Abrasive Wheel Regulations has been prominently displayed

FIGURE 1.4
An aircraft hangar fire exit clearly marked with safe condition signs

MAINTENANCE

Machines and equipment must be regularly serviced and maintained by appropriately trained and experienced personnel. This not only reduces the chance of a major breakdown leading to loss of production, it lessens the chance of a major accident caused by a plant failure. Equally important is attention to such details as regularly checking the stocking and location of First Aid cabinets and regularly checking both the condition and location of fire extinguishers. All these checks must be logged.

Safety and Warning Signs

Appropriate warning and prohibition signs should be prominently displayed. The five main types of sign, shown in Figure 1.5, are as follows:

- Prohibition signs (things that you *must not* do, for example, No Smoking)
- Mandatory signs (signs that indicate things that you *must* do, for example, Eye Protection must be used)
- Warning signs (signs that warn you about something that is dangerous, e.g. Danger High Voltage)
- Safe condition signs (signs that give you information about the safest way to go, for example, Fire Exit)
- Fire signs (signs that indicate the location of fire fighting equipment, for example, Fire Point).

Note that different colours are used to make it easy to distinguish the types of sign. For example, safe condition signs use white text on a green background, mandatory signs use white text on a blue background, and so on. It is essential that you familiarise yourself with the different types of sign and know what they mean!

> **Test your knowledge 1.5**
> Classify each of the signs shown in Figure 1.6 as either:
>
> - a prohibition sign
> - a mandatory sign
> - a warning sign
> - a safe condition sign
> - a fire sign.

> **ACTIVITY 1.4**
> Visit your engineering workshop and carry out a detailed survey of the warning signs and notices that are displayed. Make a sketch plan of the workshop area and mark on the location of each sign or notice. Classify the signs as either prohibition signs, mandatory signs, warning signs, safe condition signs, and fire signs. Also find out who is responsible for placing and maintaining the workshop signs and notices.

(a)

(b)

(c)

(d)

(e)

Prohibition signs
White text on a red background

Mandatory signs
Black text on a yellow background

Warning sings
White text on a blue background

Safe condition signs
White text on a green background

Fire signs
White text on a red background

FIGURE 1.5
A selection of different types of sign

(f)

FIGURE 1.6
See Test your knowledge 1.5

Flammable material
Oxidising agent
Caustic material
Electric shock hazard
Electric shock hazard
Compressed gas
Fork lift truck
Toxic material
Unsafe roof
Radioactive material
Explosive hazard
EX
Explosive hazard

FIGURE 1.7
A selection of caution signs

Test your knowledge 1.6
Identify each of the signs shown in Figure 1.8 and give an example of where each one might be found.

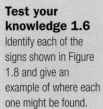

(a)
(b)
(c)
(d)
(e)

FIGURE 1.8
See Test your knowledge 1.6

DO NOT CUT GILLINER
ON THIS MA.HINE
600
DANGER
All guards must be in position before starting
Emergency stop

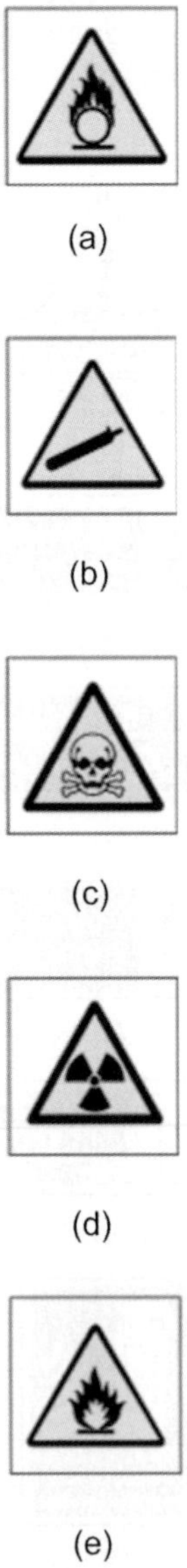

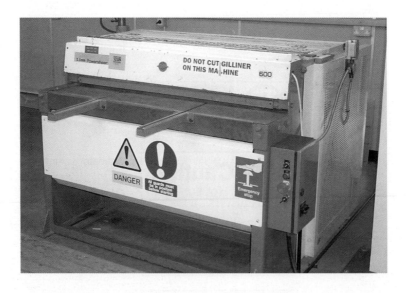

FIGURE 1.9
See Activity 1.5

ACTIVITY 1.5

Figure 1.9 shows a power guillotine suitable for cutting sheet metal. Identify the three signs displayed on the front of the machine and state the purpose of each sign. How is the guillotine stopped in an emergency?

Personal Protective Equipment (PPE)

CLOTHING AND FOOTWEAR

Suitable and unsuitable working clothing for use in an engineering workshop is shown in Figure 1.10. Overalls or protective coats should be neatly buttoned and sleeves should be tightly rolled. Safety shoes and boots should be worn (not trainers!). Overalls and protective

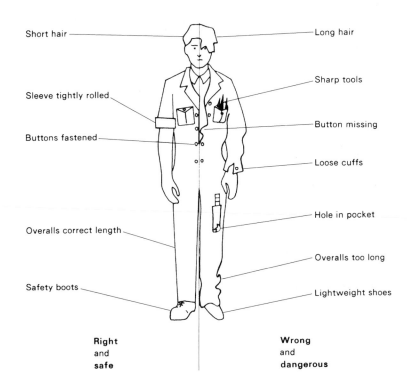

Short hair

Sleeve tightly rolled

Buttons fastened

Overalls correct length

Safety boots

Long hair

Sharp tools

Button missing

Loose cuffs

Hole in pocket

Overalls too long

Lightweight shoes

Right and safe

Wrong and dangerous

FIGURE 1.10
Correct and incorrect clothing and footwear

FIGURE 1.11
Personal protective equipment (PPE) is usually provided close to the point at which it is needed (in this case the safety goggles are housed close to the bench sander shown in Figure 1.12)

FIGURE 1.12
Example of a bench sander with PPE provided

clothing should be sufficiently loose in order to allow easy body movement but not so loose that they interfere with engineering tasks and activities.

SPECIAL EQUIPMENT

Some processes and working conditions demand even greater protection, such as safety helmets, earmuffs, respirators and eye protection worn singly or in combination. Such protective clothing must be provided by the employer when a process demands its use. Employees must, by law, make use of such equipment. Some examples are shown in Figure 1.13.

ACTIVITY 1.6

Visit your engineering workshop and identify THREE items of personal protective equipment (PPE). Explain what each is used for, where it is stored and how it is used.

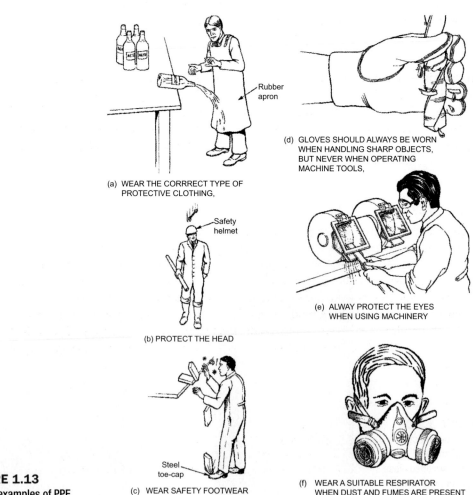

(a) WEAR THE CORRRECT TYPE OF PROTECTIVE CLOTHING,

Rubber apron

(d) GLOVES SHOULD ALWAYS BE WORN WHEN HANDLING SHARP OBJECTS, BUT NEVER WHEN OPERATING MACHINE TOOLS,

Safety helmet

(b) PROTECT THE HEAD

(e) ALWAY PROTECT THE EYES WHEN USING MACHINERY

Steel toe-cap

(c) WEAR SAFETY FOOTWEAR

(f) WEAR A SUITABLE RESPIRATOR WHEN DUST AND FUMES ARE PRESENT

FIGURE 1.13
Typical examples of PPE

Hazards and Risks

You will need to be fully aware of the hazards and health risks associated with handling and processing engineering materials. Some materials that might appear to be harmless can be hazardous if they exist in certain forms (for example, as fine airborne particles that can readily be inhaled). Some of the most common hazards include:

- Exposure to fibres (either glass or carbon) may cause skin rashes (occupational dermatitis) as well as irritation to the eyes, nose and throat
- Contact with adhesives and uncured resins may cause skin sensitisation
- Resin fumes and solvent vapours may cause irritation to the eye and nose
- Sanding dust can be an irritant if it is inhaled
- Protective coatings, paints and solvents may produce vapours that cause irritation to the eyes, nose, throat and lungs
- Etching solutions and acids can cause skin irritation and burns.
- Powdered material (and dust) may cause explosions and an increased risk of static discharge
- Use of welding, brazing and soldering equipment may result in the production of toxic fumes
- Use of welding, brazing, hot gas and soldering equipment may cause burns
- Incorrect use of tools (particularly machine tools, drills, lathes, etc.) may cause injury to operators and other personnel.

Engineering companies need to have effective procedures and training in place in order to minimise these hazards. For example, the use of a grinding wheel should be restricted to personnel who have received appropriate training and appropriate eye, hand and body protection should be provided. Furthermore, the relevant health and safety advisory notices together with any legal requirements must be prominently displayed. Figure 1.3 on page 6 shows an example of good practice.

RISK ASSESSMENTS

Engineering companies are required to carry out Risk Assessments that should be performed in order to identify hazardous activities and to inform action plans designed to improve safety and eliminate hazards. Correct storage and handling of materials. is essential and protective clothing (gloves, goggles, overalls, etc.), correct lighting and workspace organisation, adequate ventilation and efficient fume and dust extraction all have a part to play in making the workplace safe. The level of risk is determined by the probability of an event as well as the likely severity of its outcome. The evaluation of the risk involved in a given process or activity is usually based on the following questions:

1. What is the hazard?
2. How likely is it to occur?
3. Under what circumstances might it occur?
4. What controls are in place to prevent it occurring?
5. What is the likely outcome if it does occur?

> **Test your knowledge 1.7**
> List THREE substances found in an engineering workshop that could be a risk to health.

The process of identifying hazards can be a little daunting if you have never done anything like this before. To make things a little more manageable, you can divide the task into a number of smaller stages by answering questions like:

What – processes are being used?
        – plant, materials and tools are being used?
        – is the effect on people/plant/other activities?
        – are the statutory requirements?

Who   – is doing the job? (and what training have they had)
        – is exposed to risk?
        – might also be affected?
        – is supervising/monitoring the process?

Where – is the process/activity performed?
        – is the process documented?
        – do waste materials go?

When – is the process carried out?
        – were the safety procedures last reviewed/updated?

Why   – is the process being performed?
        – is the risk not being controlled?
        – is exposure not controlled at source?

How   – can the hazard be controlled?
        – can an accident occur?
        – may people be affected?

Test your knowledge 1.8
Explain briefly how the level of a particular risk is evaluated when a risk assessment is being carried out.

In order to answer all of these questions you need to observe the activity that is being performed as well as the circumstances in which it is being carried out. Further hazards may be identified by study of accident/incident reports, insurers, reports, inspection reports and other specialist survey reports. Most engineering companies use standard forms when carrying out their risk assessments. Typical guidelines and reporting forms are shown in Figures 1.14 and 1.15 respectively.

MATERIAL SAFETY DATA SHEETS

Key point
The appropriate Material Safety Date Sheet (MSDS) will provide you with essential information about the constituents, storage, use, handling, transport and disposal of materials used in an engineering workshop.

All hazardous materials should have a Material Safety Data Sheet (MSDS) supplied by, or at least available from, the material manufacturer or supplier. The sheet should be identified by the description and part number of the product and should contain comprehensive information about the product including details relating to handling, use, storage and transportation. Details of the protective clothing or equipment required to work safely with the product should also be included. Personnel working with hazardous materials should have access to the MSDS for the materials concerned.

Risk Assessment - Guidance

Since 1992 the Management of Health and Safety at Work Regulations has required a systematic review of all work activities and those thought to pose a significant risk must be subject to a suitable and sufficient assessment. The purpose of the assessment is to ensure that significant risks are reduced as far as is reasonably practicable. New activities must be risk assessed <u>before</u> they are carried out. This questionnaire and the form shown overleaf are designed to help you to carry out this risk assessment.

The risk assessment process is more effective if it is carried out by at least two persons - the person who will carry out the activity and/or is familiar with the process as well as the person who is responsible for the activity. This approach ensures that the process is better informed and more objective and it encourages ownership of the risk assessment results and recommendations.

The following steps are recommended:

1. Describe the activity (it may help to break down a complex activity into individual tasks)

2. What hazards are associated with this activity (or with each task)?

3. Estimate the risk arising from each of these hazards:

   3.1 What are the risks arising from each of these hazards? Consider both the likelihood of harm and the severity if the harm is realised

   3.2 Who is likely to be affected? What is their understanding of the risks? Consider visitors, cleaners, and all others that might be affected

   3.3 When estimating the risk, consideration should be given to the effectiveness of existing control measures:

      (a) What measures are in place to reduce the risk(s)

      (b) Are these measures sufficient?

      (c) Are there any statutory requirements relating to the activity or equipment in use? Refer to HSE and relevant industry Codes of Practice and other guidance. Your Safety Advisor or Safety Officer will be able to help locate the relevant documents

      (d) Are these requirements being met? If not, why?

   3.4 Assess each risk under the heading 'high', 'medium', or 'low' on the following basis:

      (a) 'high'   the existing risk control measures are insufficient and immediate action is needed to reduce the risk

      (b)'medium'   the existing control measures are insufficient. Addition measures are required and should be implemented by a fixed date

      (c) 'low'   existing risk control measures are sufficient. No further action is needed

4. In the light of answers to 3.4 are any additional measures required? If so:

   4.1 What additional measures are needed?

   4.2 How and when should these additional measures be implemented?

   4.3 How will you ensure that the additional and existing measures are being used?

5. How will you communicate the contents of the risk assessment to those who may be affected?

6. How often will you review this risk assessment to ensure that it remains current and effective?

7. Use the table shown overleaf as a prompt for identifying hazards and amend it to suit the circumstances of the activity you are assessing (i.e. if there is no biological hazard simply delete that row). Tick the appropriate boxes ('low', 'medium,' and 'high') and write brief notes in the other boxes.

8. It may also help to identify any sources of information (e.g. Approved Codes of Practice, guidance notes, industry guidelines, manufacturers' recommendations, etc). Finally, please ensure that you sign and data the risk assessment form and keep it safely for future reference. A copy of the form should be forwarded to your Safety Officer or Safety Advisor.

FIGURE 1.14
Typical guidance given for carrying out a risk assessment. Notice that the level of risk is assessed as being either 'high', 'medium', or 'low'

Risk Assessment (Management of Health and Safety at Work Regulations 1999)

Description and location of the Activity:

Person(s) undertaking the Activity:

Type of hazard	Examples	Details of Risk				Additional control measures needed
		Low	Med	High	Description of risk and existing control measures (include details of persons at risk)	
1. Impaired emergency access or egress	Obstructed fire exit; impaired mobility of staff/students					
2. Electric shock	Damaged cabling; poor earthing					
3. Impaired access to emergency stop control	Control obstructed or out of reach of user					
4. Slips, trips and falls	Working at height; obstructions; wet floor surfaces					
5. Fire/explosion	Flammable vapours or dusts; sparks; oxidising substances					
6. Impact, entanglement or entrapment	Moving parts of machinery; falling materials; vehicles					
7. Injury from lifting/ manual handling (see note 1)	Moving heavy items; repetitive handling; difficult postures					
8. High or low pressure	Compressed air; vacuums					
9. Hazardous substances (see note 2)	Fumes; solvents; acids; airborne dust particles					
10. Asphyxiation	Confined spaces; lack of oxygen					
11. Cuts and puncture wounds	Knives; scalpels; needles; sharp edges					
12. Radiation (see note 3)	Ionising radiation; lasers; microwave					
13. Biological (see note 2)	Bacterial or viral infection					
14. Heat/cold burns	Ovens; welding/ brazing; liquid nitrogen					
15. Loud noise and vibration (see note 4)	Machinery; engines; compressors; pumps; drilling; cutting					
16. Impaired vision	Poor lighting; obstruction of vision; strobe lighting					
17. Personal safety	Lone working; visitors and open days					
18. Training and behavioural factors	Lack of training; anti-social behaviour					
19. Other hazard (give details)						
20. Other hazard (give details)						

Notes: (1) A separate assessment is required under the Manual Handling Regulations
(2) A separate assessment is required under the Control of Substances Hazardous to Health Regulations
(3) Ionising radiation hazards require a separate assessment under the Ionising Radiations Regulations. Lasers also require a separate assessment
(4) A separate assessment is required under the Manual Handling Regulations

Further Comments:

Name(s) of person(s) carrying out this Risk Assessment:

Date of this Risk Assessment:

FIGURE 1.15
Typical risk assessment form (the types of hazard can be changed according to the type of work performed)

ACTIVITY 1.7

Use the following checklist to help you carry out a detailed and meaningful risk assessment of the electrical/electronic workshop in your school/college. You should copy the list and write brief notes against each question. Note that your risk assessment must be based on the *activities* that are carried out in the workshop so it is a good idea to start off by finding out *who* uses the workshop and *what* it is used for.

Present your findings in the form of a brief presentation (no more than 10 minutes) using appropriate visual aids and with supporting notes classifying risks as 'high', 'medium' and 'low' (see Fig. 1.15). The following questions should help you to get started:

1. Where is the electrical circuit breaker? (It should be in a prominent and immediately accessible position!)
2. Has a residual current circuit breaker (RCCB) been fitted and, if so, has the RCCB been tested lately?
3. Are all of the electrical outlets in a safe condition?
4. Does each item of test equipment have a mains lead that is in a safe condition?
5. Does each item of test equipment have a mains plug that is in a safe condition?
6. Is each item of test equipment correctly fused?
7. Have portable items of electrical equipment been tested in accordance with Portable Appliance Testing (PAT) regulations? If so, when was the last PAT test carried out?
8. Is the soldering equipment safe? Are soldering irons fitted with heatproof leads?
9. Is there any provision for solder fume extraction?
10. Are safety glasses available? What condition are they in and where are they stored?
11. Is the lighting adequate?
12. Are the safety exits properly marked and unobstructed?
13. Are appropriate fire extinguishers available and when were they last tested?
14. Is there appropriate safety information on display?
15. Is there a First Aid kit available? Where is it kept?

Your presentation should include recommendations on the safety of materials and equipment handling, use of personal protective equipment and the potential hazards that you have identified in the area. You should also suggest ways in which a work activity (such as PCB etching, bench drilling or soldering) could be improved (for example, by improving the lighting or the use of fume extraction equipment).

ACTIVITY 1.8

Use the MSDS shown in Figure 1.16 to answer the following questions:

1. When was the MSDS produced?
2. Which company produced the MSDS?
3. What are the dangers associated with inhalation?
4. What protective equipment is recommended for handling the material?

ELECTRICAL HAZARDS

All mains-operated electrical equipment is potentially dangerous. The hazards that are present can include electric shock, fire and/or smoke fumes due to the overheating of cables and equipment. Explosions set off by sparks when using unsuitable equipment when flammable vapours and gases are present.

Howard Associates

SAFETY DATA SHEET

July 2003   FDS 0703

1.   Identification of the chemical product
FERRIC CHLORIDE Ref : **AR37 - AR38 - AR371 – AR381**

Howard Associates
Brooklands Road
Weybridge
KT13 8TU

Emergency telephone number :
60080076767 (Europe) 603214989463 (Europe)
Identification of the substance or the preparation

Product name :	FERRIC CHLORIDE (Solution 37-46 %)
Chemical name :	Iran trichloride (solution 37-46 %)
Synonym(s) :	Iron chloride III (solution 37-46 %), Iron perchloride (solution 37-46 %)
Commercial Name :	SOLFLOC (R)
Formula :	$FeCl_3$
Molecular Weight :	161.5

2.   Composition information on ingredients
Ferric chloride :

- CAS Number :	7705-08-0
- EC Number (EINECS)	231-729-4
- Symbols	C
- Phrases R	34, 22, 52/53
- Concentration	37.00 – 46.00 %

3.   Hazards identification
Toxicity effects principally related to its corrosive properties.
Hazardous product for the aquatic environment.
in case of decomposition, releases dangerous products.

4.   First aid measures

Genera recommendations
-      Personal protective equipment required for rescuers (ses section 8).
-      In case of product splashing into the eyes and face, treat eyes first.
-      Submerge soiled clothing in a basin of water.

        Effects
        Main affects
Irritating to skin; corrosive to mucous membrane and eyes.
The seriousness of the lesions and the prognosis of intoxication depend directly on the concentration and duration of exposure.
Risk of liver effects.
Fatalities have been observed after a single dose of 30 grams and more taken by an adult weighing 70 kg.
Chronic exposure to the product can induce iron accumulation in tissues characterized by redbrown deposits.

        Inhalation
Severe irritation of the nose and the throat.
Cough and difficulty in breathing.
At high concentrations, risk of chemical pneumonitis, pulmonary (o)edema.
In case of repeated or prolonged exposure: risk of sore throat, nose bleeds, chronic bronchitis.
In case of repeated or prolonged exposure: risk of brown colouration of teeth.

FIGURE 1.16
Extract from a Material Safety Data Sheet for Ferric Chloride solution

Before using any electrical equipment it is advisable to carry out a number of visual checks. For example:

1. Check that the cable is not damaged or frayed
2. Check that the plug is in good condition and not cracked
3. Check that the voltage and power rating of the equipment is suitable for the supply available.

It is important to remember that bodily contact with mains or high-voltage circuits can be lethal. The most severe path for electric current within the body (i.e. the one which is most likely to stop the heart) is that which exists from one hand to the other. The hand-to-foot path is also dangerous but somewhat less dangerous than the hand-to-hand path.

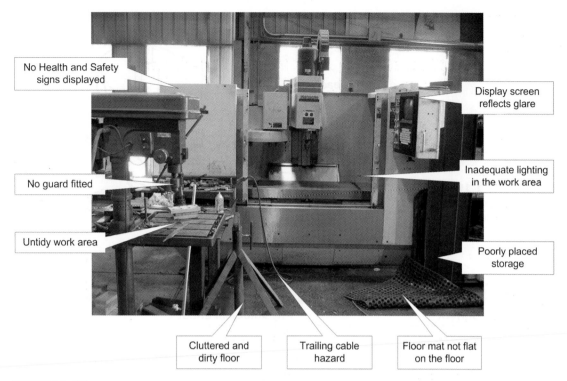

FIGURE 1.17
Example of some serious problems identified during a risk assessment of a mechanical workshop area used for CNC and manual drilling operations

Voltages in many items of electronic equipment, including all items that derive their power from the AC mains supply, are at a level which can cause sufficient current flow in the body to disrupt normal operation of the heart. The severity of such a shock will depend upon a number of factors including the magnitude of the current, whether it is alternating current (AC) or direct current (DC), and its precise path through the body. The magnitude of the current depends upon the voltage which is applied and the resistance of the body. The electrical energy developed in the body will depend upon the time for which the current flows; the duration of contact is also crucial in determining the eventual physiological effects of the shock. As a rough guide, and assuming that the voltage applied is from the 240 V AC 50 Hz mains supply, the effects shown in Table 1.1 are typical. It is important to realize that these figures are only quoted as a guide. In particular, you should note that there have been cases of lethal shocks resulting from contact with much lower voltages and at quite small values of current. Furthermore, an electric shock is often accompanied by burns to the skin at the point of contact. Such burns may be extensive and deep even when there may be little visible external damage to the skin. Burns may be particularly severe when relatively high voltages and currents are encountered.

In general *any potential in excess of 50 V* should be considered dangerous. Lesser potential may, under unusual circumstances, also be dangerous. As such, it is wise to get into the habit

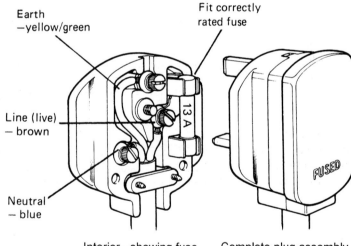

FIGURE 1.18
Wiring a mains plug

Key point
It is essential to avoid bodily contact with 'live' electrical conductors and components.

17

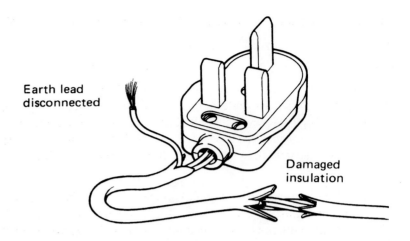

FIGURE 1.19
Some typical faults with
plug wiring

Earth lead
disconnected

Damaged
insulation

TABLE 1.1 Effect of electric current on the human body	
Current	**Physiological effect**
less than 1 mA	Not usually noticeable
1 mA to 2 mA	A slight tingle may be felt
2 mA to 4 mA	Mild shock felt
4 mA to 10 mA	Serious shock (painful)
10 mA to 20 mA	Nerve paralysis may occur (unable to let go)
20 mA to 50 mA	Loss of consciousness (breathing may stop)
more than 50 mA	Heart failure

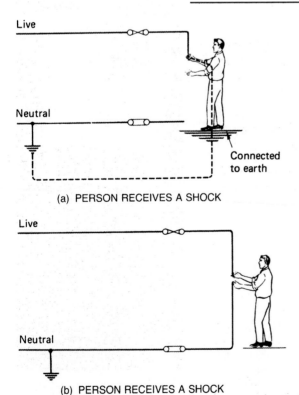

Live

Neutral

Connected
to earth

(a) PERSON RECEIVES A SHOCK

Live

Neutral

(b) PERSON RECEIVES A SHOCK

FIGURE 1.20
Ways in which an electric shock can be received

of treating all electrical and electronic circuits with great care
and avoid bodily contact at all times.

SHOCK PREVENTION

Residual current circuit breakers (RCCB), residual current devices
(RCD), or earth leakage circuit breakers (ELCB) can provide a very
worthwhile safety measure which can be instrumental in very
significantly reducing the risk of electric shock. An RCCB senses
the imbalance of current that occurs whenever current is returned
to the AC mains supply via the earth rather than the neutral wire.
This current may result from equipment failure (e.g. insulation
breakdown in a mains transformer) or from bodily contact with
the live (line) wire. Most RCCB will trip at currents of about 30 mA
within a time interval of 30 ms, sufficient to ensure that heart
failure does not occur. RCCB are not expensive and are available in
various forms (ideally they should be wired into the mains circuit
to your workshop so that all of the mains outlets are protected).

FIRE HAZARDS

Fire is the rapid oxidation (burning) of flammable materials.
For a fire to start, the following are required:

- a supply of flammable materials
- a supply of air (oxygen)
- a heat source.

Once the fire has started, the removal of one or more of the above will result in the fire going out.

FIRE PREVENTION

Fire prevention is largely a matter of 'good housekeeping'. The workplace should be kept clean and tidy. Rubbish should not be allowed to accumulate in passages and disused storerooms. Oily rags and waste materials should be put in metal bins fitted with airtight lids. Plant, machinery and heating equipment should be regularly inspected, as should fire alarm and smoke detector systems. You should also know how and where to raise the fire alarm.

Electrical installations, alterations and repairs must only be carried out by qualified electricians and must comply with the current IEE Regulations. Smoking must be banned wherever flammable substances are used or stored. The advice of the fire prevention officer of the local brigade should be sought before flammable substances, bottled gases, cylinders of compressed gases, solvents and other flammable substances are brought on site.

> **Test your knowledge 1.9**
> Describe TWO ways in which an electric shock can be received.

> **Test your knowledge 1.10**
> List THREE checks you would make before using a portable electric power tool.

> **ACTIVITY 1.9**
>
> Visit the Health and Safety Executive (HSE) website at www.hse.gov.uk and obtain information on safety in the gas welding and cutting process. Identify at least FIVE main hazards associated with the process and explain what a backfire and a flashback is. Present your answer in the form of a brief written advice notice to be displayed in the welding area of an engineering workshop.

Emergency Procedures

ELECTRIC SHOCK PROCEDURE

The following procedure should be adopted in the event of electric shock:

- Switch off the supply of current if this can be done quickly.
- If you cannot switch off the supply, do not touch the person's body with your bare hands (human flesh is a conductor and you would also receive a severe shock) but drag the affected person clear using insulating material such as dry clothing, a dry sack, or any plastic material that may be handy.
- If the affected person has stopped breathing and you have received appropriate training, commence artificial respiration immediately. Don't wait for help to come or go to seek for help—get someone else nearby to do this.

If the affected person's pulse has stopped, heart massage will also be required. If you are not confident to do this it is important to find another person to help who has received the appropriate First Aid training.

FIRE PROCEDURE

In the event of you discovering a fire, you should:

- Raise the alarm and call the fire service.
- Evacuate the premises. Regular fire drills must be held. Personnel must be familiar with normal and alternative escape routes. There must be assembly points and a roll call of personnel. A designated person must be allocated to each department or floor to ensure that evacuation is complete. There must be a central reporting point.
- Keep fire doors closed to prevent the spread of smoke. Smoke is the biggest cause of panic and accidents, particularly on staircases. Emergency exits must be kept unlocked and free from obstruction whenever the premises are in use. Lifts must not be used in the event of fire.

FIGURE 1.21
A typical fire point with CO_2 and dry powder extinguishers

● Only attempt to contain the fire until the professional brigade arrives if there is no danger to yourself or others. Always make sure you have an unrestricted means of escape. Saving lives is more important than saving property.

The order in which you perform these tasks will depend on the individual circumstances. If a fire point is nearby you should raise the alarm immediately. If, however, you have to leave the room in order to raise the alarm you should close doors and windows in order to prevent the fire spreading before you exit and sound the alarm. In all cases you should alert other people to the emergency at the earliest possible stage.

FIRE EXTINGUISHERS

Several different forms of fire extinguisher are provided in order to cope with different types of fire:

1. Class A extinguishers are for ordinary combustible materials such as wood, paper, cardboard, and most plastic materials. This type of extinguisher is based on water and may involve the use of a wall-mounted reel and hose
2. Class B fires involve flammable or combustible liquids such as fuels, solvents, oil and grease
3. Class C fires involve electrical equipment, wiring, switchgear, circuit breakers and outlets (water-based extinguishers should NEVER be used on this type of fire)
4. Class D fire extinguishers are suitable for use with chemical fires involving combustible metals such as magnesium, titanium, potassium and sodium.

It is essential to remember that, whilst water can be an effective extinguishing agent for Class A fires (paper, wood, etc.), water and air-pressurized water (APW) extinguishers MUST NOT be used on Class B or Class D fires because their use can actually cause the flames to spread and make the fire bigger! Dry powder extinguishers come in a variety of types and are suitable for a combination of class A, B and C fires. These are filled with foam or powder and pressurized with nitrogen. Carbon dioxide (CO_2) extinguishers are designed for use with Class B and C fires. CO_2 extinguishers contain carbon dioxide, a non-flammable gas, and are highly pressurized. It is worth remembering that, unlike dry chemical powder extinguishers, CO_2 extinguishers don't leave a harmful residue. This can be important when dealing with fires on expensive electrical and electronic equipment which may suffer permanent damage when dry powder extinguishers are used. Finally, it is important to have appropriate training in the operation of fire extinguishers and other fire protection equipment. If your school or college can provide you with this training you should take full advantage of it!

ACTIVITY 1.10

Produce an A3 poster that can be displayed in your school or college workshop with information on how to deal with electric shock. Make sure that the poster is easy to read and understand and include information on switching off the electricity supply together with emergency first aid contact details.

Key point
It is essential to get to know the emergency procedures for evacuation, accident, fire, etc. in your school or college. You need to know where the fire exits, alarms and fire points are and how to get assistance from a first aider or fire warden.

ACTIVITY 1.11

Produce an A3 poster that can be displayed in your school or college workshop with information on how to deal with fire. Include in your poster information on the different types of fire extinguisher that might be present and what they should (and should not) be used for. Make sure that the poster is easy to read and understand and include information on evacuation in the event of fire (including the fire assembly or fire reporting point).

ACTIVITY 1.12

Visit an engineering workshop or laboratory that is used mainly by first year students in your school/college. Prepare a safety policy for this work area and include references to relevant legislation (for example, the Health and Safety at Work Act 1974, Management of Health and Safety at Work Regulations 1999, Manual Handling Operations Regulations 1992, Control of Substances Hazardous to Health Regulations 2002). Your health and safety policy should take into account the fact that the main users of the area will be newcomers to engineering and your policy should be worded so that it can be easily understood.

You should arrange your health and safety policy under the following headings:

1. A general statement explaining what the policy is about and who needs to know about it
2. Procedure for accident and incident reporting (what needs to be reported and who should it be reported to)
3. Manual handling and lifting
4. Hazardous substances and materials
5. Hazardous equipment and processes
6. Electrical safety
7. Fire and emergency evacuation
8. Personal protective equipment
9. Safety training
10. Risk assessments and safety inspections
11. Health and safety roles and responsibilities (names and responsibilities of individuals having a specific health and safety role such as a Health and Safety Office, First Aider, Fire Warden, etc.).

Your health and safety policy should include a paragraph on each of the above. Present your findings in the form of a word-processed A4 document (no more than THREE pages).

Test your knowledge 1.11

State which type of fire extinguisher you would use in each of the following cases:

a. Paper burning in an office waste bin
b. A pan of fat burning in the kitchen of the works canteen
c. A fire in a mains voltage electrical machine.

Test your knowledge 1.12

A fire breaks out near to a store for paints, paint thinners and bottled gases. What action should be taken and in what order?

1.2 WORKING EFFICIENTLY AND EFFECTIVELY IN ENGINEERING

Engineering Work Activity

As part of this unit you will be expected to show that you can work efficiently and effectively in an engineering environment. In order to do this you will need to carry out a simple engineering activity which will usually involve the following steps:

1. Prepare the work environment ensuring that it is area free from hazards and that relevant safety procedures are implemented. You will also need to ensure that appropriate PPE and tools are selected and checked before use (ensuring that each is in a safe and usable condition).

ACTIVITY 1.13

Which of the actions listed below (and in what order) should you take in each of the following scenarios:

Scenario

A. A colleague has cut his hand on a broken light fitting.
B. A strong smell of burning appears to be coming from a small store cupboard in which flammable materials are kept.
C. Whilst operating some rotating machinery you notice that a guard has become detached and is unsafe
D. You have been taking medication for flu and have become drowsy whilst working on some equipment

E. A stranger who is not carrying any identification is found wandering in the workshop area
F. A colleague appears to have received an electric shock and is lying unconscious across some electrical trunking
G. An access way has become blocked by a pallet loaded with boxes
H. A mains connector on an arc welding set appears to have become damaged and the electrical conductors have become exposed
I. You notice that an important item of test equipment is missing from its normal place in the test lab.

Possible actions

1. Locate a fire point and sound the alarm
2. Leave the building by the nearest exit
3. Operate a fire extinguisher
4. Operate the emergency stop button
5. Summon help
6. Switch off the electrical supply
7. Close all nearby windows and doors
8. Summon a first aider
9. Stop what you are doing and take a break
10. Report the incident to site security
11. Place a 'do not use' notice on the equipment
12. Attempt to clear up
13. Inform your manager/supervisor
14. Attempt to carry out a repair before use.

Present your answer as a set of written notes with a brief explanation for each action. Choose no more than THREE actions for each scenario and ensure that you list them in the correct sequence.

2. Prepare for the work activity by ensuring that all necessary drawings, specifications, job instructions, materials and component parts are obtained and ensuring that appropriate storage arrangements are made for the work. In addition you will need to ensure that you have the necessary authorisation to carry out work from your teacher, lecturer and/or workshop supervisor.

3. Complete the work activity including all specified tasks and associated documentation. You will also need to ensure that your completed work complies with the original specification. When your work is complete you should return drawings, instructions and tools to the correct storage location and dispose of unusable tools, equipment, components and waste materials (such as oil, soiled rags, swarf, and off-cuts) following the procedures that are in place in the work area.

In addition to demonstrating that you are competent to carry out a basic engineering task, will you need to show that you can contribute to effective working relationships. This means:

1. Working with other students (including those that you may not already be familiar with), schools and college staff, specialist staff (technicians and workshop supervisors), as well as those who are 'external' to the organisation (e.g. customers, suppliers, contractors, and visitors)

2. Dealing with problems that affect engineering work and the processes that you may be using. This may involve arranging access to, and assistance with, restricted processes and materials (for example etching tanks, power guillotines, etc.) as well as access to drawings, technical libraries and quality assurance documentation

3. Conforming with institutional practices (for example, arrangements for emergency evacuation or reporting security issues) as well as contributing to organisational issues (for example, recommending and or implementing safe and effective working practices, relevant quality procedures, internal communication and teamwork).

The processes and tasks that you will carry out as part of the engineering activity will depend on the individual circumstances and, in particular, the resources and material available in your school or college. To provide you with examples of the processes that you might need to be familiar with we shall consider two representative activities, drilling and soldering.

DRILLING

Drilling is a process for producing circular holes. The holes may be cut from the solid or existing holes may be enlarged. The purpose of the drilling machine is to:

- Rotate the drill at a suitable speed for the material being cut and the diameter of the drill.
- Feed the drill into the workpiece.
- Support the workpiece being drilled; usually at right angles to the axis of the drill. On some machines the table may be tilted to allow holes to be drilled at a pre-set angle.

Drilling machines come in a variety of types and sizes, including the hand-held electrically driven power drill that you may have used at home (see Fig. 1.22). Unfortunately, when it comes to producing an accurately drilled hole, hand-held drills have a number of serious shortcomings including the difficulty of positioning the drill correctly above the workpiece and at right-angles to it. The feed force is also limited to the muscular strength of the user.

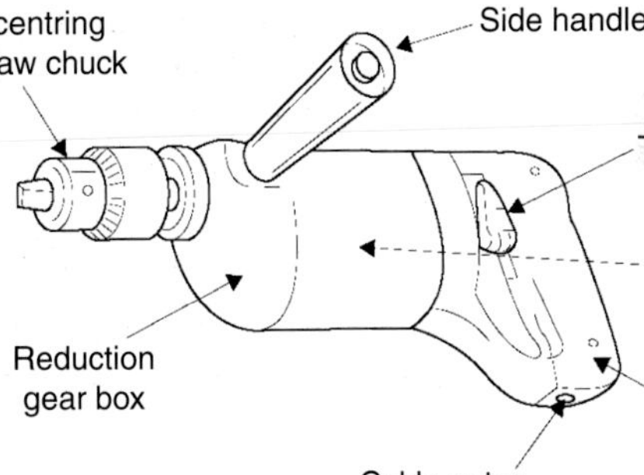

Figure 1.23 shows a more powerful, floor or bench mounted machine. The spindle rotates the drill. It can also move up and down in order to feed the drill into the workpiece and withdraw the drill at the end of the cut. Holes are generally produced with twist drills. Figure 1.24 shows a typical straight shank drill and a typical taper shank drill and names their more important features.

Large drills have taper shanks and are inserted directly into the spindle of the machine, as shown in Figure 1.25(a). They are located and driven by a taper. The tang of the drill is for extraction purposes only. It does not drive the drill. The use of a *drift* to remove the drill is shown in Figure 1.25(b).

Small drills have straight (parallel) shanks and are usually held in a self-centring chuck. Such a chuck is shown in Figure 1.25(c). The chuck is tightened with the chuck key shown. SAFETY: The chuck key must be removed before starting the machine. The drill chuck has a taper shank which is located in, and driven by, the taper bore of the drilling machine spindle.

The cutting edge of a twist drill is wedge-shaped, like all the tools we have considered so far. This is shown in Figure 1.26.

When regrinding a drill it is essential that the point angles are correct. The angles for general purpose drilling are shown in Figure 1.27(a). After grinding, the angles and lip lengths must be checked as shown in Figure 1.27(b). The point must be symmetrical. The effects of incorrect grinding are shown in Figure 1.27(c).

If the lip lengths are unequal, an oversize hole will be drilled when cutting from the solid. If the angles are unequal, then only one lip will cut and undue wear will result. The unbalanced forces will cause the drill to flex and 'wander'. The axis of the hole will become displaced as drilling proceeds. If both these faults are present at the same time, both sets of faults will be present and an inaccurate and ragged hole will result.

WORK HOLDING

It is dangerous to hold work being drilled by hand. There is always a tendency for the drill to grab the work and spin it round. Also the rapidly spinning *swarf* can produce some nasty

Key point
Drilling and soldering are typical engineering activities that you will need to be familiar with. When carrying out these activities you will need to demonstrate that you can work safely and effectively.

Self-centring three-jaw chuck
Side handle
Trigger switch
Single-phase motor
Reduction gear box
Hand grip
Cable entry

FIGURE 1.22
A hand-held electric power drill

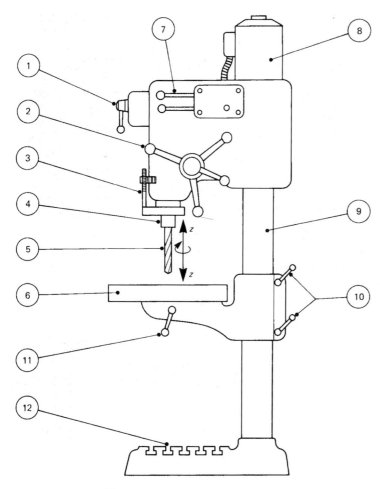

Parts of the Pillar Type Drilling Machine

1	Stop/start switch (electrics).	7	Speed change levers.
2	Hand or automatic feed lever.	8	Motor.
3	Drill depth stop.	9	Pillar.
4	Spindle.	10	Vertical table lock.
5	Drill.	11	Table lock.
6	Table.	12	Base.

FIGURE 1.23
A pillar drill

cuts to the back of your hand. Therefore the work should always be securely fastened to the machine table. Small work is usually held in a machine vice which, in turn, is securely bolted to the machine table. This is shown in Figure 1.28(a). Larger work can be clamped directly to the machine table, as shown in Figure 1.28(b). In both these latter two examples the work is supported on parallel blocks. You mount the work in this way so that when the drill 'breaks through' the workpiece it does not damage the vice or the machine table.

Figure 1.28(c) shows how an angle plate can be used when the hole axis has to be parallel to the datum surface of the work whilst Figures 1.29(a) and 1.29(b) show how cylindrical work is located and supported using *vee blocks*.

OTHER DRILLING OPERATIONS

Figure 1.30 shows some miscellaneous operations that are frequently carried out on drilling machines. These include countersinking, counterboring, and spot facing.

Countersinking

Figure 1.30(a) shows a countersink bit being used to countersink a hole to receive the heads of rivets or screws. For this reason the included angle is 90°. Lathe centre drills are unsuitable for this operation as their angle is 60°.

Counterboring

Figure 1.30(b) shows a piloted counterbore being used to counterbore a hole so that the head of a capscrew or a cheese-head screw can lie below the surface of the work. Unlike a countersink cutter, a counterbore is not self-centring. It has to have a pilot which runs in the previously drilled bolt or screw hole. This keeps the counterbore cutting concentrically with the original hole.

Spot Facing

This is similar to counterboring but the cut is not as deep. It is used to provide a flat surface on a casting or a forging for a nut and washer to 'seat' on. Sometimes, as shown in Figure 1.30(c), it is used to machine a *boss* (raised seating) to provide a flat surface for a nut and washer to 'seat' on.

SAFETY WHEN DRILLING

Before carrying out a drilling activity it is essential to observe appropriate safety precautions. These include ensuring that:

- the drill guard is serviceable and fitted correctly
- the correct speed is set for the size of drill

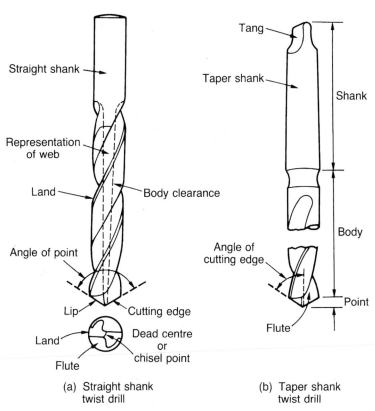

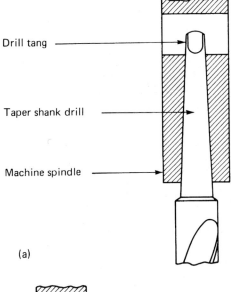

(a)

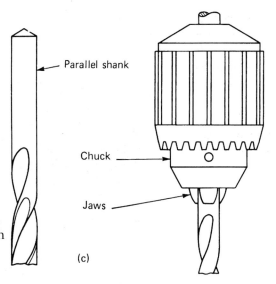

(b)

Tang

Taper shank

Shank

Body

Angle of
cutting edge

Point

Flute

(b) Taper shank
twist drill

Straight shank

Representation
of web

Land

Body clearance

Angle of point

Lip

Land

Cutting edge

Dead centre
or
chisel point

Flute

(a) Straight shank
twist drill

FIGURE 1.24
Straight shank and taper shank twist drills

Drill tang

Taper shank drill

Machine spindle

Drift

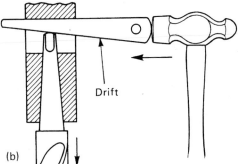

- the drill bit is in good condition and correctly fitted centrally in the chuck
- there is an appropriate means of work holding
- appropriate clothing and PPE is worn
- the work area is uncluttered and clear of swarf and other waste material
- you know where the emergency power off switch is located.

Before drilling you will need to mark out your work using a scriber where necessary to mark straight lines and a centre mark is made with a dot punch as shown in Figure 1.31(b) or with a centre punch, as shown in Figure 1.31(c). A dot punch has a fine conical point with an included angle of about 60°. A centre punch is heavier and has a less acute point angle of about 90°. It is used for making a centre mark for locating the point of a twist drill and preventing the point from wandering at the start of a cut.

Parallel shank

Chuck

Jaws

(c)

FIGURE 1.25
Methods of holding a twist drill

SOLDERING

Soldering is the process used for joining electrical conductors in order to produce a joint that is both mechanically and electrically sound. Soldering involves introducing a molten material (solder) which surrounds a mechanical joint between electrical conductors which are usually tinned copper wires, tags, pins, or printed circuit board (PCB) pads.

25

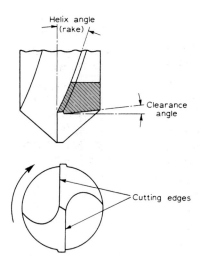

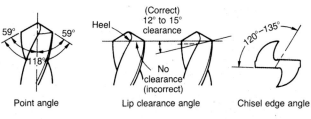

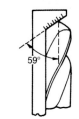

(a) Drill angles for general purpose drilling

FIGURE 1.26
Cutting edges of a twist drill

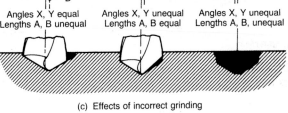

(b) Checking for correct point angle and equal lip lengths

FIGURE 1.27
Point angles for a twist drill

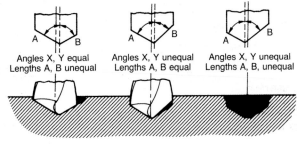

Angles X, Y equal
Lengths A, B unequal

Angles X, Y unequal
Lengths A, B equal

Angles X, Y unequal
Lengths A, B, unequal

(c) Effects of incorrect grinding

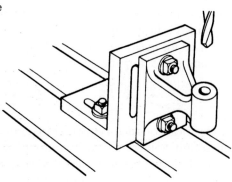

(a) Machine vice

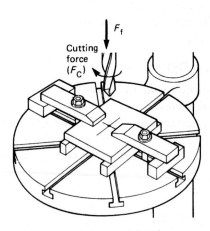

(b) Work supported on parallels and clamped to table

(c) Use of angle plate

FIGURE 1.28
Work holding

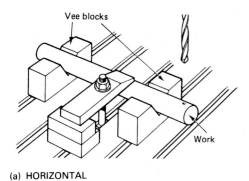

(a) HORIZONTAL

(b) VERTICAL

FIGURE 1.29
Work-holding cylindrical components

SOLDER

Solder is an alloy of two metals; tin (Sn) and lead (Pb). The proportion of tin and lead is usually expressed as a percentage or as a simple ratio. For example, the term '60–40 solder' refers to an alloy which comprises 60% tin and 40% lead by weight. The proportion of tin and lead is instrumental in determining the melting point of the solder alloy and the proportion of the constituents is usually clearly stated. Electronic solder has a low melting point temperature (typically 185°C) and is used for making joints on printed circuit boards. Joint would normally involve a single component lead and its associated copper pad and excessive temperature would have to be avoided in order to prevent damage both to the component (whose connecting lead might be of a very short length) and the PCB itself (where copper pads and tracks might lift due to the application of excessive heat).

The flux used in soldering has two roles. Firstly it helps with the removal of any oxide coating that may have formed on the surfaces that are to be soldered and secondly it helps to prevent the formation of oxides caused by the heat generated during the soldering process itself. Flux is available in both paste and liquid form.

However, in the case of solder for electronic applications, it is usually incorporated within the solder itself. In this case, the flux is contained within a number of cores that permeate through the solder material rather like the letters that appear on a piece of souvenir seaside rock.

SOLDERING TECHNIQUE

Before a soldering operation is carried out, it is vitally important that all surfaces to be soldered are clean and completely free of grease and/or oxide films. It is also important to have an adequately rated soldering iron fitted with an appropriate bit. The soldering iron bit is the all-important point of contact between the soldering iron and the joint and it should be kept scrupulously clean (using a damp sponge)

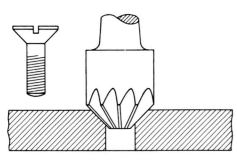

(a) COUNTERSINKING   (b) COUNTERBORING   (c) SPOTFACING

FIGURE 1.30
Countersinking, counterboring and spot facing

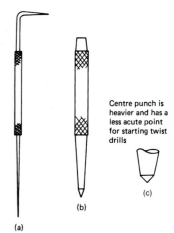

Centre punch is heavier and has a less acute point for starting twist drills

(a)

(b)

(c)

FIGURE 1.31
Scriber and centre punch

and free from oxide. To aid this process, and promote heat transfer generally, the bit should be regularly 'tinned' (i.e. given a surface coating of molten solder).

The selection of a soldering bit (see Figures 1.32 and 1.33) depends on the type of work undertaken. Figure 1.33 shows a number of popular bit profiles, the smaller bits being suitable for sub-miniature components and tightly packed boards. The procedure for making soldered joints to terminal pins and PCB pads are shown in Figures 1.35 and 1.37,

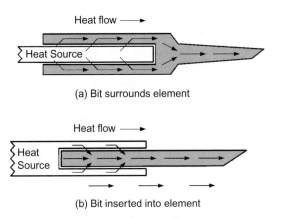

(a) Bit surrounds element

(b) Bit inserted into element

FIGURE 1.32
Heat transfer in a soldering iron bit

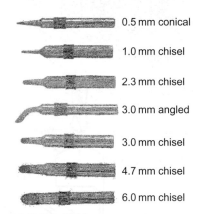

0.5 mm conical
1.0 mm chisel
2.3 mm chisel
3.0 mm angled
3.0 mm chisel
4.7 mm chisel
6.0 mm chisel

FIGURE 1.33
Typical soldering iron bits

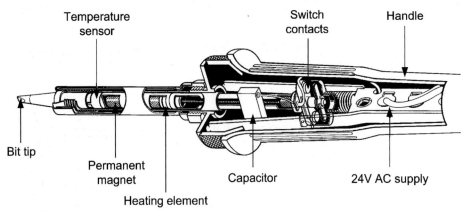

Temperature sensor — Switch contacts — Handle

Bit tip

Permanent magnet — Capacitor — 24V AC supply

Heating element

FIGURE 1.34
Construction of a temperature controlled soldering iron

Carefully insert the terminal pin and solder it to the copper strip on the underside of the board

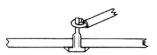

Strip the end of the wire and wrap a single turn around the pin

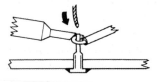

Simultaneously apply heat from the soldering iron bit to the component lead and pin. At the same time, feed solder to the joint and let it flow

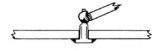

Carefully inspect the completed joint for flaws

FIGURE 1.35
Method of making a soldered connection to a terminal pin

respectively. In the case of terminal pins, the component lead or wire, should be wrapped tightly around the pin using at least one turn of wire made using a small pair of long-nosed pliers. If necessary, the wire should be trimmed using a small pair of side cutters before soldering. Next, the pin and wire should then be simultaneously heated by suitable application of the soldering iron bit and then sufficient solder should be fed on to the pin and wire (not via the bit) for it to flow evenly around the joint thus forming an airtight 'seal'. The solder should then be left to cool (taking care not to disturb the component or wire during the process). The finished joint should be carefully inspected and re-made if it suffers from any of the following faults:

1. Too little solder. The solder has failed to flow around the entire joint and some of the wire or pin remains exposed.

FIGURE 1.36
Example of a work holder which provides support for a PCB during soldering

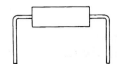

Prepare the component leads by bending to size using pliers

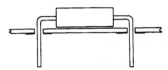

Insert the component into its correct location on the PCB

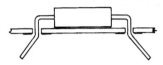

Bend the component leads through approximately 45°

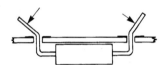

Trim the component leads using a pair of side cutters

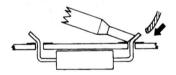

Simultaneously apply heat from the soldering iron bit to the component lead and copper pad. At the same time, feed solder to the joint and let it flow

Carefully inspect the completed joint for any flaws and rework if necessary

FIGURE 1.37
Method of making a soldered connection to a PCB mounted component

2. Too much solder. The solder has formed into a large 'blob' the majority of which is not in direct contact with either the wire or the pin.

3. The joint is 'dry'. This usually occurs if either the temperature of the joint was insufficient to permit the solder to flow adequately or if the joint was disturbed during cooling.

In the case of a joint to be made between a component and a PCB pad, a slightly different technique is used (though the requisites for cleanliness still apply). The component should be fitted to the PCB (bending its leads appropriately at right angles if it is an axial lead component) such that its leads protrude through the PCB to the copper foil side. The leads should be trimmed to within a few millimetres of the copper pad then bent slightly (so that the component does not fall out when the board is inverted) before soldering in place. Opinions differ concerning the angles through which the component leads should be bent. For easy removal, the leads should not be bent at all while, for the best mechanical joint, the leads should be bent through 90°. A good compromise, and that preferred by the author involves bending the leads through about 45°. Care should again be taken to use the minimum of solder consistent with making a sound electrical and mechanical joint.

Finally, it is important to realize that good soldering technique usually takes time to develop and the old adage 'practice makes perfect' is very apt in this respect. So, don't despair if your first efforts fail to match with those of the professional!

SAFETY WHEN SOLDERING

Whilst soldering is not a particularly hazardous operation, there are a number of essential safety and other precautions that should be observed. These are listed below:

Fumes

Solder fumes are an irritant and exposure, particularly if prolonged, can cause asthma attacks. Because of this it is essential to avoid the build up of fumes, particularly so when a soldering iron is in continuous use. The best way to do this is to use a proprietary system for fume extraction. In addition, many professional quality irons can be fitted with fume extraction facilities which are designed to clear fumes from the proximity of the soldering iron bit.

Test your knowledge 1.16
State the typical constituents of electronic solder and explain why flux is used when making a soldered joint.

Even when a soldering iron is only used intermittently it is essential to ensure that the equipment is used in a well-ventilated area where fumes cannot accumulate. Fume extraction should be fitted to all soldering equipment that is in constant use. Most good quality soldering irons that are designed for professional use can be fitted with fume extraction equipment.

Damage to Eyes

Molten solder or flux additives can cause permanent eye damage. Some means of eye protection is essential. This can take the form of safety glasses or the use of a bench magnifier. Ordinary reading or prescription glasses will also offer a measure of protection and may also be of benefit when undertaking close work.

Static Hazards

Test your knowledge 1.17
Explain what is meant by a 'dry joint' and how this can be avoided.

Many of today's electronic devices are susceptible to damage from electric fields and stray static charges. Unfortunately, simply earthing the tip of a soldering iron is not sufficient to completely eliminate static damage and additional precautions may be required to avoid any risks of electro-static damage. These may involve using a low-voltage supply to feed the heating element fitted in the soldering iron, incorporating low/zero-voltage switching, and the use of conductive (antistatic) materials in the construction of the soldering iron body and the soldering station itself.

Burn, Melt and Fire Hazards

The bit of a soldering iron is usually maintained at a temperature of between 250°C and 350°C. At this temperature, conventional plastic insulating materials will melt and many other materials (such as paper, cotton, etc.) will burn. It should also go without saying that personal contact should be avoided at all times and the use of a properly designed soldering iron stand is essential.

Shock Hazards

Mains voltage is present in the supply lead to a mains operated soldering iron and also in the supply lead to any soldering station designed for use with a low-voltage iron. If the soldering iron bit comes into contact with a mains lead there can be a danger that the insulation will melt exposing the live conductors. In such situations, heatproof insulation is highly recommended both for the supply to the iron itself and also to any power unit.

ACTIVITY 1.14

Working as a member of a small team of two or three students, you are to manufacture a simple drill gauge which will allow users to rapidly check the size of twists drills having diameters ranging from 1 mm to 5 mm in increments of 0.5 mm. Your tutor will provide you with suitable materials from which to manufacture the gauge together with a drawing showing the location and size of each of the holes that you must accurately drill. In order to complete this activity you will need tools for marking out (and centre punching) as well as access to a pillar drill.

Each member of the team should contribute to the work activity and each should produce his or her own drill gauge. In addition, you should keep a record of how your team organised the activity, who did what, and what decisions were made as the activity progressed.

You should also note down the safety precautions that you took as well as listing any PPE or other safety devices that you make use of. Finally, suggest what changes and improvements could be made as a result of your experience of carrying out this activity.

ACTIVITY 1.15

Working as a member of a small team of two or three students, you are to manufacture a simple electrical continuity tester which will allow users to rapidly check fuses, cables and electrical connectors. Your tutor will provide you with suitable materials from which to manufacture the continuity checker together with a circuit diagram and wiring layout. In order to complete this activity you will need tools for cutting and bending electrical wires and components as well as access to a soldering station.

Each member of the team should contribute to the work activity and each should produce his or her own continuity checker. In addition, you should keep a record of how your team organised the activity, who did what, and what decisions were made as the activity progressed.

You should also note down the safety precautions that you took as well as listing any PPE or other safety devices that you make use of. Finally, suggest what changes and improvements could be made as a result of your experience of carrying out this activity.

Working Relationships

In engineering, as with any other work activity, it is essential to have good working relationships with colleagues and other people that you might have contact with in your day-to-day working life. Such relationships will not only help you to be effective in doing your own job but they will also allow others to be effective in their own job roles.

Good working relationships are based on a positive attitude to work coupled with good communications skills and an ability to identify and respond to the needs of others. These skills can be easily acquired but they do involve you in being aware of what is going on around you and being willing and able to respond to the needs of others.

Few engineering activities are performed in complete isolation and all major engineering projects involve teamwork and co-operation. As you start your career in engineering you will inevitably become a member of a team and other people will depend on you just as you depend on them. A good starting point is that of gaining the respect and support of those around you by listening to other people's needs and being prepared to give advice and support whenever necessary.

ACTIVITY 1.16

Working as a member of a team of three to four students you are to design a layout for a new 'multi-workshop' designed to support a group of sixteen first year engineering students. The workshop is to provide resources for marking out, drilling (two pillar drills), turning (one small lathe), bench fitting, sheet metal cutting, folding and bending. You should create a floor plan showing the workshop layout as well as arrangements for the storage of tools, materials, and work in progress. You should also include some general purpose benches as well as a 'clean area' for paperwork and two networked PCs. Your floor plan should include appropriate dimensions and should show doors, windows, and the location of services such as electrical, gas and water supplies and the location of fire, safety and first aid equipment. At the early stages of this activity, you should ensure that you have consulted appropriate school or college staff (such as workshop technicians, health and safety officers, fire wardens, etc.). You should present your recommendations to the rest of the class in the form of a ten minute presentation with appropriate visual aids and handouts. You should also show how each member of the team contributed to the activity and indicate how effective working relationships were established and maintained between team members as the activity progressed. Finally, you should acknowledge the contribution made by others (such as specialist school or college staff).

(a)

(b)

(c)

(d)

FIGURE 1.38
See Question 3

(a)

(b)

(c)

(d)

(e)

FIGURE 1.39
See Question 10

ACTIVITY 1.17

Working as a member of a team of three to four students you are to plan and demonstrate an engineering activity (such as drilling, turning or soldering) at a school or college open-day. You should ensure that visitors are able to observe safely and that members of the team are available to provide guidance and to answer questions. You should also prepare a handout (using a single A4 page) that describes the activity that you are performing and gives details of the processes, materials, tools, resources, and safety precautions. Following the open-day, you should hold a team meeting to carry out an evaluation of your activity and indicate what worked well, what didn't work well, and why this was. You should also consider the contributions made by individual members of the team and make specific recommendations for future improvements. If the event was to be repeated in the future what changes would you make and why?

REVIEW QUESTIONS

1. List THREE duties of the employer under the Health and Safety at Work Act.
2. List THREE duties of the employee under the Health and Safety at Work Act.
3. Identify the signs shown in Figure 1.38 and explain where each would be placed.
4. Explain what is meant by a Risk Assessment. Who should be involved with carrying out a Risk Assessment?
5. List FOUR items of information that would be found in a Material Safety Data Sheet (MSDS). Where would you expect to be able to find an MSDS for a material used in an engineering laboratory?
6. Describe THREE different hazardous engineering processes and, for each process, describe ways in which the risks associated with the process can be reduced.
7. Describe TWO ways in which a person can receive an electric shock from a workshop AC mains supply.
8. Apart from causing heart failure, what else may occur when a person receives an electric shock?
9. What actions, and in what order, should be taken when a person receives an electric shock?
10. Explain the meaning of each of the signs shown in Figure 1.39.
11. What type of fire extinguisher is shown in Figure 1.40?
12. Identify the two engineering operations shown in Figure 1.41.

FIGURE 1.40
See Question 11

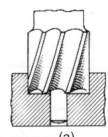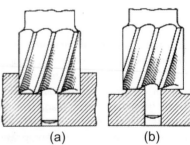

(a)   (b)

FIGURE 1.41
See Question 12

13. With the aid of a sketch, explain the construction of a vee-block work holder. What is the purpose of this device?

14. Identify an item of personal protective equipment (PPE) for use when (a) drilling and (b) soldering. In each case explain why the item should be used.

15. With the aid of a sketch, explain how heat is conducted from the element of a soldering iron to a joint that is being made.

16. Identify the fluid shown in Figure 1.42 and state the precautions that should be observed with its handling and storage.

17. Explain why good working relationships are important in engineering. Illustrate your answer with an example.

FIGURE 1.42
See Question 16

Interpreting and Using Engineering Information

SUMMARY

This unit will help you to understand how to make effective use of a wide variety of information used by engineers, including written text (such as specifications and instructions), charts and diagrams (such as flow charts, exploded views and block schematics), and drawings (such as detail and general arrangement drawings). You need to be able to extract relevant information from these sources and be able to use it to solve real engineering problems. The sources of information that you need to work with include catalogues, books and manuals, as well as electronic sources such as CD-ROM and the Internet. It's also important to know about the standards that are used for presenting information. You will learn how standards are applied and you will be expected to recognise the use of standards and be able to show that your own work conforms to them. As you work through this unit you will be expected to show that you can make appropriate use of information as a means of informing and improving your own work as an engineer. You should be able to evaluate the information that you use in order to ensure that it is accurate, current and relevant. As an engineer you will seldom be working in isolation and nearly always as a member of a team. Because of this you should also be able to present information to others in a way that is clear and unambiguous, and in a manner that conforms to accepted codes of practice and standards. Passing information on to others is another critical skill that you must learn in order to be effective as an engineer. When you have completed this unit you should be able to understand how to interpret engineering information from a wide range of drawings and documents. You should also be able to make use of this information to carry out and check your own work output.

2.1 INTERPRETING DRAWINGS AND DOCUMENTATION
Information

Engineers get the information that they need to do their jobs from a variety of different sources. Some of this information exists in printed form (in other words it consists of documents with text and diagrams) but increasingly it exists in electronic form (which can be read on a screen or printed directly from a computer). Irrespective of the form that the information actually takes, there is a need for the information to be accurate, thorough, up-to-date, and available when it is needed. Think about the information that you might

BTEC First Engineering: Mandatory and Selected Optional Units for BTEC Firsts in Engineering. DOI: 10.1016/B978-1-85617-685-9.00002-0

need in order to carry out a routine service on a motor car. You would need to obtain a service manual as well as documentation relating to the particular model of the vehicle (including year of manufacture, chassis number, engine number, etc.). You might also need details of any modifications fitted to the vehicle (for example, an immobiliser). You might also want to refer to the vehicle's 'service history' which gives details of previous services, including faults and other recommendations made by engineers. If you were doing this job in a real service facility, you would need to refer to a service schedule and job card which gives precise details of work to be carried out, detailing the recommended procedures (including tools and equipment) and quoting times for each task.

The service or maintenance schedule for a car specifies the recommended maintenance requirements for the vehicle in question. The schedule is not the same for all vehicles but depends, for example, on the type of engine and fuel, and on the type of use the vehicle will have. As an example, a typical maintenance schedule for a modern saloon car fitted with a petrol engine might run along the following lines

- Every 10,000 miles (15,000 km) or six months, whichever comes first; renew the engine oil and oil filter
- Every 20,000 miles (30,000 km) or 12 months, whichever comes first; check the front and rear brake pads and discs for wear; check and adjust the handbrake; check the steering and suspension components; check the body and under-body for corrosion; etc.
- Every 20,000 miles (30,000 km) or two years, whichever comes first; renew the element in the pollen filter
- Every two years, regardless of mileage; renew the hydraulic fluid and renew the engine coolant
- Every 40,000 miles (60,000 km) or four years, whichever comes first; renew the air cleaner element; renew the spark plugs and check the ignition system; renew the fuel filter, etc.

ACTIVITY 2.1

Refer to a service and repair manual for any current and common type of saloon or hatchback vehicle and use it to obtain the following information:

a. The engine size
b. The engine oil capacity
c. The fuel tank capacity
d. The minimum thickness of the front and rear brake pads
e. The type of spark plugs or glow plugs recommended for use in the ignition or pre-heat systems

Present your work in the form of a printed information sheet (a single A4 page). Make sure that you name the make and model of the vehicle clearly at the top of the page and list the sources of information that you've used at the bottom of the page.

Application Notes and Technical Reports

Application notes are usually brief notes (often equivalent in extent to a chapter of a book) supplied by manufacturers in order to assist engineers and designers by providing typical examples of the use of engineering components and devices.

An application note can be very useful in providing practical information that can help designers to avoid pitfalls that might occur when using a component or device for the first time. Application notes often include prototype schematics and layout diagrams.

Technical reports are somewhat similar to application notes but they focus more on the performance specification of engineering components and devices (and the tests that have been carried out on them) than the practical aspects of their use. Technical reports usually include detailed specifications, graphs, charts and tabulated data.

ACTIVITY 2.2

The following text is an extract from a technical report on the development of an electronic fuel injection system:

Electronic fuel injection systems are designed to deliver fuel to an internal combustion engine depending on the pulse width provided by a controller. Ignition systems deliver spark to each cylinder at a specific point in the engine cycle. Both of these systems are controlled by a controller that monitors feedback and changes fuel and spark accordingly.

The most important thing to balance at all times is the ratio of air to fuel, which is about 14.5:1. The air/fuel ratio affects the efficiency of an engine. If the mixture of air and fuel is too rich (or contains too much fuel) or too lean (too much air), the engine will not burn its fuel efficiently. Therefore, a certain amount of fuel is required as a function of engine speed and load, both of which need to be measured consistently. This requires a feedback system to adjust the fuel supply as engine speed changes. The ignition system is controlled similarly to the fuel injection system, in that the time at which the spark occurs is the only controlled parameter. The amount of spark is determined by the secondary ignition system. The timing of the spark is crucial so there is little room for error.

Read the above passage carefully and then answer the following questions:

1. What is the approximate ratio of air to fuel?
2. What happens if the ratio of air to fuel is not correct?
3. The amount of fuel used in the mixture is determined by two factors. What are they?
4. What is critical about the spark?
5. What determines the amount of spark?
6. What does the feedback system do?

Typical section headings used in application notes and technical reports include:

SUMMARY

A brief overview for busy readers who need to quickly find out what the application note or technical report is about.

INTRODUCTION

This sets the context and background and provides a brief description of the process or technology—why it is needed and what it does. It may also include a brief review of alternative methods and solutions.

MAIN BODY

A comprehensive description of the process or technology.

EVALUATION

A detailed evaluation of the process or technology together with details of tests applied and measured performance specifications. In appropriate cases comparative performance specifications will be provided.

RECOMMENDATIONS

This section provides information on how the process or technology should be implemented or deployed. It may include recommendations for storage or handling together with information relating to Health and Safety.

CONCLUSIONS

This section consists of a few concluding remarks.

REFERENCES

This section provides readers with a list of sources of further information relating to the process or technology, including (where appropriate) relevant standards and legislation.

Application notes explain how something is used in a particular application or how it can be used to solve a particular problem. They are intended as a guide for designers and others who may be considering using a particular process or technology for the first time.

Technical reports, on the other hand, provide information that is more to do with whether a component or device meets a particular specification or how it compares with other solutions. Technical reports are thus more useful when it comes to analysing how a process or technology performs than how it is applied.

Data Sheets and Data Books

Data sheets usually consist of abridged information on a particular engineering component or device. They usually provide maximum and minimum ratings, typical specifications, as well as information on dimensions, packaging and finish. Data sheets are usually supplied free on request from manufacturers and suppliers. Collections of data sheets for similar types of engineering components and devices are often supplied in book form. Often supplementary information is included relating to a complete family of products. An example of a data sheet is shown in Figure 2.1.

Catalogues

Test your knowledge 2.1
List each of the main sections that you would expect to find in a technical report.

Most manufacturers and suppliers provide catalogues that list their full product range. These often include part numbers, illustrations, brief specifications and prices. Whilst catalogues are often extensive documents with many hundreds or thousands of pages, short-form catalogues are usually also available. These usually just list part numbers, brief descriptions and prices but rarely include any illustrations. A brief extract from a short-form catalogue is shown in Figure 2.2.

Catalogues and data sheets are often distributed on compact disks which can provide storage for around 650 Mbytes of computer data. This is equivalent to several thousand pages of A4 text and line diagrams.

Specifications

Written specifications should take the form of a precise and comprehensive description of the product. Specifications should relate not only to the physical characteristics and appearance of a product but also to the performance of a product in terms that can be measured in order to verify its performance.

ACTIVITY 2.3

Use the data sheet shown in Figure 2.1 to answer the following questions:

1. Who is the manufacturer of the BJ284 device?
2. What is the reference number of the data sheet and when was it produced?
3. What type of device is a BJ284?
4. What is the maximum power dissipation for a BJ284?
5. State THREE applications for a BJ284.
6. Sketch the circuit symbol for a BJ284 device and label the connections.
7. What is the maximum junction temperature for a BJ284 device?
8. Under what conditions is the DC current specified?
9. What type of package is used for the BJ284?
10. A BJ284 is to be operated at a collector-emitter voltage of 6 V and a collector current of 20 A. Is this permissible? Give reasons for your answer.

Howard Associates

DATA SHEET	BJ284 NPN Silicon Power Transistor

MAIN FEATURES

- 150W max. power dissipation
- High current gain (>100 typ. at I_C = 2A)
- Large gain-bandwidth product (>6 MHz typ. at I_C = 2A)
- Rugged TO3 case
- Complementary to BJ285 PNP transistor

APPLICATIONS

- High-quality audio power amplifiers
- Linear voltage regulators
- Power switching
- Automotive ignition systems
- Power control and regulation
- Emergency lighting systems

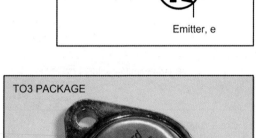

SYMBOL

TO3 PACKAGE

ABSOLUTE MAXIMUM RATINGS (T_A = +25°C)

Collector to base voltage, V_{CBO}	180V
Collector to emitter voltage, V_{CEO}	180V
Emitter to base voltage, V_{EBO}	5V
Collector current, I_C	16A
Emitter current, I_C	16A
Power dissipation, P_C	150W
Junction temperature, T_j	+150°C
Storage temperature, T_{stg}	-65°C to +150°C

ELECTRICAL CHARACTERISTICS (T_A = +25°C)

Parameter	Symbol	Condition	Min.	Typ.	Max.	Unit
Collector cut-off current	I_{CBO}	V_{CB} = 90V, I_E = 0	–	–	100	μA
Emitter cut-off current	I_{EBO}	V_{EB} = 5V, I_C = 0	–	–	100	μA
DC current gain	h_{FE}	V_{CE} = 5V, I_C = 2A	70	–	140	–
Collector-emitter saturation voltage	$V_{CE(sat)}$	I_C = 10A, I_B = 1A	–	–	3.0	V
Base to emitter voltage	V_{BE}	V_C = 5V, I_C = 10A	–	–	2.5	V
Current gain bandwidth product	f_T	V_C = 5V, I_C = 2A	–	6	–	MHz
Output capacitance	C_{ob}	V_{CB} = 10V, f = 1MHz	–	300	–	pF

FIGURE 2.1
Example of a data sheet

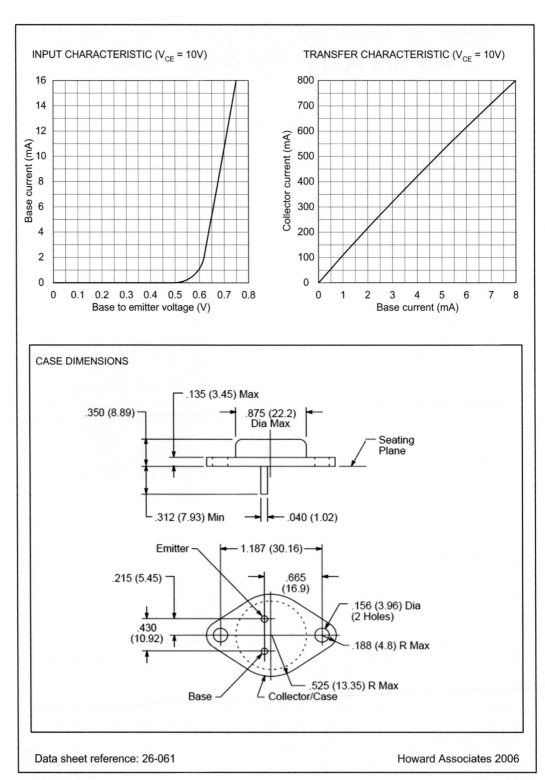

INPUT CHARACTERISTIC (V_{CE} = 10V)

TRANSFER CHARACTERISTIC (V_{CE} = 10V)

CASE DIMENSIONS

Data sheet reference: 26-061

Howard Associates 2006

FIGURE 2.1
(*Continued*)

Diecast Boxes — IP65 Sealed/Painted

A range of high-quality diecast aluminium boxes with an optional grey epoxy paint finish to RAL7001. The lid features an integral synthetic rubber sealing gasket and captive stainless steel fixing screws. Mounting holes and lid fixing screws are outside the seal, giving the enclosure protection to IP 65.

Standard supply multiple = 1   Delivery normally ex-stock

| Size | | | | | | | | Price each | | |
L	W	H	T	Finish	Manufacturer's ref:	Stock code		1-9	10-24	25+
90	45	30	3.0	none	1770-1541-21	DB65-01		£4.52	£3.95	£3.50
90	45	30	3.0	grey	1770-1542-21	DB65-01P		£5.40	£4.90	£4.45
110	50	30	4.5	none	1770-1543-22	DB65-02		£5.25	£4.50	£4.15
110	50	30	4.5	grey	1770-1544-22	DB65-02P		£6.42	£5.37	£4.95
125	85	35	5.0	none	1770-1545-23	DB65-03		£6.15	£5.17	£4.71
125	85	35	5.0	grey	1770-1546-23	DB65-03P		£7.10	£6.05	£5.65

FIGURE 2.2
Example of an extract from a short-form catalogue

Since specifications form the basis of a contract between a manufacturer or supplier and a client or customer, they need to be written in terms of what the purchaser requires and in clear, unambiguous words. There are three different types of specification:

- *General specifications:* A detailed written description of the product including its appearance, construction, and materials used.
- *Performance specification:* A list of features of the product that contribute to its ability to meet the needs of the client or end user. For example, output voltage, power, or speed.
- *Standard specification:* Describes the materials and processes (where appropriate) used in the manufacture of the product in terms of relevant standards (e.g. BS 9000).

Quality Documents

All engineering companies have *Quality Systems* in place to ensure that they produce goods and services of an appropriate quality. These systems are invariably based on documented procedures which often include one or more of the following:

- *Quality Procedures:* A detailed written description of the quality system and the controls that are in place within the company (these often form part of a Quality Manual or are documented separately as a Procedures Manual).
- *Work Instructions:* A description of a particular operation or task in terms of what must be done, who should do it, when it should be done, and what materials and processes should be used. In many cases, a series of Work Instructions are used to describe each stage in the production or manufacturing process.
- *Test Specifications:* A detailed list of characteristics and features used to verify conformance with the design specification together with details of any measurements that are to be made out and how they should be carried out.

Documents are often used to assist in the process of identification and *traceability* of products, components and materials. This is vitally important in critical sectors such as aerospace, nuclear and chemical engineering. Traceability is an essential when it becomes necessary to eliminate the causes of *non-conformance*. Traceability is achieved by coding items and maintaining records that can be updated throughout the working life of a component or record.

Test your knowledge 2.2
List THREE different types of specification. Explain why specifications are important when supplying an engineered product.

Test your knowledge 2.3
Complete the table shown in Figure 2.3 by placing a tick in the column corresponding to the most appropriate way of communicating the information (tick only one box in each row).

Application	Application note	Data sheet	Catalogue	Technical report
Summary of the precautions to be observed when handling a chemical etching fluid	✓			
Cost of diecast boxes supplied in various quantities				
Maximum working temperature for a power transistor				
Recommended printed circuit board layout for an audio amplifier				
Comparison of different types of surface finish for the interior of a domestic microwave oven				
Physical dimensions of a marine radar for fitting to the mast of a small boat				
Description of tests applied to an off-road vehicle				

FIGURE 2.3
See Test your knowledge 2.3

Test your knowledge 2.4
Give THREE examples of documents that are used as part of the quality procedures in an engineering company.

Manuals

Various types of manual are associated with engineered products including *operating manuals* that are designed to be read by the end user of the product and *service manuals* or *repair manuals* that are designed to aid the repair and/or the routine maintenance of the product. Manuals are often produced by the company that has manufactured the product but may also be produced by third-party companies and organisations who specialise in producing them.

Job cards

Job cards provide information about items of equipment and what should be done with them. Job cards are often used when equipment is sent away for service or repair, see Figure 2.4.

The Internet

The Internet and the World Wide Web are an excellent source of engineering information and many engineering companies are moving from paper-based information to electronic information that is disseminated from a website.

Test your knowledge 2.5
Give an example of the sort of information that might appear on a typical job card for the repair of an item of consumer electronics equipment.

Although the terms *Web* and *Internet* are often used synonymously, they are actually two different things. The *Internet* is the global association of computers that carries data and makes the exchange of information possible. The *World Wide Web* is a subset of the Internet—a collection of inter-linked documents that work together using a specific Internet protocol called *hypertext transfer protocol* (HTTP). In other words, the Internet exists independently of the World Wide Web, but the World Wide Web can't exist without the Internet.

The World Wide Web began in March 1989, when Tim Berners-Lee of the European Particle Physics Laboratory at CERN (the European centre for nuclear research) proposed the project as a means to better communicate research ideas among members of a widespread organization.

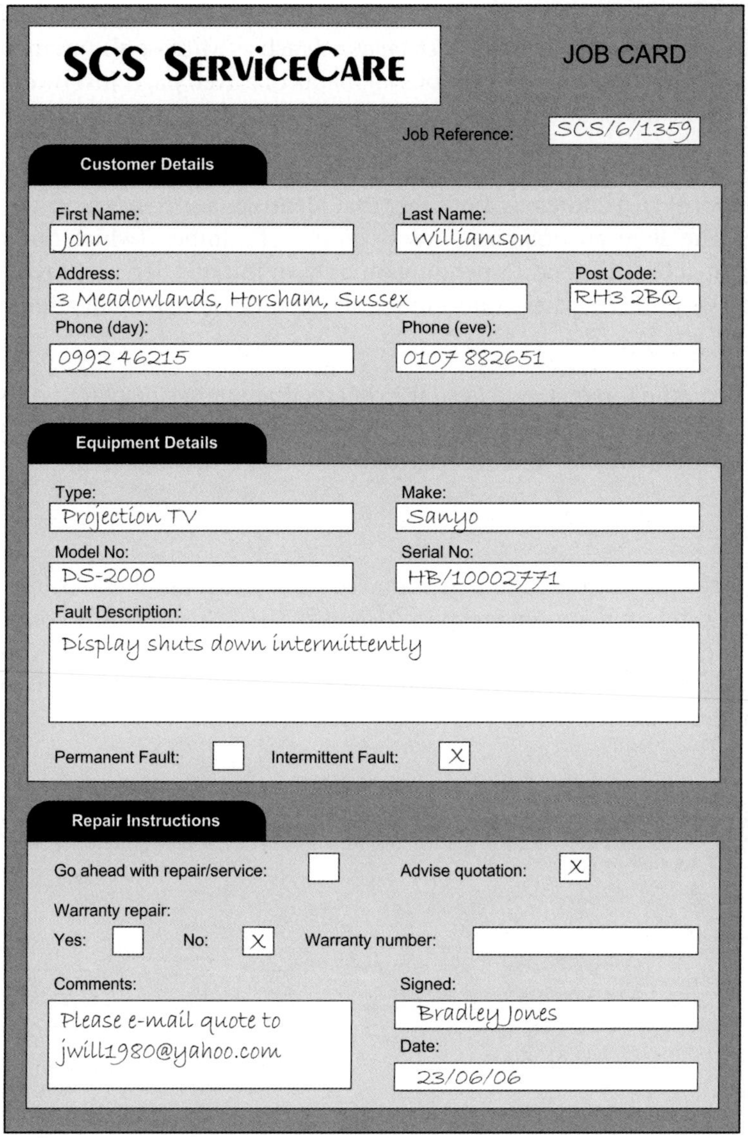

SCS SERVICECARE

JOB CARD

Job Reference: SCS/6/1359

Customer Details

First Name: John
Last Name: Williamson

Address: 3 Meadowlands, Horsham, Sussex
Post Code: RH3 2BQ

Phone (day): 0992 46215
Phone (eve): 0107 882651

Equipment Details

Type: Projection TV
Make: Sanyo

Model No: DS-2000
Serial No: HB/10002771

Fault Description: Display shuts down intermittently

Permanent Fault: ☐   Intermittent Fault: ☒

Repair Instructions

Go ahead with repair/service: ☐   Advise quotation: ☒

Warranty repair:
Yes: ☐   No: ☒   Warranty number: ☐

Comments: Please e-mail quote to jwill1980@yahoo.com
Signed: Bradley Jones
Date: 23/06/06

FIGURE 2.4
Example of a job card

Welcome to Howard Aeropsace Avionic Systems Division - Microsoft Internet Explorer

File Edit View Favorites Tools Help

Back Forward Stop Refresh Home Search Favorites Media History Mail Print

Address http://www.howardaerospace.co.uk/index.htm   Go Links

HOWARD AEROSPACE
Avionic Systems Division
Brooklands Road, Weybridge
KT13 9XA, United Kingdom

About Us   Products   Services   Site Map   What's New   Contact Us

Site last updated on 21/07/2002

My Computer

FIGURE 2.5
An example of an engineering company's website displayed in a web browser

Web sites are made up of collections of Web pages. Web pages are written in *hypertext markup language* (HTML), which tells a *Web browser* (such as Microsoft's Internet Explorer or Mozilla's Firefox) how to display the various elements of a Web page. Just by clicking on a *hyperlink*, you can be transported to a site on the other side of the world.

A set of unique addresses is used to distinguish the individual sites on the World Wide Web. An Internet Protocol (IP) address is a number that identifies a specific computer connected to the Internet. The digits are organized in four groups of numbers (which can range from 0 to 255) separated by full stops. Depending on how an Internet Service Provider (ISP) assigns IP addresses, you may have one address all the time or a different address each time you connect.

Every Web page on the Internet, and even the objects that you see displayed on Web pages, has its own unique address, known as a *uniform resource locator* (URL). The URL tells a browser exactly where to go to find the page or object that it has to display.

Search Sites

Being able to locate the information that you need from a vast number of sites scattered across the globe can be a daunting prospect. However, since this is a fairly common requirement, a special type of site, known as a *search site*, is available to help you with this task. There are three different types of search site on the Web; *search engines*, *Web directories*, and parallel and metasearch sites.

> **Test your knowledge 2.6**
> What do each of the following initials stand for?
>
> a. HTTP
> b. HTML
> c. IP
> d. ISP
> e. URL.

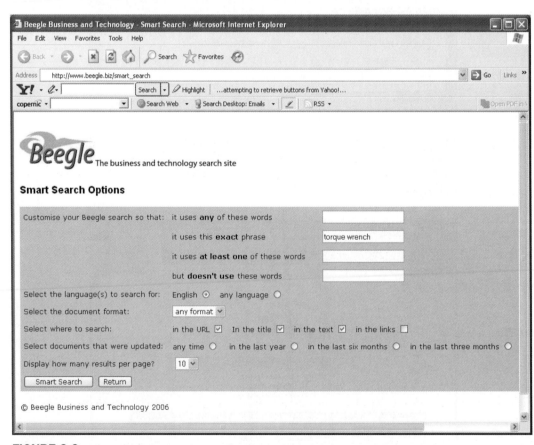

FIGURE 2.6
A typical search engine being used to search for the term 'torque wrench'. Note that the advanced search options provide you with a way of narrowing down your search to an exact phrase (see Activity 2.6)

FIGURE 2.7
A web-based reference source being used to locate and display information on fire extinguishers (see Activity 2.5)

Search engines such as Google or Excite use automated software called *Web crawlers* or *spiders*. These programs move from Web site to Web site, logging each site title, URL, and at least some of its text content. The object is to hit millions of Web sites and to stay as current with them as possible. The result is a long list of Web sites placed in a database that users search by typing in a keyword or phrase.

Web directories such as Yahoo, offer an editorially selected, topically organised list of Web sites. To accomplish that goal, these sites employ editors to find new Web sites and work with programmers to categorise them and build their links into the site's index. To make things even easier, all the major search engine sites now have built-in topical search indexes, and most Web directories have added keyword search facilities.

Intranets

Intranets work like the Web (with browsers, Web servers, and Web sites) but companies and other organisations use them internally. Companies use them because they let employees share corporate data, but they're cheaper and easier to manage than most private networks— no one needs any software more complicated or more expensive than a Web browser, for instance.

They also have the added benefit of giving employees access to the Web. Intranets are closed off from the rest of the Internet by firewall software, which lets employees surf the Web but keeps all the data on internal Web servers hidden from those outside the company.

> **Test your knowledge 2.7**
> What software is required to display a web page?

FIGURE 2.8
The Millau Viaduct bridge
nearing completion
(courtesy of Enerpac)

Extranets

One of the most recent developments has been that of the *extranet*. Extranets are several intranets linked together so that businesses can share information with their customers and suppliers. Consider, for example, the production of a European aircraft by four major aerospace companies located in different European countries. They might connect their individual company intranets (or parts of their intranets) to each other, using private leased lines or even using the public Internet. The companies may also decide to set up a set of private newsgroups so that employees from different companies can exchange ideas and share information.

Test your knowledge 2.8
Explain the difference between an intranet and the Internet.

ACTIVITY 2.4

The motorway bridge at Millau in France is the world's highest bridge. The engineering company that supplied the hydraulic system for lifting the temporary piers and pushing the bridge decks into position was Enerpac. Visit the company's website at http://www.enerpac.com and locate information on the Millau Viaduct project. Use this to answer the following questions:

1. What is the height and overall length of the bridge?
2. What valley does the bridge cross?
3. Who designed the bridge?
4. How many bridge piers were constructed?
5. How much concrete was used to build the bridge?
6. How was the bridge deck moved into position?
7. What electronic device was used to control the electro valves?

FIGURE 2.9
A temporary pillar used
to support the bridge
deck during construction
(courtesy of Enerpac)

ACTIVITY 2.5

Visit the Wikipedia website at http://www.wikipedia.com and view the entry on 'Fire Extinguishers' (similar to that shown in Fig. 2.7). Read the information and use it (together with information from other web-based resources, as appropriate) to answer the following questions:

1. Who invented the modern fire extinguisher?
2. List the main parts of a fire extinguisher and say what each part is used for.
3. What are the two main types of fire extinguisher bottle?
4. What is a Class B fire?
5. What European Standard relates to different classes of fire?
6. What name is used in the USA to describe a 'dry powder' fire extinguisher?
7. Why are halon fire extinguishers illegal in the UK?
8. What precaution should be observed when discharging a CO_2 fire extinguisher and why is this necessary?
9. In the UK, how often should a CO_2 fire extinguisher be pressure tested?
10. What is AFFF and where is it likely to be used?
11. Why do Class-D fires need special types of fire extinguisher?

ACTIVITY 2.6

Some service and repair tasks on a vehicle require the use of a torque wrench. Use information sources (such as a library or Google search) to find out what a torque wrench is and then write a short technical report (not more than one page of A4) describing a torque wrench and explaining how it is used. Include a relevant diagram, sketch or photograph in your answer.

ACTIVITY 2.7

Visit the Draper Tools website at http://www.draper.co.uk and use it to download the instruction manual for a 14.4 V Cordless Drill (Part No. CD140V). Use the information obtained from the instruction manual to answer the following questions:

1. What is the no-load speed range?
2. What is the chuck size?
3. What is the drilling capacity when used with mild steel?
4. What is the weight of the drill plus battery?
5. What is the spindle thread for the chuck?
6. What is the cell rating?
7. What is the procedure for battery disposal and why is it important to follow this recommendation?
8. What precaution should be observed when charging the battery and why is this important?

Engineering Drawings

Engineers rely heavily upon sketching and drawing as a means of communication. As an engineer you must be able to read and use working drawings as well as produce a selection of presentation drawings using both hand-drawn and computer techniques. To avoid confusion, your engineering drawings must comply with recommended standards and conventions. You will also need to be able to read electrical/electronic, pneumatic/hydraulic and mechanical engineering drawings and identify a selection of the most commonly used symbols.

Engineering drawings are sometimes referred to as *formal* or *informal*. Informal drawings (see Fig. 2.10) are usually sketches or hand-drawn diagrams that provide a quick impression of

Key point

Because they can convey a great deal of information very quickly, engineers make a great deal of use of drawings. Depending upon the way they are presented, drawings are often classified as either *formal* or *informal*.

ACTIVITY 2.8

Visit the Draper Tools website at http://www.draper.co.uk and use it to download the parts list and assembly drawing for a 230 V/400 V 160 A Turbo Arc Welder (Part No. AW180AT). Use the information obtained from the parts list and assembly drawing to answer the following questions:

1. What is the part number for the power cable and plug?
2. What is the stock number for the power regulator?
3. What is part number YS075100 and how many are fitted?
4. What is stock number 77348 and where is this fitted?
5. What controls are mounted on the front panel of the unit?
6. How many cables emerge from the front panel of the unit and what are they connected to?
7. What component is mounted in the centre of the unit?
8. Where is the fan mounted?

Key point

Engineers use hand-drawn sketches to convey their ideas quickly and without having to use a lot of words. Sketches can be 2D or 3D drawings and may, or may not, include dimensions.

what something will look like or how something will work. Formal drawings (see Fig. 2.23) generally take much longer to produce and usually contain a lot more detail. They are also much more precise and often include a scale and dimensions.

Sketches

Sketching is one of the most useful tools available to the engineer to express his or her ideas and preliminary designs. Sketches are drawn freehand and they are used to gain a quick impression of what something will look like. A sketch can be either a two dimensional (2D) representation (see Fig. 2.10) or a three dimensional (3D) representation (see Fig. 2.11). A sketch can also take the form of a block diagram or a schematic diagram (see Fig. 2.12). Where appropriate, labels, approximate dimensions and brief notes can be added to any of these types of drawing.

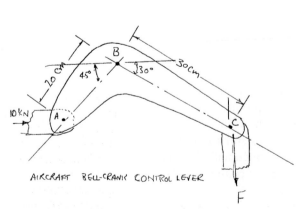

FIGURE 2.10
A 2D sketch

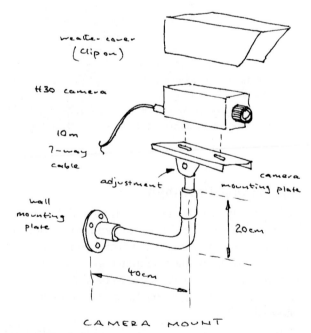

FIGURE 2.11
A 3D sketch

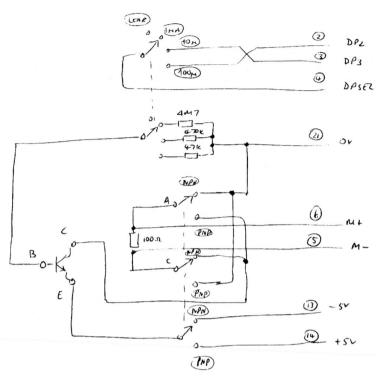

FIGURE 2.12
A schematic circuit diagram drawn as a sketch

FIGURE 2.13
An engineering component (see Activity 2.9)

Another view
Engineering drawing isn't just about using CAD to produce formal drawings. Being able to produce a quick sketch is often very useful so don't be afraid to pick up a drawing pad and put your ideas into visual form.

Key point
Block diagrams are used to show how a number of things are connected together. Block diagrams use shapes (often square or rectangular connected together with arrowed lines).

ACTIVITY 2.9
Produce a 3D sketch of an engineering component like that shown in Figure 2.13 (your tutor will supply this). Include in your sketch approximate dimensions and details of the materials used.

Block Diagrams

Block diagrams show the relationship between the various elements of a system (i.e. how they are connected together). Figure 2.14 shows the block diagram of a simple radio receiver. Diagrams like this can be very useful when carrying out fault finding.

FIGURE 2.14
Block diagram of a simple radio receiver. Note the use of arrows to show the direction of the signal and how the inputs and outputs are connected

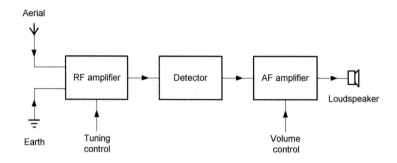

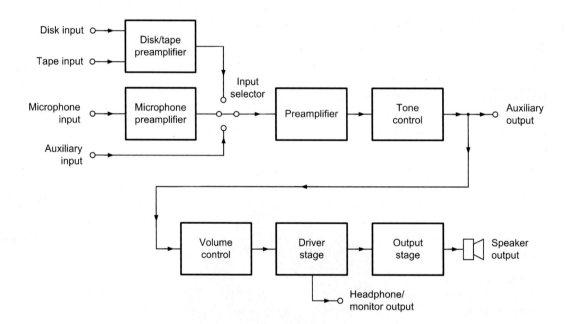

FIGURE 2.15
Block diagram of an audio system (see Activity 2.10)

> ### ACTIVITY 2.10
>
> The block diagram of an audio system is shown in Figure 2.15. Use this diagram to answer the following questions:
>
> 1. How many inputs are provided and what are they used for?
> 2. How many outputs are provided and what are they used for?
> 3. How many positions are there on the input selector switch?
> 4. Which stages appear after the volume control?
> 5. Which stage provides the auxiliary output?
> 6. If the volume control develops a fault, which outputs will be affected?
> 7. A fault has occurred in which there is no speaker output but the headphone/monitor output is normal. Which stage should be investigated?
> 8. A fault has occurred in which only the microphone and auxiliary inputs produce any output. Which stage should be investigated?

ACTIVITY 2.11

Your bicycle tyre is flat and may have a puncture or may simply need re-inflating. Draw a flow chart for checking the tyre and, if necessary, repairing or replacing it. Figure 2.18 provides you with a starting point.

Flow Diagrams

Flow diagrams or flow charts are used to illustrate a sequence of events. They are used in a wide variety of applications including the planning of engineering processes and the design of computer software. Figure 2.17 shows a flow chart for the process of drilling a hole. The shape of the symbols used in this flow chart have particular meanings as shown in Figure 2.16. For the complete set of symbols and their meanings you should refer to the appropriate British Standard, BS 4058.

Schematic Diagrams

Schematic diagrams use standard symbols to show how things are connected together. There are several types of schematic diagram including those used for electrical and electronic circuits, pneumatic (compressed air), and hydraulic (compressed fluid) circuits.

Circuit diagrams are used to show the functional relationships between the components in an electric or electronic circuit. Components (such as resistors, capacitors, and inductors) are represented by symbols and the electrical connections between the components drawn using straight lines. It is important to note that the position of a component in a circuit diagram does not represent its actual physical position in the final assembly. Circuit diagrams are sometimes also referred to as *circuit schematics*.

Figure 2.19(a) shows the *circuit schematic* for a circuit using standard electronic component symbols. Figure 2.19(b) shows the corresponding physical *layout diagram* with the components positioned on the upper (component side) of a printed circuit board (PCB). Finally, Figure 2.19(c) shows the copper *track layout* for the PCB. This layout is developed photographically as an etch-resistant pattern on the copper surface of a copper-clad board.

A *wiring diagram* is another form of schematic diagram. It shows the *physical* connections between electrical and electronic components rather than the *electrical* connections between them. Schematic diagrams are also used to represent pneumatic (compressed air) circuits and hydraulic circuits. Pneumatic circuits and hydraulic circuits share the same symbols. You can tell which circuit is which because pneumatic circuits should have open arrowheads, whilst hydraulic circuits should have solid arrowheads. Also, pneumatic circuits exhaust to the atmosphere, whilst hydraulic circuits have to have a return path to an oil reservoir. Figure 2.20 shows a simple hydraulic circuit where the components are represented by standard symbols just as the electronic components were drawn in symbolic form in the circuit schematic shown in Figure 2.19(a).

Just as electrical and electronic circuit diagrams may have corresponding layout and wiring diagrams (see Fig. 2.19), so do hydraulic, pneumatic and plumbing circuits. Only this time the wiring diagram becomes a *piping diagram*. A plumbing example is shown in Figure 2.21. As you may not be familiar with the symbols, I have named them for you. Normally this is not necessary and the symbols are recognised by their shapes.

Figure 2.22 shows the layout of a typical formal drawing. A formal engineering drawing should always have a border and a title block. So that individual features of a

FIGURE 2.16
Some flow chart symbols

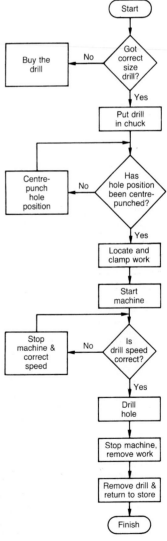

FIGURE 2.17
Flow chart for drilling a hole

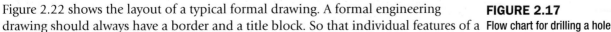

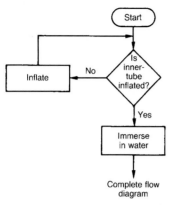

FIGURE 2.18
See Activity 2.11

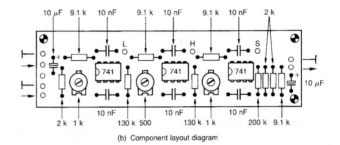

(a) Circuit diagram

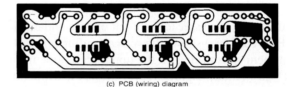

(b) Component layout diagram

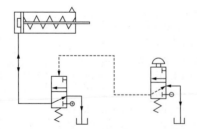

(c) PCB (wiring) diagram

FIGURE 2.19
Electronic circuit diagram together with corresponding component layout and PCB copper track layout diagrams

Key point

Schematic diagrams are used to show how components are connected together in electrical, pneumatic and hydraulic circuits. Schematic diagrams use standard symbols for components and the links between them are shown with.

Key point

Just as electrical and electronic circuit diagrams have corresponding layout and wiring diagrams, so do hydraulic, pneumatic and plumbing circuits. Physical layout diagrams show how components and parts look when they are assembled whilst the wiring and piping diagrams show how they are connected together.

FIGURE 2.20
A simple hydraulic circuit

drawing can be easily located, the border often has letters along one axis and numbers along the other. It is thus possible to identify a particular *drawing zone*, for example C4 has been shaded in Figure 2.22. To save time, formal drawing sheets are usually printed to a standard layout for use in a particular company ready for the draftsperson to add the drawing and complete the boxes and tables. A selection of formal drawing sheets are also provided as *templates* in most CAD programs.

The title block can be expanded horizontally to accommodate any written information that is required. The title block should contain:

- The drawing number (which is often repeated in the top left-hand corner of the drawing)
- The drawing name or title
- The scale used for the drawing
- The projection used (first or third angle)

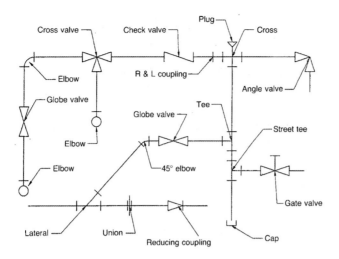

(a) Circuit diagram (schematic)

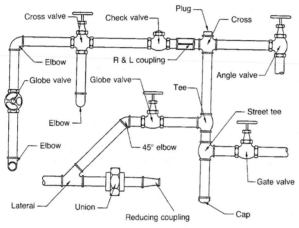

(b) Piping diagram

FIGURE 2.21
A typical plumbing circuit with corresponding piping diagram

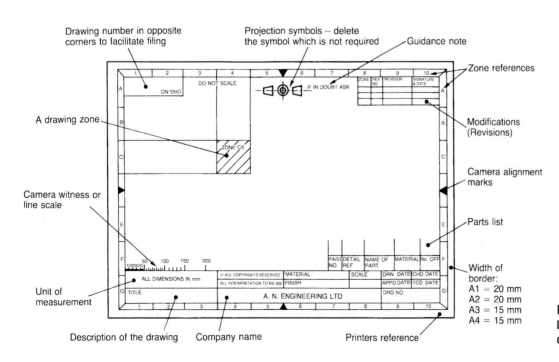

FIGURE 2.22
Layout of a typical formal drawing

53

- The name and signature of the person who made the drawing (i.e. the *originator*)
- The name and signature of the person who checks and/or approves the drawing, together with the date
- The drawing issue number and its release date
- Any other information as required.

In addition, a list of component parts may be provided (together with numbered references shown on the drawing), the materials that are to be used, the finish that is to be applied, the units used for measurement and tolerances, reference to appropriate standards (e.g. BS 308), and guidance notes (such as 'do not scale').

The scale should be stated on the drawing as a ratio. The recommended scales are as follows:

- Full size, i.e. 1:1
- Reduced scales (smaller than full size), e.g.:
  1:2 1:5 1:10 1:20 1:50 1:100 1:200 1:500 1:1,000 Don't use the words full-size, half-size, quarter-size, etc!
- Enlarged scales (larger than full size), e.g.:
  2:1 5:1 10:1 20:1 50:1 100:1

General Arrangement Drawings

Figure 2.23 shows a typical *general arrangement* (GA) drawing. This shows as many of the features listed above as are appropriate for this drawing. It shows all the components correctly assembled together. Dimensions are not usually given on GA drawings although, sometimes, overall dimensions will be given for reference when the GA drawing is of a large assembly drawn to a reduced scale.

The GA drawing shows all the parts used in an *assembly*. These are often listed in a table together with the quantities required. Manufacturers' catalogue references are also given for

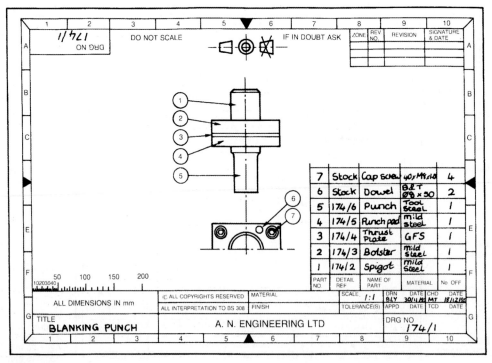

FIGURE 2.23
A General Arrangement (GA) drawing

any components that are not actually being manufactured. The parts are usually 'bought-in' as 'off-the-shelf' parts from other suppliers. The detail drawing numbers are also included for components that have to be manufactured as special items.

Detail Drawings

As the name implies, detail drawings provide all the details required to make the component shown on the drawing. If you look back at Figure 2.23 you will see from the table that the detail drawing for the punch has the reference number 174/6. Figure 2.24 shows this detail drawing. In this instance, the drawing provides the following information:

- The shape of the punch
- The dimensions of the punch and the manufacturing tolerances
- The material from which the punch is to be made and its subsequent heat treatment
- The unit of measurement (millimetre)
- The projection (first angle)
- The finish
- The guidance note 'Do not scale drawing'
- The name of the company
- The name of the draftsperson
- The name of the person checking the drawing.

It should go without saying that the amount of information given will depend upon the nature of the job. Drawings for a critical aircraft component, for example, will be much more fully detailed than those for a wheelbarrow component!

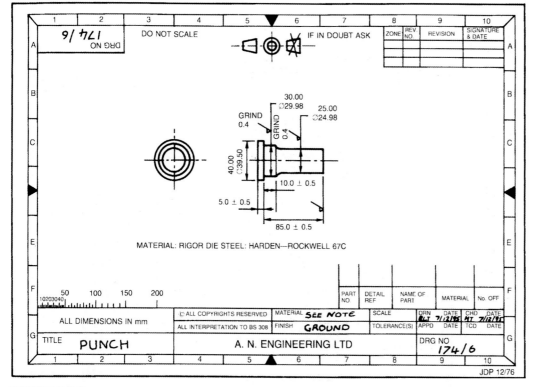

FIGURE 2.24
A typical detail drawing

Test your knowledge 2.10
Refer to the detail drawing shown in Figure 2.24.

a. What units are used for the dimensions?
b. What is the overall length of the component?
c. What is the diameter of the component?
d. What finish is applied to the component?
e. What material is used for the component?

Using CAD

Computer aided design (CAD) has now largely replaced manual methods used for producing formal engineering drawing. CAD software is used in conjunction with a computer and the drawing produced on the computer screen is saved in a computer file on disk. Networked CAD/CAM and computer aided engineering systems (CAE) have made it possible to share data and drawings over a network and also make them available to computer numerically controlled (CNC) machine tools that carry out automated manufacturing operations. Your school or college will be able to provide you with access to CAD equipment as part of your BTEC course. Figures 2.25 and 2.26 show screen shots of two popular CAD packages used in engineering.

Figure 2.25 shows a general arrangement (GA) drawing produced using AutoSketch. Some dimensions have been included. Figure 2.26 shows a 3D assembly diagram produced using AutoCAD. Drawings like this show how a complex product is assembled from its component parts. Product design involves a great deal of 3D work like this in the initial stages of designing a product before any of the parts or components are manufactured. Modern CAD packages are often linked into more complex computer aided manufacturing (CAM) systems which allow drawing data to be passed on electronically to the software and machine tools used in the manufacturing process.

Exploded Views

The final type of specialised drawing that we shall be looking at is called an *exploded view*. Quite simply, an exploded view is a pictorial representation of a product that is taken apart. By drawing the individual component parts separately but in approximately the same physical relationship as when assembled, you can gain a very good idea of how something is put together. Exploded views can be extremely useful when a product has to be serviced or maintained. A service or maintenance engineer has only to take a look at an exploded diagram to see how the various parts fit together.

Key point
Computer aided design (CAD) refers to the use of a computer to design a part and to produce engineering drawings. 2D CAD is used to show schematics, GA and detail drawings. 3D CAD is often used for assembly diagrams and in product design.

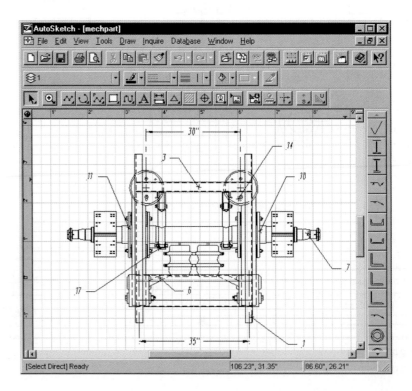

FIGURE 2.25
A GA drawing using
56 AutoSketch

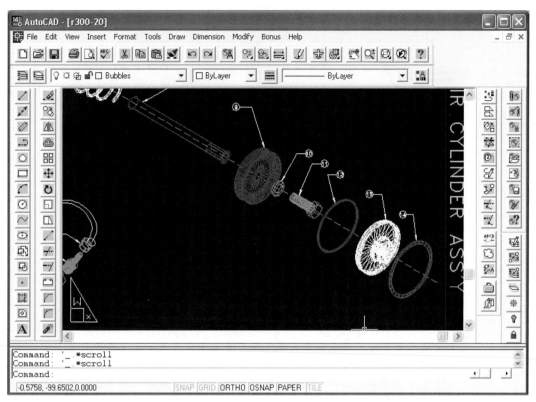

FIGURE 2.26
A 3D assembly diagram using AutoCAD

A typical exploded diagram for the tailstock assembly of a lathe is shown in Figure 2.27. Note that *part numbers* are included on the diagram.

Production Documentation

A variety of different types of document (including charts, graphs, tables and drawings) are used in conjunction with the manufacturing processes. They include production plans, job cards (or work instructions), test reports and quality control documentation.

PRODUCTION PLANS

Production plans normally describe the sequence of tasks and processes that must be carried out when manufacturing an engineered product or delivering an engineering service. Production plans are necessarily quite detailed and usually include:

- resources required (e.g. machine tools, assembly jigs, personnel)
- materials, parts and components to be used
- processes to be used
- sequence of production (and timing, where appropriate)
- arrangements for inspection and/or quality control procedures
- Health and Safety factors.

JOB CARDS AND WORK INSTRUCTIONS

Job cards and work instructions are normally used to describe a particular process within the overall sequence of manufacturing processes described in a production plan. Examples of processes that might require a job card or work instruction are balancing the wheels of a car or fitting an electrical connector to a wiring harness.

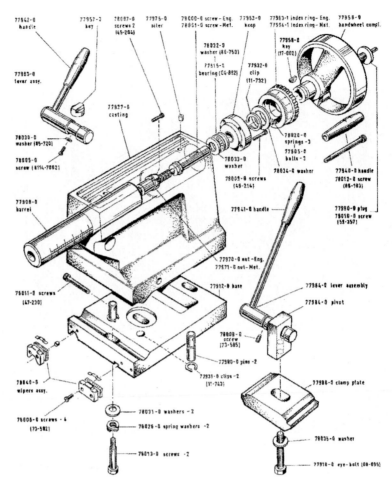

FIGURE 2.27
An exploded view of a tailstock assembly

ACTIVITY 2.12

Prepare a production plan for a simple balsawood flying model aircraft. Your production plan should consider all of the processes that are involved (e.g. marking out the balsawood sheet and cutting the individual parts) and the equipment required (e.g. modelling knife, pins, balsa cement, etc). Your production plan should also indicate the required sequence of operations and the time allowed for each.

ACTIVITY 2.13

Prepare a work instruction for balancing an assembled balsawood model aircraft (see Activity 2.12) so that the centre of gravity is correctly located. Make sure that your job card can be used by someone with no previous knowledge. Include all necessary diagrams.

Charts and Graphs

In order to make it easier to understand (or when presenting information to a non-technical audience), numerical engineering data is often presented in the form of a chart or a graph. Some typical examples are shown in Figures 2.28 and 2.32. Notice how it is easy to see a *trend* when data is presented in this way.

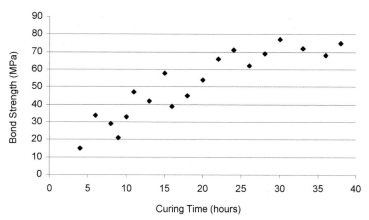

FIGURE 2.28
Engineering data represented in the form of a scatter diagram

Test your knowledge 2.11
Refer to the scatter diagram shown in Figure 2.28.

1. What was the maximum bond strength observed and at what curing time was this achieved?
2. What was the minimum bond strength observed and at what time was this achieved?
3. What appears to happen after 25 hours' curing time?

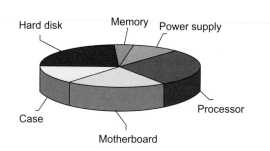

FIGURE 2.29
Engineering data represented in the form of a pie chart

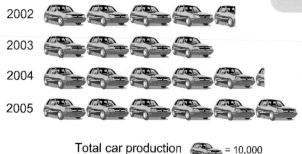

Total car production = 10,000

FIGURE 2.30
Engineering data represented in the form of a pictogram

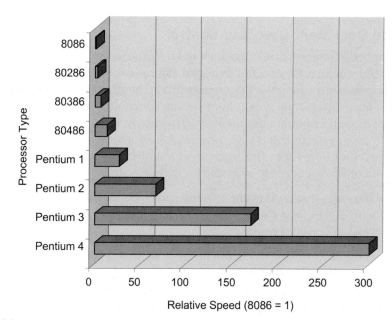

Relative Speed (8086 = 1)

FIGURE 2.31
Engineering data represented in the form of a bar chart

Test your knowledge 2.12
Refer to the pie chart shown in Figure 2.29 which shows the proportion of costs associated with the manufacture of a PC.

1. Which component costs the least?
2. Which component costs the most?
3. If the total cost of production is £200 what is the approximate cost of:
   a. the motherboard, and
   b. the case.

59

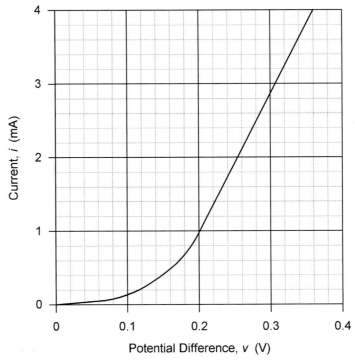

FIGURE 2.32
Engineering data represented in the form of a line graph

Tables

Numerical engineering data is frequently shown in the form of tables of data. Typical examples are those that show the dimensions of screw threads, limits and fits, welding data, tapping data, wire gauges and current capacities, etc. You will find several of these in use in your school/college workshop and you should get into the habit of using them when the need arises.

Drawing and Document Care and Control

Finally, it is important to ensure that drawings and documents are stored and used in such a way as to ensure that they are not damaged as a result of physical handling or the environment in which they are used. Documents should be stored in appropriate cabinets well away from dust, dirt, oil and grease. Some means for controlling the issue (and return!) of documents is also an important requirement, particularly where a number of people need to have access to a document. Many companies have a technical library and a person appointed to be responsible for it. This operates in a similar way to that of a public library, where books and other materials are available for loan to borrowers.

Like the formal drawings that we introduced earlier, engineering documents should be clearly marked with the originator's name, the issue number and/or the date of issue, and the date of any subsequent revisions or modifications.

Many companies are now moving to the electronic storage of drawings and documents. This has the obvious advantage that the original drawings or documents are stored in a form that is very secure and also that they can be updated very easily whenever the need arises. *Document control* is an important aspect of the operation of all large engineering companies and the use of computerised systems has made this much easier and more effective.

Material	Carbon steel	Stainless steel	Aluminium alloy	Copper	Brass	Concrete
Density (kg/m³)	7850	8000	2720	8960	8450	2400
Young's Modulus (GN/m²)	207	213	68.9	104	105	13.8
Linear Expansion Coefficient (μm/mK)	11	18	23	11.2	19	
Yield Stress (MN/m²)	230 to 460	200 to 585	30 to 280	47 to 320	62 to 430	
Ultimate Stress (MN/m²)	400 to 770	500 to 800	90 to 300	200 to 350	330 to 530	27 to 55

FIGURE 2.33
Engineering data presented in the form of a table (see Test your knowledge 2.16)

Another view
Being able to communicate information to other people quickly, clearly and accurately is an important engineering skill. As engineer you will be rarely working alone!

ACTIVITY 2.14

Danridge Engineering is a structural steelwork company involved in the design, fabrication and erection of steel framed structures. The company uses the document types listed in the table below:

Document format	Electronic file type	Application
Portable document files	PDF	Data sheets, application notes, catalogues
AutoCAD drawing	DWG and DXF	General assembly and detail drawings
Raster scanned documents	BMP	Notes, sketches and diagrams
MS Excel files	XLS	Production control data, quality control data
MS Word files	DOC and DOCX	Reports, production plans and work instructions
Web pages	HTM and HTML	On-line documentation accessible via the company's website

All of the above information is held on the company's Intranet server and is accessible to company personnel both on and off-site. The electronic system replaces an earlier paper/file based system which was both time-consuming and prone to error due to the use of outdated paper documents retained in offices and workshops. All documents are now controlled and marked with the originator's name, version number and issue date. Daily backups of all stored documents are taken by the network manager and backup copies are stored in a secure off-site location.

Danridge's goals for their electronic system were to:

1. Reduce volume of paperwork and eliminate transmittal bottlenecks
2. Have one safe searchable storage place for all company information
3. Implement version control of edited files
4. Be able to search on drawing and contract number
5. Enforce company standards for data filing and information sharing
6. Reduce paper flow for environmental cost savings
7. Eliminate duplication of outdated documents in paper files and on local hard drives
8. Improve security of data.

Explain how each of these goals has been achieved.

Test your knowledge 2.16
The table shown in Figure 2.33 refers to some common engineering materials. Use this information to answer the following questions:

1. What are the units of density?
2. Which material is the most dense?
3. Which metal has the lowest yield stress?
4. Which material has the highest ultimate stress?
5. Which material has the least coefficient of linear

Test your knowledge 2.17
Give TWO reasons why most engineering companies are moving to electronic storage of drawings and documents.

Test your knowledge 2.18
List FOUR items of information that are needed in order to control the documents used in an engineering company.

REVIEW QUESTIONS

1. Give an example of (a) a formal and (b) an informal engineering drawing. Explain how they differ.
2. The drawing in Figure 2.34 shows the parts of an engineered component. What materials are used and what processes are applied to them?

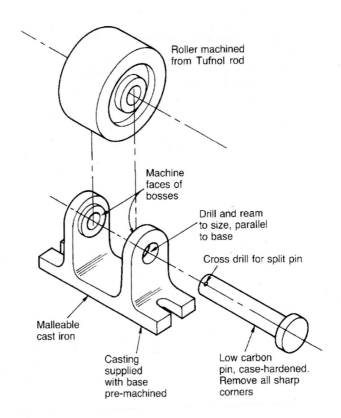

Roller machined from Tufnol rod

Machine faces of bosses

Drill and ream to size, parallel to base

Cross drill for split pin

Malleable cast iron

Casting supplied with base pre-machined

Low carbon pin, case-hardened. Remove all sharp corners

FIGURE 2.34
See Question 2

3. Explain the need for document and drawing control in a large engineering company.
4. An engineering company wishes to set up an intranet. Explain what this is and how it differs from the Internet.
5. Describe four different ways in which engineering information can be presented. Suggest a typical application for each method.
6. List the main sections used in a technical report and explain what each section is used for.
7. Figure 2.35 shows how the output of a semiconductor manufacturing plant has varied over the years from 1997 to 2001.
   a. How many logic devices were produced in 1998?
   b. In which year did the production fall?
   c. In which year was the greatest production achieved?
8. Describe the information that might appear on a typical job card for the repair of an item of consumer electronic equipment.
9. What type of drawing is shown in (a) Figure 2.36 and (b) Figure 2.37.
10. Sketch a block diagram showing the relationship between the various parts of a PC system. Your diagram should include the following components; keyboard, display, system unit, mouse, network or modem connection. Label your diagram clearly.
11. Sketch a typical flow chart to illustrate the process of transferring images from a digital camera to a PC. Label your diagram clearly.
12. List and briefly describe FIVE items of information that normally appears in the title block of a formal engineering drawing.
13. Explain THREE advantages of using CAD to generate engineering drawings.

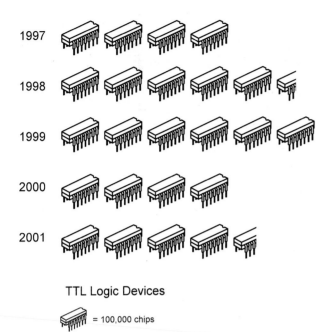

1997

1998

1999

2000

2001

TTL Logic Devices

= 100,000 chips

FIGURE 2.35
See Question 7

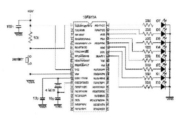

FIGURE 2.36
See Question 9

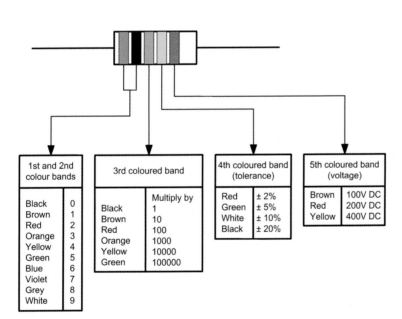

FIGURE 2.37
See Question 9

1st and 2nd colour bands		3rd coloured band		4th coloured band (tolerance)		5th coloured band (voltage)	
Black	0		Multiply by	Red	± 2%	Brown	100V DC
Brown	1	Black	1	Green	± 5%	Red	200V DC
Red	2	Brown	10	White	± 10%	Yellow	400V DC
Orange	3	Red	100	Black	± 20%		
Yellow	4	Orange	1000				
Green	5	Yellow	10000				
Blue	6	Green	100000				
Violet	7						
Grey	8						
White	9						

FIGURE 2.38
See Question 14

14. The chart shown in Figure 2.38 shows how the colour code of a tubular capacitor is interpreted. Values are in picofarad (pF).
   a. What does the fourth coloured band indicate?
   b. If the fifth colour band is yellow what does this indicate?
   c. A capacitor is marked with the following bands; yellow, violet, brown, green, red. What is the value, tolerance, and working voltage of this component?
   d. What colour coding should appear on a capacitor having a value of 2,200 pF, ±2%, 400 V d.c. working?

Mathematics for Engineering Technicians

SUMMARY

Mathematics is a tool that engineers use to solve problems. Being able to use mathematics is therefore an essential skill that you must develop. This unit will provide you with sufficient mathematical knowledge for you to solve real engineering problems that you will meet in the other core units as well as the specialist units that follow. It will also provide you with a firm foundation that will enable you to progress to the core units of the BTEC National award. In this book, we have adopted a 'topic-based' approach in which the unit has been divided into fifteen sections. Each section is designed to introduce you to a particular mathematical skill. The topics are introduced in a logical sequence and each section builds on those that have gone before. The sections deal with topics such as numbers, notation, electronic calculators, approximation, variables and constants, proportionality, laws of indices, standard form, precedence, equations, length and area, volume, graphs, trigonometry, ratios and percentages. We've also included a large number of worked examples and Test your knowledge questions to provide you with plenty of practice in using mathematics to solve typical engineering problems. Finally, it's important that you have access to an electronic calculator throughout your study of this unit. Further details can be found in Appendix 2.

3.1 ARITHMETIC, ALGEBRA AND GRAPHICAL METHODS

Let's begin with numbers. Engineers tend to use a lot of numbers for the simple reason that they are *precise*. For example, instead of saying that we need a 'large storage tank' we would convey a lot more meaning (and be much more *specific*) if we said that we need 'a tank with a capacity of 4,500 litres'. In fact, when engineers draw up a *specification* they do so using numbers and drawings in preference to written descriptions.

Whole numbers that have a positive (+) sign, such as 1, 2, 3, 4, ... , are known as *positive integers*. Negative numbers (which have a − sign), such as −1, −2, −3, −4, ... , are known as *negative integers*. Note that, if a number is positive, we don't usually include a positive (+) sign to show that it's positive. Instead, we simply assume that it's there!

The number of units that a number is from zero (regardless of its direction or sign) is known as its *absolute value*. Positive numbers are conventionally shown to the right of the *number line* (see Fig. 3.1) whilst negative numbers are shown to the left. When the sign is shown these

65

BTEC First Engineering: Mandatory and Selected Optional Units for BTEC Firsts in Engineering. DOI: 10.1016/B978-1-85617-685-9.00003-2

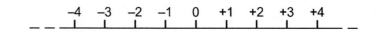

FIGURE 3.1

The number line (showing positive and negative integers)

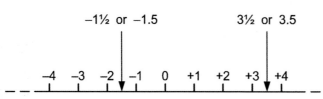

FIGURE 3.2

Decimal and fractional numbers shown on the number line

TABLE 3.1 Some common fractions and their decimal values	
Fraction	**Decimal value**
½	0.5
¼	0.25
1/8	0.125
1/16	0.0625
1/100	0.01
1/1,000	0.001

numbers are said to be *signed*. For example, the number +5 has an absolute value of 5 and its sign is + (positive). Similarly, the number −3 has an absolute value of 3 and its sign is − (negative). Note that the number zero (0) is unique in that it is neither a positive integer nor is it a negative integer.

When we are performing arithmetic using numbers we do need to be careful when we show the signs. For example, if we need to find the sum of the first three positive integers (1, 2, and 3) we would write this as follows:

Sum of first three positive integers:

$$1 + 2 + 3 = 6$$

However, if we are asked to find the sum of the first three negative integers (−1, −2, and −3) we would write:

Sum of first three negative integers:

$$(-1) + (-2) + (-3) = -1 - 2 - 3 = -6$$

Notice how we have used brackets to help to clarify the arithmetic. Brackets can be very useful as we shall see a little later on.

Because we frequently have to deal with numbers that lie between two integer numbers, integers are often not precise enough for use in engineering applications. We can get over this problem in two ways; using *fractions* and using a *decimal point*. For example, the number that sits mid-way between the positive integers 3 and 4 can be expressed as 3½ or 3.5. Similarly, the number that sits equally between −1 and −2 can be expressed as −1½ or −1.5 (see Fig. 3.2). A table of some common fractions and their corresponding decimal values is shown in Table 3.1.

Laws of Signs

There are four basic laws for using signs. They are:

FIRST LAW

To add two numbers with like signs, add their absolute values and prefix their common sign to the result.

Example: $2 + 3 = (+2) + (+3) = +5$

SECOND LAW

To add two signed numbers with unlike signs, subtract the smaller absolute number from the larger and prefix the sign of the number with the larger absolute value to the result.

$$Example: 2 + (-3) = (+2) + (-3) = -(3 - 2) = -1$$

THIRD LAW

To subtract one number from another, change the sign of the number to be subtracted and follow the rules for addition.

$$Example: 2 - 3 = (+2) - (+3) = -(3 - 2) = -1$$

FOURTH LAW

To multiply (or divide) one signed number by another, multiply (or divide) their absolute values; then, if the numbers have like signs, prefix a plus sign to the result; if they have unlike signs, prefix a minus sign to the result.

$$Example: 2 \times 3 = (+2) \times (+3) = +6$$
$$Example: 2 \times -3 = (+2) \times (-3) = -6$$
$$Example: -2 \times -3 = (-2) \times (-3) = +6$$

Units and Symbols

You will find that quite a large number of units and symbols are used in science and engineering so let's get started by introducing some of them. In fact, it's important to get to know these units and also to be able to recognize their abbreviations and symbols before you actually need to use them. Later we will go on to perform calculations and solve equations using these units but for now we will simply list some of the most common units and symbols so that at least you can begin to get to know something about them.

The units shown in Table 3.2 are called *fundamental units* (or *base units*) and they are part of the International System (known as 'SI') of units. Other units can be derived from the seven fundamental units. These are called *derived units* and a selection of some of them is shown in Table 3.3.

TABLE 3.2 Fundamental SI units			
Name	**Symbol**	**Unit**	**Abbreviation**
Mass	M	kilogram	kg
Length	L	metre	m
Time	T	second	s
Electric Current	I	Ampere	A
Temperature	θ	Kelvin	K
Amount of Substance	N	mole	mol
Luminous Intensity	J	candela	cd

Key point

When a positive number is divided by a negative number (or vice versa) the result is negative but when the two numbers are *both* negative the result is positive.

Key point

When a positive number is multiplied by a negative number the result is negative but when two negative numbers are multiplied together the result is positive.

Test your knowledge 3.1

Find the value of each of the following expressions:

a. $2 - 13$
b. $-7 + 19$
c. 5×-9
d. -11×-7
e. $15 \div -3$

Test your knowledge 3.2

Which of the following statements is correct?

a. $1 + (-1) = 0$
b. $1 - (-1) = 0$
c. $-1 \times +1 = 0$
d. $-1 \times -1 = 0$
e. $-1 + +1 = 0$

TABLE 3.3 Some common derived units

Quantity	Unit	Abbreviation	Derivation
Energy (or *work*)	Joule	J	1 J of energy (or 'work done') is used when 1 N of force moves through 1 m in the direction of the force
Force	Newton	N	Unit of force (a force of 1 N gives a mass of 1 kg an acceleration of 1 m/s^2)
Pressure	Pascal	Pa	A pressure of 1 Pa exists when a force of 1 N is exerted over an area of 1 m^2
Power	Watt	W	A power of 1 W is developed when 1 J of energy is used in 1 s
Electric Charge	Coulomb	C	An electric charge of 1 C is transferred when a current of 1 A flows for 1 s
Frequency	Hertz	Hz	A wave has a frequency of 1 Hz if one complete cycle occurs in 1 s
Velocity	metre/second	m/s (or m s^{-1})	A body travelling at a velocity of 1 m/s moves through 1 m every second in the direction of travel
Acceleration	metre/second/second	m/s/s (or m/s^2 or m s^{-2})	A body travelling with an acceleration of 1 m/s^2 increases its velocity by 1 m/s every second in the direction of travel

Key point

All derived units are derived in terms of the seven fundamental (or *base)* units.

Key point

Symbols used for electrical and other quantities are normally shown in italic font whilst units are shown in normal (non-italic) font. Thus *M* and *L* are symbols whilst J and W are units.

Key point

To multiply a number by 1,000 (one thousand) we move the decimal point three places to the right. To divide a number by 1,000 we move the decimal point three places to the left.

TABLE 3.4 Some common prefixes and multipliers

Prefix	Abbrev.	Multiplier	
tera	T	10^{12}	(=1,000,000,000,000)
giga	G	10^{9}	(=1,000,000,000)
mega	M	10^{6}	(=1,000,000)
kilo	k	10^{3}	(=1,000)
(none)	(none)	10^{0}	(=1)
centi	c	10^{-2}	(=0.01)
milli	m	10^{-3}	(=0.001)
micro	μ	10^{-6}	(=0.000,001)
nano	n	10^{-9}	(=0.000,000,001)
pico	p	10^{-12}	(=0.000,000,000,001)

Multiples and Sub-multiples

Unfortunately, the numbers that we meet in engineering can sometimes be very large or very small. For example, the voltage present at the antenna input of a VHF radio could be as little as 0.0000015 V. At the other extreme, the resistance present in an amplifier stage could be as high as 10,000,000 Ω. Having to take into account all of these zeros can be a bit of a problem but we can make life a lot easier by using a standard range of multiples and sub-multiples. These use a *prefix* letter that adds a *multiplier* to the quoted value, as shown in Table 3.4.

It is important to note that multiplying by 1,000 is equivalent to moving the decimal point *three* places to the *right*. Dividing by 1,000, on the other hand, is equivalent to moving the decimal point *three* places to the *left*. Similarly, multiplying by 1,000,000 is equivalent to moving the decimal point *six* places to the *right* whilst dividing by 1,000,000 is equivalent to moving the decimal point *six* places to the *left*.

EXAMPLE 3.1

A high speed vehicle test track has a total distance of 3.75 km. Express this in metres.

To convert from km to m we need to apply a multiplier of 10^3 or 1,000. Thus to convert 3.75 km to m we multiply 3.75 by 1,000, as follows:

$$3.75 \text{ km} = 3.75 \times 1,000 = 3,750 \text{ m}$$

EXAMPLE 3.2

An LED requires a current of 15 mA. Express this in A.

To convert mA to A, we apply a multiplier of 10^{-3} or 0.001. Thus to convert 15 mA to A we multiply 15 by 0.001, as follows:

$$15 \text{ mA} = 15 \times 0.001 = 0.015 \text{ A}$$

Note that multiplying by 0.001 is equivalent to moving the decimal point three places to the *left*.

EXAMPLE 3.3

An insulation tester produces a voltage of 2,750 V. Express this in kV.

To convert V to kV we apply a multiplier of 10^{-3} or 0.001. Thus we can convert 2,750 V to kV as follows:

$$2,750 \text{ V} = 2,750 \times 0.001 = 2.75 \text{ kV}$$

Here again, multiplying by 0.001 is equivalent to moving the decimal point three places to the *left*.

EXAMPLE 3.4

A capacitor has a value of 27,000 pF. Express this in μF.

There are 1,000,000 pF in 1 μF. Thus, to express 27,000 pF in μF we need to multiply by 0.000,001. The easiest way of doing this is simply to move the decimal point six places to the left. Hence 27,000 pF is equivalent to 0.027 μF (note that we have had to introduce an extra zero before the 2 and after the decimal point).

Unit Conversion

Unfortunately, the standard SI units are not used universally and several other types of unit are in common use. To overcome this problem we can use *conversion factors* to convert from one type of unit to another. Table 3.5 shows some useful conversion factors.

EXAMPLE 3.5

An aircraft flies at an altitude of 37,000 ft. Express this in metres.

To covert from feet (ft.) to metres (m) we need to multiply by a conversion factor (see Table 3.5) of 0.3048.

$$\text{So } 37,000 \text{ ft.} = 37,000 \times 0.3048 = 11,277.6 \text{ m}.$$

Note that we can also express this result in km by simply moving the decimal point three places to the left (i.e. 11.2776 km).

Key point

To multiply a number by 1,000,000 (one million) we move the decimal point six places to the right. To divide a number by 1,000,000 we move the decimal point six places to the left.

Test your knowledge 3.3

1. State the SI unit for frequency.
2. State the SI unit for electric charge.
3. State the SI unit for force.
4. State the abbreviation for the SI unit of acceleration.
5. State the symbol used for electric current.
6. Twenty Joules of energy are used in half a second. What power does this correspond to?
7. A military aircraft increases its speed from 250 m/s to 300 m/s in 5 s. What acceleration is this?
8. An electric current of 5 A flows for half a minute. How much charge is moved?

EXAMPLE 3.6

A race car travels at an average speed of 175 miles per hour. Express this in metres per second.

To convert from miles per hour (mph) to metres per second (m/s) we need to multiply by a conversion factor (see table) of 0.4470.

$$\text{So } 175 \text{ mph} = 175 \times 0.4470 = 78.225 \text{ m/s}.$$

TABLE 3.5 Conversion factors for some common non-SI units

From:	To:	Conversion factor (multiply by):
miles per hour (mph)	metres per second (m/s)	0.4470
metres per second (m/s)	miles per hour (mph)	2.2369
miles per hour (mph)	kilometres per hour (km/h)	1.6093
kilometres per hour (km/h)	miles per hour (mph)	0.6214
litres (l)	cubic metres (m³)	0.001
cubic metres (m³)	litres (l)	1,000
feet (ft.)	metres (m)	0.3048
metres (m)	feet (ft.)	3.2808
miles (mi.)	kilometres (km)	1.6093
kilometres (km)	miles (mi.)	0.6214
yards (yd.)	metres (m)	0.9144
metres (m)	yards (yd.)	1.0936
kilograms (kg)	tonne (metric)	0.001
tonne (metric)	kilograms (kg)	1,000
ton (UK)	tonne (metric)	1.01605
tonne (metric)	ton (UK)	0.9842
pounds per square inch (p.s.i.)	Pascal (Pa)	6894.8
Pascal (Pa)	pounds per square inch (p.s.i.)	1.4504×10^{-4}
litres (l)	gallons (gal. UK)	0.21997
gallons (gal. UK)	litres (l)	4.5461

Notation

Standard notation is used in mathematics to simplify the writing of mathematical expressions. This notation is based on the use of symbols that you will already recognise including: = (equal), + (addition), − (subtraction), × (multiplication), and ÷ (division). Other symbols that you may not be so familiar with include < (less than), > (greater than), ∝ (proportional to) and √ (square root). You need to understand what each of these symbols means and how each of them is used so we shall take a brief look at some with which you might not already be familiar, such as indices, reciprocals (or negative indices), and square roots.

INDICES

The number 4 is the same as 2×2, that is, 2 multiplied by itself. We can write (2×2) as 2^2. In words, we would call this 'two raised to the power two' or simply 'two squared'. Thus:

$$2 \times 2 = 2^2$$

By similar reasoning we can say that:

$$2 \times 2 \times 2 = 2^3$$

and

$$2 \times 2 \times 2 \times 2 = 2^4$$

In these examples, the number that we have used (i.e. 2) is known as the *base* whilst the number that we have raised it to is known as an *index*. Thus, 2^4 is called 'two to the power of four', and it consists of a base of 2 and an index of 4. Similarly, 5^3 is called 'five to the power of 3' and has a base of 5 and an index of 3.

Special names are used when the indices are 2 and 3, these being called 'squared' and 'cubed', respectively. Thus 7^2 is called 'seven squared' and 9^3 is called 'nine cubed'. When no index is shown, the power is 1, i.e. 2^1 means 2. Also, note that *any* number raised to the power 0, is 1. Hence, $2^0 = 1$, $3^0 = 1$, $4^0 = 1$, and so on.

EXAMPLE 3.7

Find the value of $2^5 + 3^3$

Now $2^5 = 2 \times 2 \times 2 \times 2 \times 2 = 32$ and $3^3 = 3 \times 3 \times 3 = 27$

So $2^5 + 3^3 = 32 + 27 = 59$

EXAMPLE 3.8

Find the value of $10^2 - 5^3$

Now $10^2 = 10 \times 10 = 100$ and $5^3 = 5 \times 5 \times 5 = 125$

So $10^2 - 5^3 = 100 - 125 = -25$

RECIPROCALS

The *reciprocal* of a number is when the index is -1 and its value is given by 1 divided by the base. Thus the reciprocal of 2 is 2^{-1} and its value is $\frac{1}{2}$ or 0.5. Similarly, the reciprocal of 4 is 4^{-1} which means $\frac{1}{4}$ or 0.25.

EXAMPLE 3.9

Find the value of $3^2 + 2^{-1}$

Now $3^2 = 3 \times 3 = 9$ and $2^{-1} = \frac{1}{2}$ or 0.5

So $3^2 + 2^{-1} = 9 + 0.5 = 9.5$

EXAMPLE 3.10

Find the value of $\dfrac{1}{2} + \dfrac{1}{4} + \dfrac{1}{8}$

Now, $\dfrac{1}{2} = 0.5, \dfrac{1}{4} = 0.25$, and $\dfrac{1}{8} = 0.25$

So $\dfrac{1}{2} + \dfrac{1}{4} + \dfrac{1}{8} = 0.5 + 0.25 + 0.125 = 0.875$

Test your knowledge 3.5

1. A military vehicle weighs 13,500 kg. Express this in metric tonnes.
2. A train travels at 120 km/h. Express this in miles per hour.
3. A fuel tank has a capacity of 1,200 UK gallons. Express this in litres.
4. A tyre is inflated to 28 pounds per square inch. Express this in Pascal.

Key point

When a number is multiplied by itself we say that the number is squared and we show this by writing the base number with an index of 2. For example, $3 \times 3 = 3^2 = 9$.

NEGATIVE INDICES

We have already said that the reciprocal of a number is the same as that number raised to the power -1. If the reciprocal happens to be a number raised to a power other than 1 then this is the same as the number raised to the same but *negative* power.

This is probably sounding a lot more complex than it really is so here are a few examples:

$$\frac{1}{2} = 2^{-1}, \frac{1}{2^2} = 2^{-2}, \text{and } \frac{1}{2^3} = 2^{-3}$$

Test your knowledge 3.6

Which of the following statements are correct:

a. $3^4 = 12$

b. $0.5^{-1} = 2$

c. $\dfrac{1}{50} = 0.02$

d. $3^3 - 2^{-1} = 8.5$

e. $\dfrac{1}{5} + \dfrac{1}{4} = \dfrac{1}{9}$

f. $0.2^{-1} + 2^{-1} = 5.5$

g. $\sqrt{10000} = 10^2$

h. $\sqrt{0.4} = 0.2$

i. $\dfrac{10^2}{10} = 10$

j. $2^0 = 2$

EXAMPLE 3.11

Find the value of 2^{-3}

Now $2^{-3} = \dfrac{1}{2^3} = \dfrac{1}{2 \times 2 \times 2} = \dfrac{1}{8} = 0.125$

SQUARE ROOTS

The *square root* of a number is when the index is ½. The square root of 2 is written as $2^{\frac{1}{2}}$ or $\sqrt{2}$. The value of a square root is the value of the base which when multiplied by itself gives the number. Since $3 \times 3 = 9$, then $\sqrt{9} = 3$. However, $(-3) \times (-3) = 9$, so we have a second possibility, i.e. $\sqrt{9} = \pm 3$. There are always two answers when finding the square root of a number and we can indicate this is by placing a $\pm$ sign in front of the result meaning 'plus or minus'. Thus:

$$4^{\frac{1}{2}} = \sqrt{4} = \pm 2$$

and

$$9^{\frac{1}{2}} = \sqrt{9} = \pm 3.$$

EXAMPLE 3.12

Find the value of $\sqrt{25}$

Now $25 = 5 \times 5$ (or -5×-5)

So $\sqrt{25} = \pm 5$

EXAMPLE 3.13

Find the value of $\sqrt{100} + \sqrt{4}$

Now $\sqrt{100} = \pm 10$ and $\sqrt{4} = \pm 2$

So $\sqrt{100} + \sqrt{4} = \pm 10 \pm 2$

Thus we have four potential answers;

$+10 +2 = 12, +10 -2 = 8, -10 +2 = -8, \text{and} -10 -2 = -12$

EXAMPLE 3.14

Find the value of $\dfrac{64^{\frac{1}{2}}}{16^{\frac{1}{2}}}$

Now $64^{\frac{1}{2}} = \sqrt{64} = 8$ and $16^{\frac{1}{2}} = \sqrt{16} = 4$

So $\dfrac{64^{\frac{1}{2}}}{16^{\frac{1}{2}}} = \dfrac{8}{4} = 2$

Electronic Calculators

You will find that an electronic calculator (see Fig. 3.3) can be extremely useful when it comes to solving engineering problems. You will need to ensure that the calculator has a full range of mathematical functions ($+, -, \div, \times, \sqrt{}$, etc) as well the ability to use engineering notation (see later). Such calculators are often referred to as *scientific calculators*. The problems that follow all require the use of a calculator.

SIGNIFICANT FIGURES

If you have tried Test your knowledge 3.7 you will probably have noticed that your answers were not *exactly* the same as mine! The reason for this is that I may have been working with more or less figures in my answer. In fact, I have not included more than five digits in any of my answers. Instead, I've *rounded up* or *rounded down* to the nearest digit in the least significant (right-most) position. The rule is that, starting with the rightmost digit, if the number in a particular digit position is 5, or greater, then 1 is added to the number in the next (more significant) position. However, if the number in a particular digit position is less than 5, then the number in the next position remains unchanged. The number is then truncated at the point at which it has the correct number of digits (figures). This sounds a lot more complicated than it really is so I've given a few examples in Table 3.6.

FIGURE 3.3
The Casio fx-83ES scientific calculator recommended for BTEC engineering students

Test your knowledge 3.7
Use your calculator to evaluate the following expressions:

a. 3.17^2

b. 0.45^{-1}

c. $2.5^2 + 4.2^{-1}$

d. $9.1^3 - 0.4^{-1}$

e. $\dfrac{1}{1.25} + \dfrac{1}{4.5} + \dfrac{1}{8.2}$

f. $0.245^{-1} - 0.55^{-2}$

g. $\sqrt{1255}$

h. $\sqrt{19.5 + 2.2^2}$

i. $\dfrac{6.75^2}{2.5^2}$

j. $\dfrac{\sqrt{226}}{3.2^2}$

Key point
Numbers are often expressed to a given number of significant figures in which case numbers are rounded up or rounded down (starting with the least significant digit position of the result) before discarding the remaining rightmost digits.

Key point

Numbers are often expressed to a given number of decimal places in which case numbers are rounded up or rounded down (starting with the least significant decimal digit position of the result) and only the correct number of digits to the right of the decimal point are retained.

TABLE 3.6 Examples of rounding up and rounding down

Number	To four significant figures	To three significant figures	To two significant figures	To one significant figure
2.3333	2.333	2.33	2.3	2
3.6666	3.667	3.67	3.7	4
1.4923	1.592	1.59	1.6	2
6.8744	6.874	6.87	6.9	7
1,638.4	1,638	1,640	1,600	2,000

Notice in this last example how we have replaced some of the original digits with zeros!

Test your knowledge 3.8

1. Express 9.81 to two significant figures.
2. Express 1.177 to three significant figures.
3. Express 26.666 to four significant figures.
4. Express 8,192 to two significant figures.
5. Express 41.575 to two decimal places.

EXAMPLE 3.15

Find, to six significant figures, the square root of 251.

Using a calculator I find that:

$$\sqrt{251} = 15.84297952$$

The first six figures of this result are 15.8429. The next (seventh) digit is 7. This is larger than 5 so we must round up the last digit of our result by adding one to it. This makes our result (to six significant figures), 15.8430.

DECIMAL PLACES

Sometimes we are more interested in the number of digits after (i.e. to the right) of the decimal point rather than the total number of digits (figures) in an answer. In this case we simply round up or round down (as before) but ensure that we have the correct number of digits to the right of the decimal point. Any number to the left of the decimal point remains unchanged.

EXAMPLE 3.16

Find, to three decimal places, the square root of 200.

Using a calculator I find that:

$$\sqrt{200} = 14.142315$$

The digits to the right of the decimal point must be reduced to just three. The fourth digit after the decimal point is 3 thus we don't need to round up the number to the left. This makes our result (to three decimal places), 14.142.

Approximation

Because precise calculations can take time (and may not always be required) engineers frequently need to make quick estimates based on *approximate values*. It's also quite useful to make a rough estimate of an answer to a problem or calculation before you arrive at a more accurate answer using a calculator. If there is a big discrepancy between the approximate result and the calculated answer you will know that something must have gone wrong and you will need to check your working again!

You may sometimes hear approximate values referred to as 'ballpark figures'. Approximation is frequently used when we don't actually need an exact figure straight away (we can always work this out later). Instead, we just need to get a 'feel' for what the exact value would be.

Let's suppose that we need to estimate the cost of a rectangular sheet of metal with sides 2.9 metres and 4.1 metres if we know that the metal costs £1.51 per square metre. To arrive at an approximate estimate of the cost we could simply round the numbers up or down and then multiply them together in order to find the area (in square metres) and then multiply the result by 1.5 (instead of 1.51) in order to arrive at an approximate cost in pounds.

Hence, estimated area of metal = 3 × 4 = 12 square metres

estimated cost = 12 × £1.5 = £18

We can do all of this by applying a little mental arithmetic—there's actually no need to use a calculator!

Now, if we needed to arrive at an *exact* value, we could use the same reasoning but enter the values into a calculator (or use long multiplication) as follows:

Actual area of metal = 2.9 × 4.1 = 11.89 square metres

estimated cost = 11.89 × £1.51 = £17.954

As you can see, our estimate of the cost was actually quite close to the real value but we got there much more quickly!

Variables and Constants

Unfortunately, we don't always know the value of a particular quantity that we need to use in a calculation. In some cases the value might actually change, in which case we refer to it as a *variable*. In other cases, the value might be fixed (and we might actually prefer not to actually quote its value). In this case we refer to the value as a *constant*.

In either case, we can use a *symbol* to represent the quantity. The symbol itself (often a single letter) is a form of shorthand notation. For example, in the case of the voltage produced by a battery we would probably use v to represent *voltage* whereas, in the case of the time taken to travel a certain distance, we might use t.

An example of a *variable* quantity is the outside temperature. On a hot summer's afternoon the temperature may well exceed 30°C whilst, on a cold winter's morning it might be as little as −4°C. An example of a *constant* quantity might be the temperature at which water freezes and becomes ice; i.e. 0°C.

Let's take another example. Figure 3.4 shows a lorry carrying several different loads. In Figure 3.4(a) the lorry is not carrying any load at all. The weight of the lorry in this condition (known as its *unladen weight)* is 4 tonnes. In Figure 3.4(b) a load of 1 tonne is being carried. If you add this to the unladen weight of the lorry the total weight of the lorry is (4 + 1) or 5 tonnes. In Figures 3.4(c), (d) and (e) loads of 2.5, 4 and 8 tonnes are being carried and this results in total weights of 6.5, 8 and 10 tonnes respectively.

It should be obvious that the unladen weight is a *constant* and the load weight is a *variable*. We can express the relationship between the unladen weight, load weight, and total weight using a simple formula, as follows:

Total weight = unladen weight + load weight

$$W_t = W_u + W_l$$

where W_t represents the total weight, W_u represents the unladen weight, and W_l represents the load weight.

Key point

Approximation is used whenever we need a rough estimate of a value. Approximation is also a good way of checking the validity of a complicated calculation using a calculator!

Test your knowledge 3.9

1. A race car travels at 205 km/h around a test circuit which has a total length of 7.9 km. Estimate (without using a calculator) how many circuits the car will complete in one hour?
2. Determine the actual number of circuits completed by the race car in Question 1.

Key point

Constants are fixed values that do not change when performing calculations. Variables are values that do change and may also rely on the values of other variable quantities. Both constants and variables can be represented by symbols.

Test your knowledge 3.10

The recommended tyre pressure for a commercial vehicle is 30 lb per square inch (psi) plus 2 psi for every 100 kg of load carried. What should the tyre pressure be when the vehicle carries a load of 750 kg?

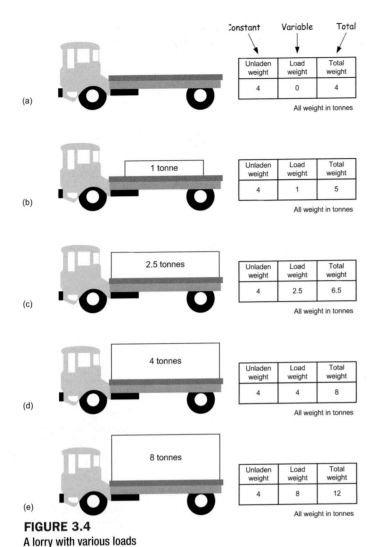

FIGURE 3.4
A lorry with various loads

(a)

Constant	Variable	Total
Unladen weight	Load weight	Total weight
4	0	4

All weight in tonnes

(b) 1 tonne

Unladen weight	Load weight	Total weight
4	1	5

All weight in tonnes

(c) 2.5 tonnes

Unladen weight	Load weight	Total weight
4	2.5	6.5

All weight in tonnes

(d) 4 tonnes

Unladen weight	Load weight	Total weight
4	4	8

All weight in tonnes

(e) 8 tonnes

Unladen weight	Load weight	Total weight
4	8	12

All weight in tonnes

The formula can be quite useful. For example, suppose the lorry manufacturer has specified a maximum total weight of 12.5 tonnes. We might want to check that we don't exceed this value. We can calculate the maximum load weight by re-arranging the formula as follows:

$$W_l = W_t - W_u$$

where $W_t = 12.5$ tonnes and $W_u = 4$ tonnes.

From which:

$$W_l = 12.5 - 4 = 8.5 \text{ tonnes}$$

Hence the maximum load that the lorry can carry (without exceeding the manufacturer's specification) is 8.5 tonnes.

Proportionality

In engineering applications, when one quantity changes it normally affects a number of other quantities. For example, if the engine speed of a car increases its road speed will also increase (see Fig. 3.5). To put this in a mathematical way we can say that:

road speed is *directly proportional* to engine speed

Using mathematical notation and symbols to represent the quantities, we would write this as follows:

$$v \propto N$$

where v represents road speed and N represents the engine speed.

In some cases, an increase in one quantity produces a *reduction* in another quantity. For example, if the road speed of a car increases the time taken for it to travel a measured distance will decrease (see Fig. 3.6). To put this in a mathematical way we would say that:

time taken to travel a measured distance is *inversely proportional* road speed.

Using mathematical notation and symbols to represent the quantities, we would write this as follows:

$$t \propto \frac{1}{v}$$

where t represents the time taken and v represents the road speed.

EXAMPLE 3.17

The current in an electric circuit is directly proportional to the voltage applied to it and inversely proportional to the resistance of the circuit. Using I to represent current, V to represent voltage, and R to represent resistance we can say that:

$$I \propto V$$

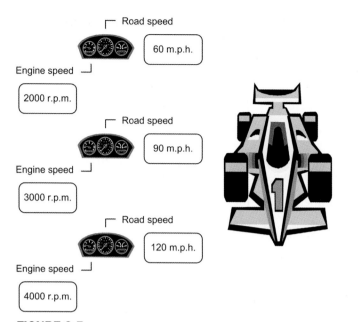

FIGURE 3.5

Relationship between engine speed and road speed

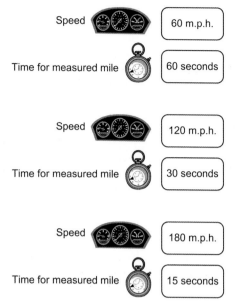

FIGURE 3.6

Relationship between road speed and time taken to travel a measured distance

(Current, *I*, is proportional to voltage, *V*)

and

$$I \propto \frac{1}{R}$$

(Current, *I*, is inversely proportional to resistance, *V*)

We can combine these two relationships to obtain an *equation* which involves all three variables, *I*, *V* and *R*:

$$I = \frac{V}{R}$$

EXAMPLE 3.18

The power delivered to a loudspeaker is proportional to the square of the voltage applied to it and inversely proportional to the impedance of the speaker. Determine the power that would be delivered to a 4 ohm (Ω) loudspeaker when connected to an amplifier that delivers 20 V.

From the information given, and using *P*, *V* and *Z* to represent power, voltage and impedance, we can obtain the following relationships:

$$P \propto V^2$$

(power, *P*, is proportional to the square of the voltage, *V*) and

$$P \propto \frac{1}{Z}$$

(power, *P*, is inversely proportional to the impedance, *Z*)

We can combine these two relationships to obtain an *equation:*

$$P = \frac{V^2}{Z}$$

Test your knowledge 3.11

1. The density of a body is directly proportional to its mass and inversely proportional to its volume. Use the symbols, ρ, *m*, and *V*, to write down an expression for density in terms of mass and volume.

2. A block of polyurethane foam has a mass of 0.155 kg and a volume of 1.2 m³. Determine the density of the alloy (in kg/m³) using the relationship that you obtain in Question 1. Express your answer correct to four decimal places.

Key point

If the value of one variable increases when the value of another variable increases we say that they are *proportional* to one another. If the value of one variable increases when the value of another variable decreases we say that they are *inversely proportional* to one another.

Key point

When a number has an index of 0 its value is 1 (regardless of the actual value of the number). Thus 2^0, 3^0, 4^0 and so on are all equal to 1.

Test your knowledge 3.12

Which of the following expressions are true?

a. $\dfrac{2^2}{2} = 2$

b. $\dfrac{1}{2^{-1}} = 2$

c. $4^{\frac{1}{2}} = 2$

d. $\dfrac{(2^2)^2}{2^3} = 2$

e. $\dfrac{2^{10} \times 2^2}{2^{11}} = 2$

Key point

When multiplying two numbers having the same base the indices are added, thus: $3^3 \times 3^2 = 3^{3+2} = 3^5$. If one number is divided by another of the same base, the indices are subtracted, thus: $3^3/3^2 = 3^{3-2} = 3^1 = 3$.

Now we know that $V = 20\,V$ and $Z = 4\,\Omega$. Replacing the symbols that we have been using by the values that we know gives:

$$P = \frac{V^2}{Z} = \frac{20^2}{4} = \frac{400}{4} = 100\ \text{W}$$

Laws of Indices

When simplifying calculations involving indices, certain basic rules or laws can be applied, called the *laws of indices*. These are listed below:

- when multiplying two or more numbers having the same base, the indices are added. Thus $2^2 \times 2^4 = 2^{2+4} = 2^6$.
- when a number is divided by a number having the same base, the indices are subtracted. Thus $2^5/2^2 = 2^{5-2} = 2^3$.
- when a number which is raised to a power is raised to a further power, the indices are multiplied. Thus $(2^5)^2 = 2^{5\times2} = 2^{10}$.
- when a number has an index of 0, its value is 1. Thus $2^0 = 1$.
- when a number is raised to a negative power, the number is the reciprocal of that number raised to a positive power. Thus $2^{-4} = 1/2^4$. Similarly, $1/2^{-3} = 2^3$.
- when a number is raised to a fractional power the denominator of the fraction is the root of the number and the numerator is the power. Thus $4^{\frac{3}{4}} = \sqrt[4]{4^2} = (2)^2 = 4$ and $25^{\frac{1}{2}} = \sqrt{25^1} = \pm5$.

EXAMPLE 3.19

Find the value of $\dfrac{3^3 \times 3^4}{3^5}$

Now $\dfrac{3^3 \times 3^4}{3^5} = \dfrac{3^{3+4}}{3^5} = \dfrac{3^7}{3^5} = 3^{7-5} = 3^2 = 9$

Standard Form

A number written with one digit to the left of the decimal point and multiplied by 10 raised to some power is said to be written in *standard form* (this is also sometimes referred to as *scientific notation*). Thus: 1234 is written as 1.234×10^3 in standard form, and 0.0456 is written as 4.56×10^{-2} in standard form.

When a number is written in standard form, the first factor is called the *mantissa* and the second factor is called the *exponent*. Thus the number 6.8×10^3 has a mantissa of 6.8 and an exponent of 10^3.

Numbers having the same exponent can be added or subtracted in standard form by adding or subtracting the mantissae and keeping the exponent the same. Thus:

$$2.3 \times 10^4 + 3.7 \times 10^4 = (2.3 + 3.7) \times 10^4 = 6.0 \times 10^4,$$

and

$$5.7 \times 10^{-2} - 4.6 \times 10^{-2} = (5.7 - 4.6) \times 10^{-2}$$
$$= 1.1 \times 10^{-2}$$

When adding or subtracting numbers it is quite acceptable to express one of the numbers in non-standard form, so that both numbers have the same exponent. This makes things much easier as the following example shows:

$$2.3 \times 10^4 + 3.7 \times 10^3 = 2.3 \times 10^4 + 0.37 \times 10^4$$
$$= (2.3 + 0.37) \times 10^4$$
$$= 2.67 \times 10^4$$

Alternatively,

$$2.3 \times 10^4 + 3.7 \times 10^3 = 23000 + 3700 = 26700 = 2.67 \times 10^4$$

The laws of indices are used when multiplying or dividing numbers given in standard form. For example,

$$(2.5 \times 10^3) \times (5 \times 10^2) = (2.5 \times 5) \times (10^{3+2})$$
$$= 12.5 \times 10^5 \text{ or } 1.25 \times 10^6$$

To round this section off, here are some examples of engineering quantities expressed in standard form:

$$221 \text{ N} = 2.21 \times 10^2 \text{ N}$$
$$0.035 \text{ V} = 3.5 \times 10^{-2} \text{ V}$$
$$454 \text{ kHz} = 4.54 \times 10^5 \text{ Hz}$$
$$65.5 \text{ μC} = 6.55 \times 10^{-5} \text{ C}$$

ENGINEERING NOTATION

Engineering notation is very similar to standard form (scientific notation) but uses a mantissa that lies in the range 1 to 999 and an index that is a multiple of three. Engineering notation is useful because it is very easy to convert to the standard multiples and sub-multiples of the standard units (which are all power of three). Here are some examples of using engineering notation:

$$3,495 \text{ m} = 3.495 \times 10^3 \text{ m} (= 3.495 \text{ km})$$
$$0.075 \text{ V} = 75 \times 10^{-3} \text{ V} (= 75 \text{ mV})$$
$$12,576 \text{ N} = 12.576 \times 10^3 \text{ N} (= 12.576 \text{ kN})$$
$$67,625,000 \text{ Hz} = 67.625 \times 10^6 \text{ Hz} (= 67.625 \text{ MHz})$$
$$0.00025 \text{ A} = 250 \times 10^{-6} \text{ A} (= 250 \text{ μA})$$

EXPONENT NOTATION

Unfortunately, some computers and electronic calculators are not well suited to the entry of numbers expressed using either standard form or using engineering notation. To overcome this problem, scientific and engineering calculators often make use of a special mode of data entry using a key which is often labelled 'EXP' and a display which uses an 'E' symbol to indicate that *exponent notation* is being used.

In exponent notation, the 'E' means 'multiply by 10 raised to a power given by the number that follows'. So, '1E1' means $1 \times 10^1 = 10$, '1E2' means $1 \times 10^2 = 100$, '1E3' means $1 \times 10^3 = 1,000$ and so on. Note that, although we can usually omit the sign of the exponent when the exponent is positive, '1E3' actually means '1E+3', '1E5' actually means '1E+5', and so on. So, instead of writing (or entering) 10,000, we can simply write '1E4' (in other words 1×10^4). Similarly, we would write '1E5' (meaning 1×10^5) for 100,000, and so on.

Test your knowledge 3.13

Evaluate the following:

a. $\dfrac{4^3}{4^4}$

b. $10^3 \times 10^2$

c. $\dfrac{6^4 \times 6^2}{6^5}$

Test your knowledge 3.14

Express the following in standard form:

a. 166 kN
b. 515 m/s
c. 377 nF
d. 0.52 μA
e. 47.5 mW
f. 0.022 MΩ.

Test your knowledge 3.15

Express the following using engineering notation:

a. 2,650 N
b. 0.525 V
c. 0.022 Ω
d. 65,000 m
e. 825,500 Hz
f. 0.00650 A.

Test your knowledge 3.16

Express the answers to Test your knowledge 3.15 using standard multiples and sub-multiples.

Test your knowledge 3.17

1. Express (a) 19,950 and (b) 0.0075 using exponent notation
2. Express (a) 1.59125E2 and (b) 1.915E − 6 using standard form

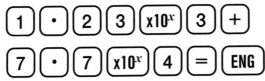

FIGURE 3.8
Key entry for Example 3.20

FIGURE 3.7
Exponent entry using the Casio fx-83ES calculator

M   □D   ▲

1.23×10З+7.7Ю4

78.23×10³

FIGURE 3.9
Calculator display for Example 3.20

Here are some examples of using exponent notation:

$$2,251 = 2.251\text{E}3$$
$$68,295 = 6.8295\text{E}4$$
$$1,577,625 = 1.577625\text{E}6$$
$$0.005491 = 5.491\text{E}{-3}$$
$$0.00001221 = 1.221\text{E}{-5}$$

The Casio *fx*-83ES calculator recommended for BTEC Engineering students does not have an exponent key but has a better method for entering exponents using a key that is labelled '$\times 10^x$'. It's worth getting into the habit of using this key in your calculations so here's an example to show you how it works:

Test your knowledge 3.18
Use your calculator to solve the following using exponents and displaying your results using engineering notation:

a. $0.6 \times 10^6 + 1.9 \times 10^5$
b. $5.1 \times 10^{-3} - 3.6 \times 10^{-2}$
c. $27 \times 10^5 \times 0.15 \times 10^{-3}$
d. $0.45 \times 10^5 \div 17 \times 10^7$.

EXAMPLE 3.20

Find the result of adding 1.23×10^3 and 7.7×10^4.

The required key entry using the *fx*-83ES is shown in Figure 3.8 and the calculator display is shown in Figure 3.9. Note how we have used the 'ENG' key in order to display the result using engineering notation—for small or large numbers it's good to get in the habit of using this notation!

Precedence

The order in which arithmetic operations (such as addition, subtraction, multiplication and division) are performed is important. Consider the following:

$$2 + 6 \times 3 - 8 \div 4$$

If the arithmetic is performed in the order that it appears in the expression (i.e. addition, multiplication, subtraction, and division) then we would arrive at the answer 4. If we carry out the arithmetic in the reverse order (i.e. division, subtraction, multiplication, and addition) this might suggest that the answer is 8. Clearly, what we need is a rule that establishes the correct answer in which arithmetic operations are performed, we call this *precedence*.

Key point
To multiply a number by 1,000,000 (one million) we move the decimal point six places to the right. To divide a number by 1,000,000 we move the decimal point six places to the left.

The rule of precedence for arithmetic operations is that they should be performed in the following order: brackets, of, division, multiplication, addition, and subtraction. The best way to remember this law is to use the acronym, BODMAS. The result of the example that we mentioned earlier would then (correctly) be 18. But how do we arrive at this?

There are no brackets in the expression (if there had been we would need to evaluate these first) so the first operation that we need to worry about is division. To make what we are

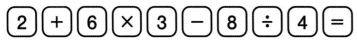

FIGURE 3.10
Key entry for the precedence example

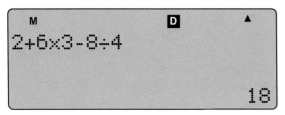

FIGURE 3.11
Calculator display for the precedence example

Test your knowledge 3.19
Without using a calculator, determine the value of:

a. $27 - 9 \times 2 + 12 \div 3$
b. $2 + 4 \div 0.5 - 1.5 \times 2$
c. $15 \div (9 - 4) + 1.5 \times 2 - 5$
d. $2 \times 0.4 - 10 \div (3.5 - 1.5)$.

doing a little more obvious we will introduce some brackets to each of the operations in turn. Now, starting with the division that we just mentioned:

$$2 + 6 \times 3 - 8 \div 4 = 2 + 6 \times 3 - (8 \div 4) \quad \text{division first}$$
$$2 + 6 \times 3 - 2 \quad = 2 + (6 \times 3) - 2 \text{ multiplication next}$$
$$2 + 18 - 2 \quad = (2 + 18) - 2 \quad \text{addition next}$$
$$20 - 2 \quad = 18 \quad \text{and finally subtraction.}$$

It might be interesting to see if your calculator produces the same result! The key entry and calculator displays for the *fx*-83ES are shown respectively in Figures 3.10 and 3.11.

Equations

When we need to understand the relationship between different quantities we often express this in the form of an *equation*. To save time and effort we write an equation using symbols rather than words, for example, instead of writing 'distance travelled is the same as speed divided by time' we would simply write:

$$s = v \times t$$

where s represents distance (in metres), v represents speed (or velocity) (in metres per second), and t represents time (in seconds).

So, if we need to find the distance travelled by a car travelling at 45 m/s in 20 s we could substitute values in the equation, using $v = 10$ and $t = 20$, as follows:

$$s = v \times t = 45 \times 20 = 900 \text{ m}$$

In fact, engineers frequently need to solve equations in order to find the value of an unknown quantity, as the following examples show:

EXAMPLE 3.21

A current of 0.5 A flows in a 56 Ω resistor. Given that $V = I R$ determine the voltage that appears across the resistor.

It's a good idea to get into the habit of writing down what you know before attempting to solve an equation. In this case:

$$I = 0.5 \text{ A, } R = 56 \, \Omega, \, V = ?$$

Now $V = I R = 0.5 \times 56 = 28 \text{ V}$

Test your knowledge 3.20
Use your calculator to check your answers to Test your knowledge 3.16.

Another view
At some stage in this Unit (and in order to meet Edexcel's assessment requirements) you will need to be able to demonstrate your proficiency with using a calculator. This includes being able to enter and solve calculations *in one single expression* (this is sometimes referred to as a *chained calculation*). It's therefore important to know how to get the best from your calculator and how to enter (and correct) complex expressions involving the use of brackets and fractions as well as the basic arithmetic operations.

> **Key point**
> Any arithmetic operation may be applied to an equation as long as the same operation is applied to both the left-hand side and the right-hand side.

EXAMPLE 3.22

Resistors of 3.9 kΩ, 5.6 kΩ and 10 kΩ are connected in parallel. Calculate the effective resistance of the circuit.

The resistance of the parallel circuit is given by the equation:

$$\frac{1}{R} = \frac{1}{R1} + \frac{1}{R2} + \frac{1}{R3}$$

We know that:

$$R1 = 3.9\ \text{k}\Omega = 3.9 \times 10^3\Omega$$
$$R2 = 5.6\ \text{k}\Omega = 5.6 \times 10^3\Omega$$
$$R3 = 10\ \text{k}\Omega = 1 \times 10^4\Omega = 10 \times 10^3\Omega$$

Hence:

$$\frac{1}{R} = \frac{1}{3.9 \times 10^3} + \frac{1}{5.6 \times 10^3} + \frac{1}{10 \times 10^3}$$
$$= \frac{10^{-3}}{3.9} + \frac{10^{-3}}{5.6} + \frac{10^{-3}}{10} = \left(\frac{1}{3.9} + \frac{1}{5.6} + \frac{1}{10}\right) \times 10^{-3}$$
$$= (0.256 + 0.179 + 0.1) \times 10^{-3} = 0.535 \times 10^{-3}$$

Now since $1/R = 0.535 \times 10^3$ we can invert both sides of the equation so that:

$$R = \frac{1}{0.535 \times 10^{-3}} = 1.87 \times 10^3 = 1.87\ \text{k}\Omega$$

Note that any arithmetic operation may be applied to an equation as long as the equality of the equation is maintained. In other words, the *same* operation *must* be applied to both the left-hand side (LHS) and the right-hand side (RHS) of the equation.

> **Another view**
> It's worth noting from Example 3.22 that we have could have used the laws of indices to simplify the powers of ten before attempting to use the calculator to determine the final result. However, in this case we've made use of the calculator's built-in exponent facility. Whichever technique you use it's important to be confident that you are correctly using the exponent notation since it's not unknown for students to produce answers that are incorrect by a factor of 1,000 or even 1,000,000 and an undetected error of this magnitude could be totally disastrous! Finally, before attempting to substitute values into an equation, it's important to be clear about what you know (and what you don't know) and always make sure that you have the correct units.

EXAMPLE 3.23

A copper wire has a length l of 1.5 km, a resistance R of 5 Ω and a resistivity ρ of 17.2×10^{-6} Ω mm. Find the cross-sectional area, a, of the wire, given that:

$$R = \frac{\rho l}{a}$$

Once again, it is worth getting into the habit of summarising what you know from the question and what you need to find (don't forget to include the units):

$$R = 5\ \Omega,\ \rho = 17.2 \times 10^{-6}\Omega\ \text{mm},\ l = 1500 \times 10^3\text{mm},\ a = ?$$

Substituting the values that we know into the equation, $R = \rho l/a$ gives:

$$5 = \frac{(17.2 \times 10^{-6}) \times (1500 \times 10^3)}{a}$$

Cross multiplying (i.e. exchanging the '5' for the 'a') gives:

$$a = \frac{(17.2 \times 10^{-6}) \times (1500 \times 10^3)}{5}$$

Note that 'cross multiplying' here produces the same as multiplying both sides by a and dividing both sides by 5. Thus we have maintained the equality of the equation. Finally, we can evaluate the result using a calculator. Using the key entry shown in Figure 3.12 we arrive at the result 5.16. Finally, note that, since we have been working in mm, the result, a, will be in mm^2. So, $a = 5.16\,\text{mm}^2$

(1 7 • 2 x10^x − 6) ×

(1 5 0 0 x10^x 3) ÷ 5 =

FIGURE 3.12
Key entry for Example 3.22

Transposition

Being able to change the subject of an equation (as we did in Example 3.23) is quite important. We call this *transposition* and it allows us to make the quantity that we are trying to find the subject of the equation. All we need to remember is that whatever we do to one side of an equation we must do to the other side, as shown in the next few examples.

■ EXAMPLE 3.24

Make t the subject of the equation, $v = u + a\,t$.

Let's start by subtracting u from both sides of the equation, thus:

$$v - u = u + a\,t - u$$

Re-grouping the terms on the right-hand side (RHS) gives:

$$v - u = u - u + a\,t$$

But $u - u = 0$ so we can reduce the RHS to:

$$v - u = 0 + a\,t = a\,t$$

Next we must divide both sides by a:

$$\frac{v - u}{a} = \frac{a\,t}{a}$$

Now $a/a = 1$ so:

$$\frac{v - u}{a} = 1 \times t = t$$

Finally, exchanging the LHS and RHS gives:

$$t = \frac{v - u}{a}$$

Note that it's not necessary to show all of the intermediate stages when working through the transposition (as we have done in this first example).

■ EXAMPLE 3.25

Make v the subject of the equation, $h = v^2/2g$

Start by multiplying both sides by *2g*, as follows:

$$h \times 2g = \frac{v^2}{2g} \times 2g$$

Now $2g/2g = 1$ so the RHS becomes just v^2, as follows:

$$h \times 2g = v^2$$

Re-arranging this expression, exchanging the LHS and RHS gives:

$$v^2 = 2gh$$

Finally, taking the square root of both sides gives:

$$\sqrt{v^2} = \sqrt{2gh}$$

Now $\sqrt{v^2} = v$ so the final version of our transposed equation is:

$$v = \sqrt{2gh}$$

Test your knowledge 3.21

1. Make C the subject of the equation:

$$X = \frac{1}{2\pi fC}$$

2. Make u the subject of the equation:

$$s = ut + \frac{at^2}{2}$$

3. Make L the subject of the equation:

$$f = \frac{1}{2\pi\sqrt{LC}}$$

Test your knowledge 3.22

1. A force of 50 N is applied to an oil seal having an area of 0.0045 m². Given that pressure is equal to force divided by area, determine the pressure acting on the seal. Express your answer in Pascal (Pa) correct to three significant figures.

2. A car travelling at 50 m/s accelerates uniformly to reach a speed of 75 m/s in a time of 6.8 seconds. Given that v = u + a t, where v is the final velocity, u is the initial velocity, a is the acceleration and t is the time, make a the subject of the equation and then determine the acceleration. Express your answer in m/s² correct to three significant figures.

ACTIVITY 3.1

A manufacturer of satellite launch vehicles has developed a new rocket motor that is capable of ejecting gas at a rate of 1,500 kg/s with an exhaust velocity of 50 km/s. The manufacturer needs to know the maximum mass of launcher plus fuel and payload that the engine is capable of lifting. The manufacturer knows that:

a. $F = v \times M_g$

where F is the thrust exerted (in Newton), v is the velocity of the expelled gas, and M_g is the mass of gas expelled every second, and

b. $M_r = F/g$

where M_r is the mass of the rocket (in kg), F is the thrust in Newton, and g is the gravitational constant (9.81 m/s^2).

Use the above information to determine the maximum mass of the rocket (including fuel and payload) that can be lifted by the rocket engine. Show all working and state any assumptions made.

Test your knowledge 3.23

1. When a batch of 20 pilot lamps are tested, 15% are found to be faulty. How many lamps are working?
2. An electronic solder consists of a 60% tin and 40% lead by weight. How much lead is present in a 2 kg reel of solder?

Ratios and Percentages

Ratio tells us how a whole number is divided into different parts. Note that we are often not concerned with the actual amount but more about the proportions into which it is divided. This may sound a whole lot more complex than it is so here are a few examples to show you how it works.

- A two-stroke engine requires a fuel mixture of 40:1. This means that one part of two-stroke oil must be mixed with every 40 parts of petrol. The ratio should be the same for any quantity of fuel.
- A gearbox has a ratio of 4 to 1. This means that, for every four turns of the input shaft (from the engine) the output shaft (to the transmission) will just make one turn. The ratio is not affected by engine speed.

Percent simply means 'per hundred'. So 10% means 'ten parts in a hundred', 20% means 'twenty parts in a hundred', and so on. Like ratios, percentages are used without reference to the actual value of the quantity concerned. Here are some typical examples of how we use percentages:

- A heat exchanger has an efficiency of 90%. This means that, for every 100 Joules of input, only 90 Joules will be usefully converted and the remaining 10 Joules will be lost.
- A manufacturer of spare parts offers a 20% discount on all orders over £200. This means that, if a purchaser places an order for £1,000 worth of spare parts, the discount will amount to £200 and the purchaser only needs to pay £800.

Graphs

Graphs provide us with a visual way of representing data. They can also be used to show, in a simple pictorial way, how one variable affects another variable. Many different types of graph are used in engineering. We shall start by looking at the most basic type, the straight line graph.

STRAIGHT LINE GRAPHS

Earlier in Example 3.17 we introduced the idea of *proportionality*. In particular, we showed that the current flowing in a circuit was directly proportional to the voltage applied to it (from Ohm's Law). We expressed this using the following mathematical notation:

$$I \propto V$$

where I represents the current and V represents output voltage.

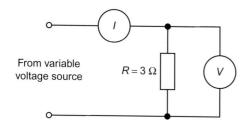

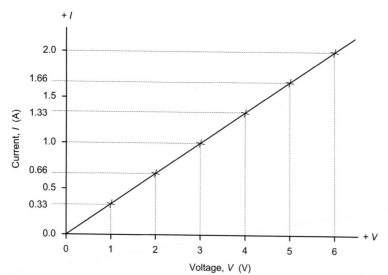

FIGURE 3.13
Circuit showing experimental set-up to obtain readings of current and voltage

FIGURE 3.14
Graph of current plotted against voltage

We can illustrate this relationship using a simple graph showing current, I, plotted against voltage, V. Let's assume that the voltage applied to the circuit is varied in 1 V steps over the range 1 V to 6 V and the circuit has a resistance of 3 Ω. By taking a set of measurements of V and I (see Fig. 3.13) we would obtain the following table of corresponding values shown below:

Voltage, V (V)	1	2	3	4	5	6
Current, I (A)	0.33	0.66	1.0	1.33	1.66	2.0

The resulting graph is shown in Figure 3.14. To obtain the graph, a point is plotted for each pair of corresponding values for V and I. When all the points have been drawn they are connected together by drawing a line. Notice that, in this case, the line that connects the points together takes the form of a straight line. This is *always* the case when two variables are directly proportional to one another.

It is conventional to show the *dependent variable* (in this case it is current, I) plotted on the vertical axis and the *independent variable* (in this case it is voltage, V) plotted on the horizontal axis. If you find these terms a little confusing, just remember that, what you know is usually plotted on the horizontal scale whilst what you don't know (and may be trying to find) is usually plotted on the vertical scale. In fact, the graph contains the same information regardless of which way round it is drawn!

EXAMPLE 3.31

The following measurements are made on an electronic component:

Temperature, θ (°C)	10	20	30	40	50	60
Resistance, R (Ω)	105	110	115	120	125	130

Plot the graph showing how resistance varies with temperature. Determine the resistance of the component at 0° C and suggest the relationship that exists between resistance and temperature.

The results of the experiment are shown plotted in graphical form in Figure 3.15. Note that the graph consists of a straight line but that it does not pass through the *origin* of the graph (i.e. the point at which θ and V are 0° C and 0 V respectively). The second most important feature to note (after having noticed that the graph is a straight line) is that, when $\theta = 0°$ C, $R = 100\ \Omega$.

By looking at the graph we could suggest a relationship (i.e. an *equation*) that will allow us to find the resistance, R, of the component at any given temperature, θ. The relationship is simply:

$$R = 100 + \frac{\theta}{2}\,\Omega$$

If you need to check that this works, just try inserting a few pairs of values from those given in the table. You should find that the equation balances every time!

STRAIGHT LINE GRAPHS

The shape of a graph is dictated by the equation that connects its two variables. For example, the general equation for a straight line (like those shown earlier) takes the form:

$$y = m\,x + c$$

where y is the *dependent variable* (plotted on the vertical or *y-axis*), x is the *independent variable* (plotted on the horizontal or *x-axis*), m is the slope (or *gradient*) of the graph and c is the *intercept* on the *y*-axis. Figure 3.16 shows this information plotted on a graph.

The values of m (the gradient) and c (the y-axis intercept) are useful when quoting the specifications for electronic components. In the case of Example 3.15, the electronic component being tested (in this case a *thermistor*) has:

- a resistance of 100 Ω at 0° C (thus $c = 100\ \Omega$)
- a characteristic that exhibits an increase in resistance of 0.5 Ω per °C (thus $m = 0.5\ \Omega/°C$).

SQUARE LAW GRAPHS

Of course, not all graphs have a straight line shape. In the previous example we saw a graph that, whilst substantially linear, became distinctly curved at one end. Many graphs are curved rather than linear. One common type of curve is the square law. To put this into context, consider the relationship between the power developed in a load resistor and the current

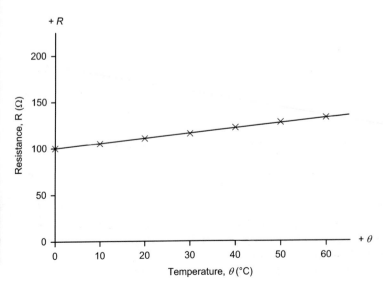

FIGURE 3.15
Graph of resistance plotted against temperature

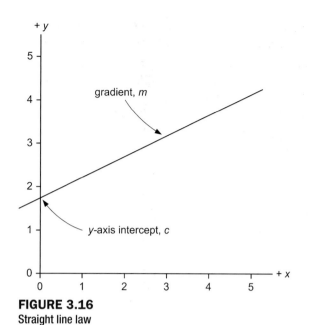

FIGURE 3.16
Straight line law

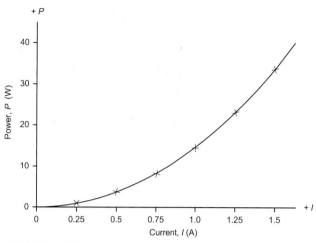

FIGURE 3.17
A square law graph

applied to it. Assuming that the load has a resistance of 15 Ω we could easily construct a table showing corresponding values of power and current, as shown below:

Current, I (A)	0.25	0.5	0.75	1	1.25	1.5
Power, P (W)	0.94	3.75	8.44	15	23.44	33.75

We can plot this information on a graph showing power, P, on the vertical axis plotted against current, I, on the horizontal axis. In this case, P is the *dependent variable* and I is the *independent variable*. The graph is shown in Figure 3.17.

It can be seen that the relationship between P and I in Figure 3.17 is far from linear. The relationship is, in fact, a *square law relationship*. We can actually deduce this from what we know about the power dissipated in a circuit and the current flowing in the circuit. You may recall that:

$$P = I^2 R$$

where P represents power in Watts, I is current in Amps, and R is resistance in ohms.

Since R remains constant, we can deduce that:

$$P \propto I^2$$

In words, we would say that 'power is proportional to current squared'. Many other examples of square law relationships are found in engineering.

MORE COMPLEX GRAPHS

Many more complex graphs exist and Figure 3.18 shows some of the most common types. Note that these graphs have all been plotted over the range $x = \pm 4$, $y = \pm 4$. Each graph consists of four quadrants. These are defined as follows (see Fig. 3.19):

First quadrant	Values of x and y are both positive
Second quadrant	Values of x are negative whilst those for y are positive
Third quadrant	Values of x and y are both negative
Fourth quadrant	Values of x are positive whilst those for y are negative.

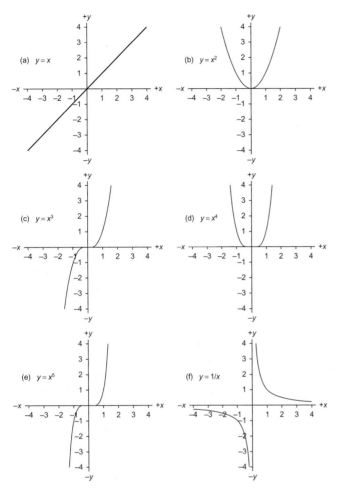

FIGURE 3.18
Some complex graphs

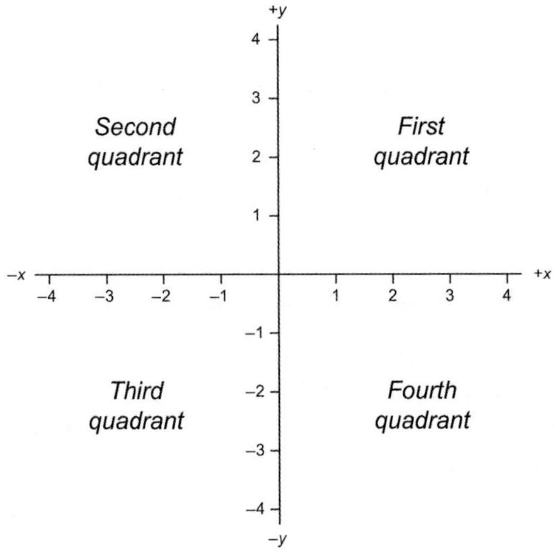

FIGURE 3.19
The four quadrants

The straight line relationship, $y = x$, is shown in Figure 3.18(a). This graph consists of a straight line with a gradient of 1 that passes through the *origin* (i.e. the point where $x = 0$ and $y = 0$). The graph has values in the first and third quadrants.

The relationship $y = x^2$ is shown in Figure 3.18(b). This graph also passes through the origin but its gradient changes, becoming steeper for larger values of x. As you can see, the graph has values in the first and second quadrants.

The graph of $y = x^3$ is shown in Figure 3.18(c). This *cubic law* graph is steeper than the square law of Figure 3.18(b) and it has values in the first and third quadrants.

Figure 3.18(d) shows the graph of $y = x^4$. This graph is even steeper than those in Figures 3.18(b) and 3.18(c). Like the square law graph of Figure 3.18(b), this graph has values in the first and second quadrants.

The graph of $y = x^5$ is shown in Figure 3.18(e). Like the cubic law graph of Figure 3.18(c), this graph has values in the first and third quadrants.

Finally, Figure 3.18(f) shows the graph of $y = 1/x$ (or $y = x^{-1}$). Note how the y values are very large for small values of x and very small for very large values of x. This graph has values in the first and third quadrants.

If you take a careful look at Figure 3.18 you should notice that, for odd powers of x (i.e. x^1, x^3, x^5, and x^{-1}) the graph will have values in the first and third quadrant whilst for even powers of x (i.e. x^2 and x^4) the graph will have values in the first and second quadrants.

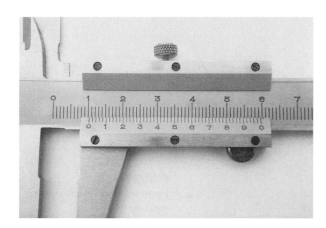

FIGURE 3.20
A vernier caliper used for accurate measurement of small components. The reading indicated is 1.04

3.2 MENSURATION AND TRIGONOMETRY

Length, area and volume are extremely important in engineering and engineers frequently have to carry out calculations involving these quantities. Measurement of length is often carried out with a rule but when we need more accurate measurements, particularly of small components, we often use a vernier caliper (see Fig. 3.20).

Area and volume can be calculated from measurements of length but, as we shall see later in this section, the way that we calculate area or volume depends on the particular shape that we are dealing with. Furthermore, recognising how more complex shapes can be made from basic shapes is a particular skill that engineers need.

Shapes

Figure 3.21 shows several common shapes which have different numbers of straight sides. Notice that the triangle has three sides, the square has four sides, the pentagon has five sides, and so on. And, although the circle does not have straight lines, it can actually be considered to be an object with an infinite number of sides of equal length!

Angular and Linear Measure

Being able to measure angles as well as length is important in many engineering applications. The essential difference between *angular measure* and *linear measure* is illustrated in Figure 3.22.

In Figure 3.22(a) we are concerned with the distance between points A and B measured along a straight line which joins the two points. In Figure 3.22(b), we are concerned with the amount of rotation between lines A and B (which can be of any length) which meet at point O.

One complete rotation, starting at point X and returning to point X in Figure 3.23(a), is equivalent to an angle of 360°. Some other angles are illustrated in Figures 3.23(b) to 3.23(d). When lines are perpendicular to one another (i.e. at right angles) the angle between them is 90°, as shown in Figure 3.24.

Area

The formulae for determining areas of various shapes are shown in Figure 3.25. Some simple rectangular shapes are shown in Figure 3.26. In Figure 3.26(a) the shape is a square measuring 1 m × 1 m. The area of the shape is thus $1 \times 1 = 1\,\text{m}^2$. In Figure 3.26(b) the

Shape	Name	Number of sides
	Triangle	3
	Square	4
	Pentagon	5
	Hexagon	6
	Septagon	7
	Octagon	8
	Nonagon	9
	Circle	infinite

FIGURE 3.21
Some common shapes

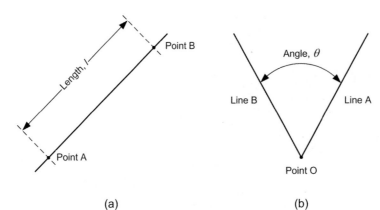

FIGURE 3.22
Linear measure (a) and angular measure (b)

(a)

(b)

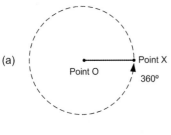

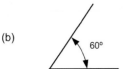

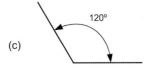

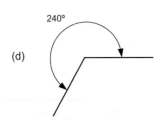

FIGURE 3.23
One complete rotation is equivalent to moving through an angle of 360°

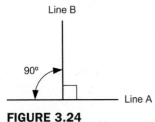

Name	Shape and dimensions	Area
Square		$A = a^2$
Rectangle		$A = a\,b$
Right-angled triangle		$A = \frac{1}{2}\,b\,h$
Circle	$r = \dfrac{d}{2}$	$A = \pi r^2$

FIGURE 3.25
Formulae for determining the area of various common engineering shapes

FIGURE 3.24
Perpendicular lines

shape is a square measuring 2 m × 2 m. The area of the shape is thus 2 × 2 = 4 m². In Figure 3.26(c) the shape is a square measuring 3 m × 3 m. The area of the shape is thus 3 × 3 = 9 m².

In Figure 3.26(d) we are dealing with a rectangle (rather than a perfect square). The rectangle has dimensions 4 m × 3 m and its area is 4 × 3 = 12 m². Finally, in Figure 3.26(e) we have a shape that can be divided into two rectangles. The dimensions of one rectangle is 2 m × 2 m whilst the other is a perfect square measuring 2 m × 2 m. The area of the shape is the sum of the areas of the two rectangles, i.e. (2 × 3) + (2 × 2) = 6 + 4 = 12 m².

Triangles

As you saw in Figure 3.25, a triangle is an object that has three sides. The sum of the interior angles of a triangle is 180°. Thus, if we know two of the angles of a triangle we can find the remaining (third) angle.

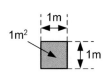

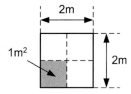

Total area = 1 x 1 = 1m²

(a)

Total area = 2 x 2 = 4m²

(b)

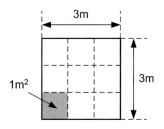

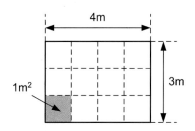

Total area = 3 x 3 = 9m²

(c)

Total area = 4 x 3 = 12m²

(d)

Total area = (2 x 3) + (2 x 2) = 6 + 4 = 10m²

(e)

FIGURE 3.26
Area of some rectangular
shapes

EXAMPLE 3.26

Determine the third angle of the triangle shown in Figure 3.27.

Now the sum of the angles in a triangle is 180°. Hence:

$$\theta + 45° + 60° = 180°$$

$$\text{So } \theta = 180° - (45° + 60°) = 180° - 105° = 75°$$

Some shapes involving triangles are shown in Figure 3.28. In Figure 3.28(a) the shape is a triangle which is half of a perfect square having sides 1 m × 1 m. The area of the triangle is thus $0.5 \times (1 \times 1) = 0.5\,\text{m}^2$.

In Figure 3.28(b) the area is made up from three perfect squares (each of area 1 m²) and one triangle having an area of 0.5 m². The total area is thus 3.5 m².

Finally, the shape in Figure 3.28(c) can be divided into a rectangle with an area of 12 m² and a triangle having an area of 3 m². The total area is thus (12 + 3) = 15 m.

Key point

The sum of the angles of a triangle is 180°. Thus, if two of the angles are known the third can easily be found.

Test your knowledge 3.27

Find the area of each shape shown in Figure 3.29.

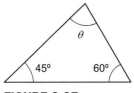

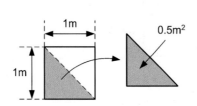

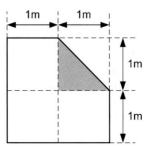

FIGURE 3.27
See Example 3.26

Area of triangle = 0.5m²

(a)

Total area = 3 + 0.5 = 3.5m²

(b)

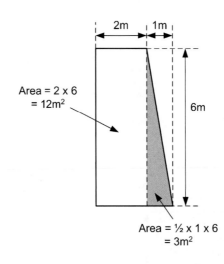

Area = 2 x 6
= 12m²

6m

Area = ½ x 1 x 6
= 3m²

Total area = 12 + 3 = 15m²

(c)

FIGURE 3.28
Some shapes involving triangles

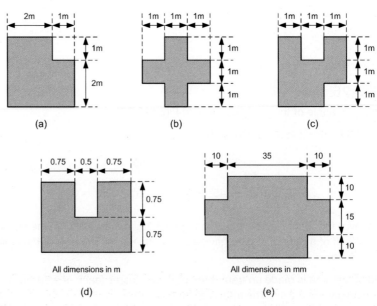

(a)

(b)

(c)

All dimensions in m

(d)

All dimensions in mm

(e)

FIGURE 3.29
See Test your knowledge 3.27

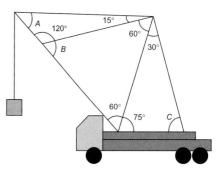

FIGURE 3.30
See Test your knowledge 3.28

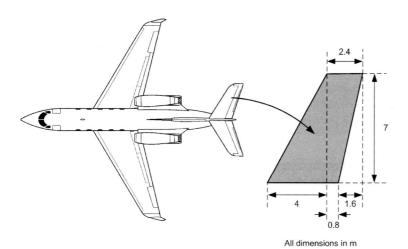

All dimensions in m

FIGURE 3.31
See Activity 3.2

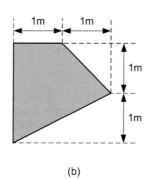

(a)   (b)

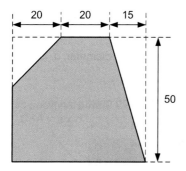

All dimensions in mm

(c)

FIGURE 3.32
See Test your knowledge 3.29

Test your knowledge 3.28
Figure 3.30 shows the arrangement used in a mobile crane. Find the three unknown angles; *A, B,* and *C.*

Test your knowledge 3.29
Find the area of each shape shown in Figure 3.32.

ACTIVITY 3.2

Figure 3.31 shows the dimensions of the tail plane of a *Gulfstream* aircraft. Determine the total area of the tail plane to three significant figures (for the purpose of this exercise you can assume that the tail plane is perfectly flat). Show all working in your answer and include notes and diagrams to show how the calculation was performed using your electronic calculator.

Test your knowledge 3.31
Determine the volume of the concrete pillar shown in Figure 3.37.

The volume can now be calculated as follows:

$$V = \pi r^2 h = 3.142 \times (12)^2 \times 120 = 54{,}293.8 \text{ mm}^3$$

To convert this to cubic metres (m³) we need to multiply by 10^{-9} as follows:

Hence $V = 54{,}293.8 \times 10^{-9} = 54.2938 \times 10^{-6} \text{ m}^3$

Trigonometry

Test your knowledge 3.32
Determine the internal volume of the equipment cabinet shown in Figure 3.38.

Trigonometrical ratios are to do with the way in which we measure angles. Take a look at the right-angled triangle shown in Figure 3.40. The angle that we are interested in (we have used the Greek symbol 6 to denote this angle) is *adjacent* to one side and *opposite* to the other side. The third side (called the *hypotenuse*) is the longest side of the triangle.

PYTHAGORAS THEOREM

In Figure 3.41, the theorem of Pythagoras states that, for a right-angled triangle, 'the square on the hypotenuse is equal to the sum of the squares on the other two sides'. Writing this as an equation we arrive at:

Test your knowledge 3.33
Determine the volume of metal used to make the right-angled flange shown in Figure 3.39.

$$c^2 = a^2 + b^2$$

where c is the hypotenuse and a and b are the other two sides.

Taking square roots of both sides of the equation we can see that:

$$c = \sqrt{a^2 + b^2}$$

Thus if we know two of the sides (for example, a and b) of a right angled triangle we can easily find the third side (c).

Test your knowledge 3.34
A vertical cylindrical gas storage tank has two hemispherical ends (i.e. each end is half of a sphere). Determine the volume of the tank if it has a total height of 3 m and a diameter of 1 m.

EXAMPLE 3.32

A triangular brace made from sheet metal is shown in Figure 3.42. Determine the length of the third side.

From Figure 3.42 the two perpendicular sides of the triangle have lengths of 50 mm and 120 mm. The remaining side can be calculated from:

$$c = \sqrt{50^2 + 120^2} = \sqrt{2500 + 14400} = \sqrt{16900} = 130 \text{ mm}$$

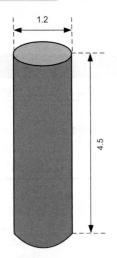

All dimensions in m

FIGURE 3.37
See Test your knowledge 3.31

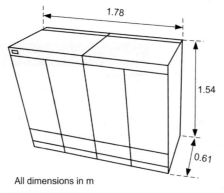

All dimensions in m

FIGURE 3.38
See Test your knowledge 3.32

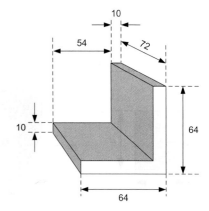

All dimensions in mm

FIGURE 3.39
See Test your knowledge 3.33

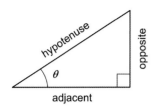

FIGURE 3.40
A right-angled triangle

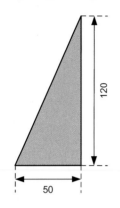

All dimensions in mm

FIGURE 3.42
See Example 3.32

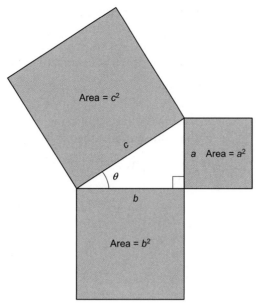

FIGURE 3.41
Pythagoras' theorem

TRIGONOMETRICAL RATIOS

The ratios a/c, b/c, and a/b are known as the basic *trigonometric ratios.* They are known as sine *(sin)*, cosine *(cos)* and tangent *(tan)* of angle θ respectively (see Fig. 3.40). Thus:

$$\sin\theta = \frac{opposite}{hypotenuse} = \frac{a}{c}, \quad \cos\theta = \frac{adjacent}{hypotenuse} = \frac{b}{c},$$

$$\text{and } \tan\theta = \frac{opposite}{adjacent} = \frac{a}{b}.$$

TRIGONOMETRICAL EQUATIONS

Equations that involve trigonometrical expressions are known as trigonometrical equations. Fortunately they are not quite so difficult to understand as they sound! Consider the equation:

$$\sin\theta = 0.5$$

This equation can be solved quite easily using a calculator. However, before doing so, you need to be sure to select the correct mode for expressing angles on your calculator. If you are using a 'scientific calculator' you will find that you can set the angular mode to either *radian* measure or *degrees.* A little later we will explain the difference between these two angular measures but for the time being we shall just use degrees.

If you solve the equation (by keying in 0.5 and pressing the *inverse sine* function keys) you should see the result 30° displayed on your calculator. Hence we can conclude that:

$$\sin 30° = 0.5$$

Actually, a number of other angles will give the same result! Try pressing the sine function key and entering the following angles in turn:

$$30°, 210°, 390° \text{ and } 570°$$

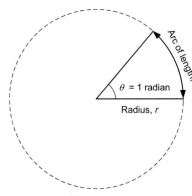

FIGURE 3.43
Definition of the radian

They should all produce the same result, 0.5! This should suggest to you that the graph of the sine function repeats itself (i.e. the shape of the graph is *periodic*). In the next section we shall plot the sine function but, before we do we shall take a look at using radian measure to specify angles.

RADIANS

The *radian* is defined as the angle subtended by an arc of a circle equal in length to the radius of the circle. This relationship is illustrated in Figure 3.43.

The circumference, l, of a circle is related to its radius, r, according to the formula:

$$l = 2\pi r$$

Thus

$$r = \frac{l}{2\pi}$$

Now, since there are 360° in one complete revolution we can deduce that one radian is the same as $360°/2\pi = 57.3°$. On other words, to convert:

● degrees to radians divide by 57.3
● radians to degrees multiply by 57.3

It is important to note that one complete cycle of a periodic function (i.e. a waveform) occurs in a time, T. This is known as the *periodic time* or just the *period*. In a time interval equal to T, the angle will have changed by 360°.

The relationship between time and angle expressed in degrees is thus:

$$\theta = \frac{T}{t} \times 360° \quad \text{and} \quad t = \frac{T}{\theta} \times 360°$$

Thus, if one complete cycle (360°) is completed in 0.02 s (i.e. $T = 20$ ms) an angle of 180° will correspond to a time of 0.01 s (i.e. $t = 10$ ms).

Conversely, if we wish to express angles in radians:

$$\theta = \frac{T}{t} \times 2\pi \quad \text{and} \quad t = \frac{T}{\theta} \times 2\pi$$

Thus, if one complete cycle (2π radians) is completed in 0.02 s (i.e. $T = 20$ ms) an angle of π radians will correspond to a time of 0.01 (i.e. $t = 10$ ms).

Graphs of Trigonometrical Functions

To plot a graph of $y = \sin \theta$ we can construct a table of values of $\sin \theta$ as θ is varied from 0° to 360° in suitable steps (say, every 30°). This exercise (carried out using a scientific calculator) will produce a table that looks something like this:

Angle, θ	0°	30°	60°	90°	120°	150°	180°
$\sin \theta$	0	0.5	0.866	1	0.866	0.5	0
Angle, θ	210°	30°	60°	90°	120°	150°	
$\sin \theta$	−0.5	−0.866	−1	−0.866	−0.5	0	

Plotting the values in the table reveals the graph shown in Figure 3.44. We can use the same technique to produce graphs of $\cos \theta$ and $\tan \theta$, as shown in Figures 3.45 and 3.46.

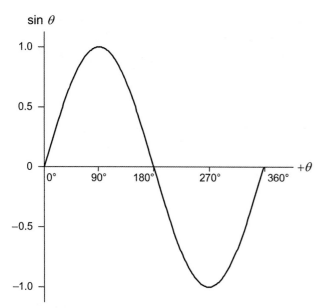

FIGURE 3.44
Graph of $y = \sin \theta$

cos θ

FIGURE 3.45
Graph of $y = \cos \theta$

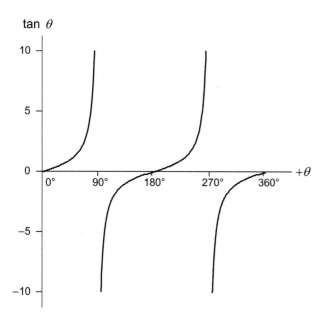

FIGURE 3.46
Graph of $y = \tan \theta$

Test your knowledge 3.39
The following data was obtained from measurements made on a crank arm:

Angle, θ	Distance, d (m)
0°	0
30°	1.1
60°	1.9
90°	2.2
120°	1.9
150°	1.1
180°	0
210°	−1.1
240°	−1.9
270°	−2.2
300°	−1.9
330°	−1.1
360°	0

Plot a graph showing how distance varies with angle and use it to find the values of θ that correspond to a distance of 1 m. Label your axes clearly. Also determine the maximum positive and negative displacements of the crank.

REVIEW QUESTIONS

1. A wire has a diameter of 0.0875 mm. Express this in metres using engineering notation.
2. Evaluate the following expression correct to three significant figures:

$$\frac{1}{2} + \frac{1}{3} + \frac{1}{9}$$

3. Given that $F = ma$, find F when $m = 22.5$ kg and $a = 1.75$ m/s.
4. Make I the subject of the equation $P = I^2R$.
5. Make u the subject of the equation $v^2 = u^2 + 2as$.
6. Make h the subject of the equation $v = \sqrt{2gh}$.

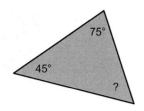

FIGURE 3.47
See Question 12

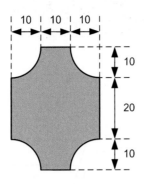

All dimensions in mm
FIGURE 3.48
See Question 13

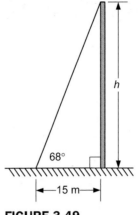

FIGURE 3.49
See Question 14

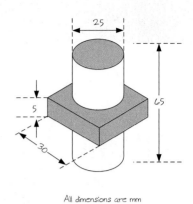

All dimensions are mm
FIGURE 3.50
See Question 15

7. A radio beacon operates on a wavelength of 6 m. Determine the frequency of the station given that $v = f \lambda$ where v is the speed of light (3×10^8 m/s), f is the frequency (in Hz), and λ is the wavelength (in m).

8. If $R = \dfrac{\rho \, l}{A}$, find ρ if $R = 2\,\Omega$, $l = 0.8 \times 10^3$ m, $a = 4.4 \times 10^{-6}$ m^2.

9. Determine the value of $\sin 60° \times (1 - \cos 60°)$.

10. Without using your calculator evaluate the following:
   a. $3 + 7 \times 2 - 35 \div 5$
   b. $144 \div 12 + 7 - 3 \times 3$

11. Use your calculator to evaluate, to three significant figures, the values of:
   a. $\sqrt{52} + 14^2$
   b. $\dfrac{22}{7} - \dfrac{1}{3}$
   c. $\dfrac{\sqrt{3}}{2} - 2\sqrt{2}$
   d. $4 + \left(1 + \dfrac{5}{9}\right)^3$

12. Find the unknown angle in Figure 3.47.

13. Find the area of the metal plate shown in Figure 3.48.

14. Determine the height of the aerial mast shown in Figure 3.49.

15. Figure 3.50hows a sketch of a cast metal component. Determine, correct to three significant figures, the total volume of metal in the component (express your answer in cubic mm).

16. A fuel tank is 70% full. If the tank has a total capacity of 1,100 litres, determine the amount of additional fuel required to fill the tank.

17. A right-angled triangle has two perpendicular sides measuring 150 mm and 200 mm. Determine the length of the third side and the area of the triangle.

18. The following data was obtained in an experiment:

Time, t (s)	0	1	2	3	4	5
Velocity, v (m/s)	0	2.5	10	22.5	40	62.5

Plot a graph showing how velocity, v, varies with time, t. Label the axes clearly and use the graph to determine:

   a. the velocity when $t = 2.5$ s, and
   b. the time when $v = 50$ m/s.

Applied Electrical and Mechanical Science

SUMMARY

Science underpins all aspects of engineering and this unit will provide you with an understanding of the principles that form the basis of engineering systems. The unit includes definitions of some important electrical and mechanical quantities and the units used for measurement. The unit will also introduce you to essential concepts such as power being the rate at which work is done and electric current being the rate at which charge is conveyed. Fundamental concepts like these underpin the operation of all engineering systems and it's important that you should have a good grasp of them at an early stage in your career as an engineer. The unit is doubly important because it will provide you with the essential knowledge for the specialist units that follow. Without a good understanding of basic scientific principles it can be very difficult to make good progress with these units.

When you complete this unit you should be able to define and apply concepts and principles relating to both electrical and mechanical science. There are many parallels between these two disciplines and many scientific concepts apply equally to both. Finally, the unit will provide you with plenty of scope for practical work and measurements carried out in the laboratory or engineering workshop.

4.1 ELECTRICAL SCIENCE

Electrical and electronic components and systems are used in a huge range of engineered products. An aircraft, for example, would simply not get off the ground without the electrical and electronic systems that maintain the flight instruments, operate the flight controls, manage the engines, provide navigation and communication with the ground. The term 'fly by wire' provides us with a clue as to just how important the role of electricity is in the operation of a modern aircraft!

Conductors and Insulators

Electric current is the name given to the flow of electrons (or negative charge carriers). Electrons orbit around the nucleus of atoms just as the earth orbits around the sun (see Fig. 4.1). Electrons are held in one or more shells, constrained to their orbital paths by virtue of a

BTEC First Engineering: Mandatory and Selected Optional Units for BTEC Firsts in Engineering. DOI: 10.1016/B978-1-85617-685-9.00004-4

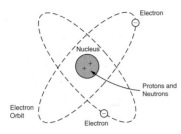

FIGURE 4.1

A single atom of helium (H_e) showing its two electrons in orbit around its nucleus

Key point

Charges with the same polarity (i.e. both positive or both negative) will repel one another whilst charges with opposite polarity (i.e. one positive and the other positive) will attract one another.

Key point

Current is the rate of flow of charge. Thus, if more charge moves in a given time, more current will be flowing. If no charge moves then no current is flowing.

Key point

A current of one Ampere (1 A) is equal to one coulomb (1 C) per second.

force of attraction towards the nucleus which contains an equal number of protons (positive charge carriers).

Since like charges repel and unlike charges attract, negatively charged electrons are attracted to the positively charged nucleus. A similar principle can be demonstrated by observing the attraction between two permanent magnets; the two North poles of the magnets will repel each other, while a North and South pole will attract. In the same way, the unlike charges of the negative electron and the positive proton experience a force of mutual attraction.

The outer shell electrons of a conductor can be reasonably easily interchanged between adjacent atoms within the lattice of atoms of which the substance is composed. This makes it possible for the material to conduct electricity. Typical examples of conductors are metals such as copper, silver, iron and aluminium. By contrast, the outer shell electrons of an insulator are firmly bound to their parent atoms and virtually no interchange of electrons is possible. Typical examples of insulators are plastics, rubber and ceramic materials.

Electric Charge and Current

All electrons and protons carry a tiny electric *charge* but its value is so small that a more convenient unit of charge is used called the *Coulomb*. One coulomb is a total charge, Q, equivalent to that of 6.21×10^{18} electrons. In other words, a single electron has a charge of 1.61×10^{-19} C!

Current, I, is defined as the rate of flow of charge and its unit is the Ampere, A. Hence:

$$I = \frac{Q}{t}$$

where Q is in Coulombs and t is in seconds.

So, for example: if a steady current of 3 A flows for two minutes, then the amount of charge transferred will be:

$$Q = I \times t = 3\,A \times 120\,s = 360 \text{ Coulombs}$$

Voltage and Resistance

The ability of an energy source (e.g. a battery) to produce a current within a conductor may be expressed in terms of electromotive force (e.m.f.). Whenever an e.m.f. is applied to a circuit a potential difference (p.d.) exists. Both e.m.f. and p.d. are measured in volts (V).

In many practical circuits there is only one e.m.f. present (the battery or supply) whereas a p.d. will be developed across *each* component present in the circuit.

The conventional flow of current in a circuit is from the point of more positive potential to the point of greatest negative potential (note that electrons move in the *opposite* direction!). Direct current results from the application of a direct e.m.f. (derived from batteries or a d.c. power supply). An essential characteristic of these supplies is that the applied e.m.f. does not change its polarity (even though its value might be subject to some fluctuation).

For any conductor, the current flowing is directly proportional to the e.m.f. applied. The current flowing will also be dependent on the physical dimensions (length and cross-sectional area) and material of which the conductor is composed.

The amount of current that will flow in a conductor when a given e.m.f. is applied is inversely proportional to its resistance. Resistance, therefore, may be thought of as an

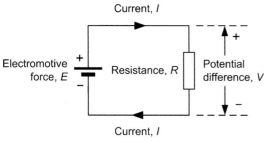

FIGURE 4.2
A simple circuit to illustrate the relationship between voltage (*V*), current (*I*) and resistance (*R*). Note that the direction of conventional current flow is from positive to negative

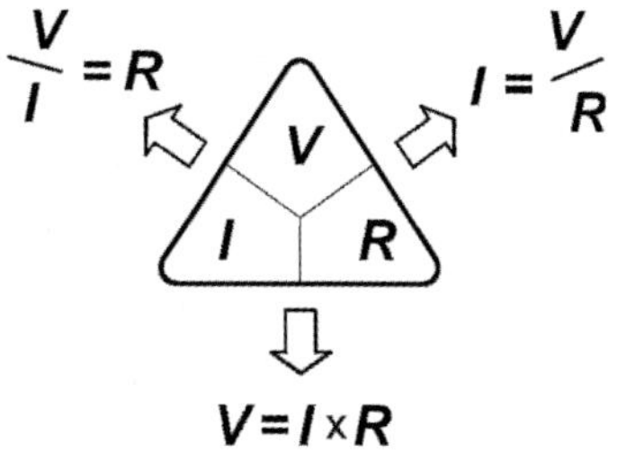

$$\frac{V}{I} = R \qquad I = \frac{V}{R}$$

$$V = I \times R$$

FIGURE 4.3
Triangle showing the relationship between *V*, *I* and *R*

opposition to current flow; the higher the resistance the lower the current that will flow (assuming that the applied e.m.f. remains constant).

Ohm's Law

Provided that temperature does not vary, the ratio of p.d. across the ends of a conductor to the current flowing in the conductor is a constant. This relationship is known as Ohm's Law and it leads to the relationship:

$$\frac{V}{I} = \text{a constant} = R$$

where *V* is the potential difference (or voltage drop) in Volts (V), *I* is the current in Amperes (A), and R is the resistance in Ohms (Ω) (see Fig. 4.2).

The formula may be arranged to make *V*, *I* or *R* the subject, as follows:

$$V = I \times R, \; I = \frac{V}{R} \text{ and } R = \frac{V}{I}$$

The triangle shown in Figure 4.3 should help you remember these three important relationships. However, it's worth noting that, when performing calculations of currents, voltages and resistances in practical circuits it is seldom necessary to work with an accuracy of better than $\pm 1\%$ simply because component tolerances are usually greater than this. For example, a resistor marked $100\,\Omega$ having a tolerance of 5% can have a value of anything between $95\,\Omega$ and $105\,\Omega$.

In calculations involving Ohm's Law, it can sometimes be convenient to work in units of $k\Omega$ and mA (or $M\Omega$ and μA) in which case potential differences will be expressed directly in V. Note that 1,000 mA is the same as 1 A, $1,000\,\Omega$ is the same as $1\,k\Omega$, 0.001 A is the same as 1 mA, 0.001 V is the same as 1 mV, and so on.

EXAMPLE 4.1

A $12\,\Omega$ resistor is connected to a 6 V battery. What current will flow in the resistor?

Here we must use the relationship, *I* = *V/R* (where *V* = 6 V and *R* = $12\,\Omega$):

$$I = \frac{V}{R} = \frac{6}{12} = 0.5 \text{ A or } 500\,\text{mA}$$

Hence a current of 500 mA will flow in the resistor.

Test your knowledge 4.3
A voltage drop of 15V appears across a resistor when a current of 400 mA flows in it. What is the value of the resistor?

EXAMPLE 4.2

A current of 100mA flows in a 56 Ω resistor. What voltage drop (potential difference) will be developed across the resistor?

Here we must use $V = I \times R$ and ensure that we work in units of Volts (V), Amperes (A) and Ohms (Ω).

$$V = I \times R = 0.1\,A \times 56\,\Omega = 5.6\,V$$

Note that 100mA is the same as 0.1A and this calculation shows that a p.d. of 5.6V will be developed across the resistor.

Test your knowledge 4.4
A power supply is designed to supply an output of 15V, 0.75A. What value of load resistor would be required to test the power supply at its full rated output?

EXAMPLE 4.3

A voltage drop of 15V appears across a resistor in which a current of 1mA flows. What is the value of the resistance?

$$R = \frac{V}{I} = \frac{15}{1 \times 10^{-3}} = 15 \times 10^3 = 15k\Omega$$

Note that if you work in units of mA and V the calculation will produce an answer in kΩ, i.e.

$$R = \frac{V}{I} = \frac{15}{1} = 15mA$$

Resistance and Resistivity

The resistance of a metal conductor is directly proportional to its length and inversely proportional to its area. The resistance is also directly proportional to the resistivity (or specific resistance) of the material. Resistivity is defined as the resistance measured between the opposite faces of a cube having sides of 1 cm.

The resistance, R, of a conductor is thus given by the formula:

$$R = \frac{\rho l}{A}$$

where R is the resistance (ft), ρ is the resistivity (Ωm), l is the length (m), and A is the area (m^2).

Table 4.1 shows the electrical properties of some common metals. Note that we have included a column for 'relative conductivity'. This shows how good a material is relative to copper.

TABLE 4.1 Properties of some common metals

Metal	Resistivity (at 20°C) (Ωm)	Relative conductivity (copper = 1)	Temperature coefficient of resistance (per °C)
Silver	1.626×10^{-8}	1.06	0.0041
Copper (annealed)	1.724×10^{-8}	1.00	0.0039
Copper (hard drawn)	1.777×10^{-8}	0.97	0.0039
Aluminium	2.803×10^{-8}	0.61	0.0040
Mild steel	1.38×10^{-7}	0.12	0.0045
Lead	2.14×10^{-7}	0.08	0.0040
Nickel	8.0×10^{-8}	0.22	0.0062

EXAMPLE 4.4

A coil consists of an 8m length of annealed copper wire having a cross-sectional area of l mm². Determine the resistance of the coil.

We will find the resistance by using the formula,

$$R = \frac{\rho l}{A}$$

The value of ρ for annealed copper given in Table 4.1 (above) is $1.724 \times 10^{-8}\,\Omega m$.

The length of the wire, l, is 4 m while the area, A, is 1 mm² or 1×10^{-6} m² (note that it is important to be consistent in using units of metres for length and square metres for area).

Hence the resistance of the coil will be given by:

$$R = \frac{1.724 \times 10^{-8} \times 8}{1 \times 10^{-6}} = 13.724 \times 10^{(-8+6)}$$

Thus $R = 13.792 \times 10^{-2}$ or $0.13792\,\Omega$

EXAMPLE 4.5

A wire having a resistivity of $1.724 \times 10^{-8}\,\Omega m$, length 20 m and cross-sectional area 1 mm² carries a current of 5 A. Determine the voltage drop between the ends of the wire.

First we must find the resistance of the wire (as in Example 4.4):

$$R = \frac{\rho l}{A} = \frac{1.6 \times 10^{-8} \times 20}{1 \times 10^{-6}} = 32 \times 10^{-2} = 0.32\,\Omega$$

The voltage drop can now be calculated using Ohm's Law:

$$V = I \times R = 5A \times 0.32\,\Omega = 1.6\,V$$

This calculation shows that a potential difference of 1.6 V will be dropped between the ends of the wire.

Energy and Power

At first you may be a little confused about the difference between energy and power. Put simply, energy is the ability to do work while power is the rate at which work is done. In electrical circuits, energy is supplied by batteries or generators. It may also be stored in components such as capacitors and inductors. Electrical energy is converted into various other forms of energy by components such as resistors (producing heat), loudspeakers (producing sound) and light emitting diodes (producing light).

The unit of energy is the Joule (J). Power is the rate of use of energy and it is measured in Watts (W). A power of 1 W results from energy being used at the rate of 1 J per second. Thus:

$$P = \frac{W}{t}$$

where P is the power in Watts (W), W is the energy in Joules (J), and t is the time in seconds (s).

The power in a circuit is equivalent to the product of voltage and current. Hence:

$$P = I \times V$$

where P is the power in Watts (W), I is the current in Amperes (A), and V is the voltage in Volts (V). The formula may be arranged to make P, I or V the subject, as follows:

$$P = I \times P, I = \frac{P}{V} \text{ and } V = \frac{P}{I}$$

Key point

Metals, like copper and silver are good conductors of electricity. Good conductors have low resistance whilst poor conductors have high resistance.

Another view

Potential difference and voltage are essentially the same thing and engineers tend to use these two terms interchangeably. Even though we also measure it in terms of 'volts', electromotive force is a little different. You can think of this as the 'force' that causes current to move around a circuit, producing potential differences (or voltage drops) across the resistances that are present in the circuit.

Test your knowledge 4.5

A wirewound resistor is made from a 4 m length of aluminium wire ($\rho = 2.18 \times 10^{-8}\,\Omega m$). Determine the resistance of the wire if it has a cross-sectional area of 0.2 mm².

Test your knowledge 4.6

A current of 2 A flows in a 22 Ω resistor. What power is dissipated in the resistor?

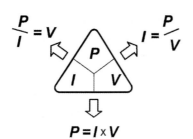

$$P = I \times V$$

FIGURE 4.4
Triangle showing the relationship between P, I and V

The triangle shown in Figure 4.4 should help you remember these relationships. The relationship, $P = I \times V$, may be combined with that which results from Ohm's Law ($V = I \times R$) to produce two further relationships. First, substituting for V gives:

$$P = I \times (I \times R) = I^2R$$

Secondly, substituting for I gives:

$$P = \left(\frac{V}{R}\right) \times V = \frac{V^2}{R}$$

EXAMPLE 4.6

A current of 1.5 A is drawn from a 3 V battery. What power is supplied?

Here we must use $P = I \times V$ (where I = 1.5 A and V = 3 V).

$$P = I \times V = 1.5 \times 3 = 4.5\,W$$

Hence a power of 4.5 W is supplied.

EXAMPLE 4.7

A voltage drop of 4 V appears across a resistor of 100 Ω. What power is dissipated in the resistor?

Here we use $P = V^2/R$ (where V = 4 V and R = 100 Ω).

$$P = \frac{V^2}{R} = \frac{4^2}{100} = 0.16\,W$$

Hence the resistor dissipates a power of 0.16 W (or 160 mW).

EXAMPLE 4.8

A current of 20 mA flows in a 1 kΩ resistor. What power is dissipated in the resistor?

Here we use $P = I^2 \times R$ but, to make life a little easier, we will work in mA and kΩ (in which case the answer will be in mW).

$$P = I^2 \times R = (20\,mA \times 20\,mA) \times 1k\Omega = 400\,mW$$

Thus a power of 400 mW is dissipated in the 1 kΩ resistor.

Electrostatics

If a conductor has a deficit of electrons, it will exhibit a net positive charge. If, on the other hand, it has a surplus of electrons, it will exhibit a net negative charge. An imbalance in charge can be produced by friction (removing or depositing electrons using materials such as silk and fur, respectively) or induction (by attracting or repelling electrons using a second body which is, respectively, positively or negatively charged).

Force Between Charges

Coulomb's Law states that, if charged bodies exist at two points, the force of attraction (if the charges are of opposite polarity) or repulsion (if the charges have the same polarity) will be proportional to the product of the magnitude of the charges divided by the square of their distance apart. Thus:

$$F = \frac{kQ_1Q_2}{r^2}$$

where Q_1 and Q_2 are the charges present at the two points (in Coulombs), r the distance separating the two points (in metres), F is the force (in Newtons), and k is a constant depending upon the medium in which the charges exist.

In vacuum or 'free space',

$$k = \frac{1}{4\pi\varepsilon_0}$$

where ε_0 is the *permittivity of free space* ($8.854 \times 10^{-12}\,C/Nm^2$).

Combining the two previous equations gives:

$$F = \frac{kQ_1Q_2}{4\pi \times 8.854 \times 10^{-12}r^2} \;\; \text{Newtons}$$

Electric Fields

The force exerted on a charged particle is a manifestation of the existence of an electric field. The electric field defines the direction and magnitude of a force on a charged object. The field itself is invisible to the human eye but can be drawn by constructing lines which indicate the motion of a free positive charge within the field; the number of field lines in a particular region being used to indicate the relative strength of the field at the point in question. Figures 4.5 and 4.6 show the electric fields between charges of the same and opposite polarity while Figure 4.7 shows the field which exists between two charged parallel plates (an arrangement known as a *capacitor*).

Electromagnetism

When a current flows through a conductor a magnetic field is produced in the vicinity of the conductor. The magnetic field is invisible but its presence can be detected using a compass needle (which will deflect from its normal North-South position).

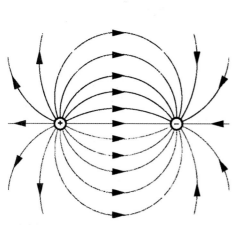

FIGURE 4.5
Electric field between two unlike electric charges

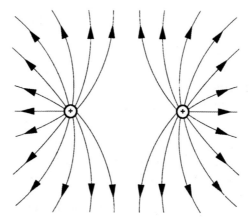

FIGURE 4.6
Electric field between two like electric charges (in this case both positive)

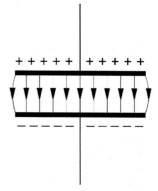

FIGURE 4.7
Electric field between two parallel plates

107

If two current-carrying conductors are placed in the vicinity of one another, the fields will interact with one another and the conductors will experience a force of attraction or repulsion (depending upon the relative direction of the two currents).

Force Between Two Current-Carrying Conductors

The mutual force which exists between two parallel current-carrying conductors will be proportional to the product of the currents in the two conductors and the length of the conductors but inversely proportional to their separation. Thus:

$$F = \frac{k I_1 I_2 l}{d}$$

where I_1 and I_2 are the currents in the two conductors (in Amps), l is the parallel length of the conductors (in metres), d is the distance separating the two conductors (in metres), F is the force (in Newtons), and k is a constant depending upon the medium in which the charges exist.

In vacuum or 'free space',

$$k = \frac{\mu_0}{2\pi}$$

where μ_0 is a constant known as the *permeability of free space* ($4\pi \times 10^{-7}$ or $12.57 \times 10^{-7}\,\mathrm{H/m}$).

Combining the two previous equations gives:

$$F = \frac{\mu_0 I_1 I_2 l}{2\pi d}$$

or

$$F = \frac{4\pi \times 10^{-7} I_1 I_2 l}{2\pi d}$$

or

$$F = \frac{2 \times 10^{-7} I_1 I_2 l}{d} \text{ Newtons}$$

Magnetic Fields

The field surrounding a straight current-carrying conductor is shown in Figure 4.8. The magnetic field defines the direction of motion of a free North pole within the field. In the case of Figure 4.8, the lines of flux are concentric and the direction of the field (determined by the direction of current flow) is given by the right-hand rule where the thumb represents the direction of current and the fingers (in grasp position) show the direction of magnetic flux.

Magnetic Field Strength

The strength of a magnetic field is a measure of the density of the flux at any particular point. In the case of Figure 4.8, the field strength will be proportional to the applied current and inversely proportional to the perpendicular distance from the conductor. Thus:

$$B = \frac{k I}{d}$$

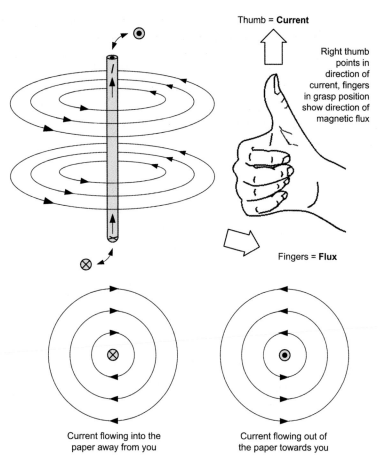

FIGURE 4.8
Magnetic field surrounding a straight conductor

where B is the magnetic flux density (in Tesla), I is the current (in Amperes), d is the distance from the conductor (in metres), and k is a constant.

Assuming that the medium is vacuum or 'free space', the density of the magnetic flux will be given by:

$$B = \frac{\mu_0 I}{2\pi d}$$

where B is the *flux density* (in Tesla), μ_0 is the permeability of 'free space' ($4\pi \times 10^{-7}$ or 12.57×10^{-7}), I is the current (in Amperes), and d is the distance from the centre of the conductor (in metres).

The flux density is also equal to the total flux divided by the area of the field. Thus:

$$B = \frac{\Phi}{A}$$

where Φ is the flux (in Webers) and A is the area of the field (in square metres).

In order to increase the strength of the field, a conductor may be shaped into a loop (Fig. 4.9) or coiled to form a *solenoid* (Fig. 4.10). Note, in the latter case, how the field pattern is exactly the same as that which surrounds a bar magnet.

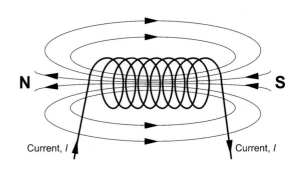

(a) Magnetic field around a solenoid

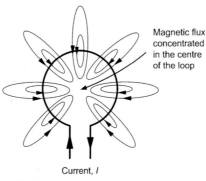

Magnetic flux concentrated in the centre of the loop

Current, I

FIGURE 4.9
Forming a conductor into a loop increases the strength of the magnetic field in the centre of the loop

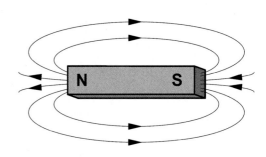

(b) Magnetic field around a permanent magnet

FIGURE 4.10
The magnetic field surrounding a solenoid coil resembles that of a permanent magnet

Test your knowledge 4.9
A flux density of 1.8 mT is developed in air over an area of 6 cm². Determine the total flux.

EXAMPLE 4.9

Determine the flux density produced at a distance of 50 mm from a straight wire carrying a current of 20 A.

Applying the formula $B = \mu_0\, I/2\pi\, d$ gives:

$$B = \frac{12.57 \times 10^{-7} \times 20}{2 \times 3.142 \times 50 \times 10^{-3}} = \frac{251.4 \times 10^{-7}}{314.2 \times 10^{-3}}$$

from which:

$$B = 0.8 \times 10^{-4} \text{ Tesla}$$

Thus $B = 80 \times 10^{-6}$ T or $B = 80\,\mu$T.

Test your knowledge 4.10
Determine the flux density produced at a distance of 2 cm from a straight wire carrying a current of 5 A.

Test your knowledge 4.11
Determine the current that must be applied to a straight wire conductor in order to produce a flux density of 200 μT at a distance of 12 mm in free space.

EXAMPLE 4.10

A flux density of 2.5 mT is developed in free space over an area of 20 cm². Determine the total flux.

Re-arranging the formula $B = \Phi/A$ to make Φ the subject gives $\Phi = B \times A$ thus:

$$\Phi = (2.5 \times 10^{-3}) \times (20 \times 10^{-4}) = 50 \times 10^{-7} \text{ Webers}$$

from which $B = 5\,\mu$Wb

Magnetic Circuits

Materials such as iron and steel possess considerably enhanced magnetic properties. Hence they are employed in applications where it is necessary to increase the flux density produced by an electric current. In effect, magnetic materials allow us to channel the electric flux into a 'magnetic circuit', as shown in Figure 4.11.

In the circuit of Figure 4.11(b) the *reluctance* of the magnetic core is analogous to the *resistance* present in the electric circuit shown in Figure 4.11(a). We can make the following comparisons between the two types of circuit (see Table 4.2).

In practice, not all of the magnetic flux produced in a magnetic circuit will be concentrated within the core and some 'leakage flux' will appear in the surrounding free space (as shown in Fig. 4.13). Similarly, if a gap appears within the magnetic circuit, the flux will tend to spread out as shown in Figure 4.12. This effect is known as 'fringing'.

B-H Curves

Figure 4.14 shows flux density, B, plotted against magnetising force (or *magnetic field intensity*), H, for some common magnetic materials. Because these materials all support the existence of a magnetic flux they are referred to as *ferromagnetic materials.*

The slope of the *B-H* curves indicates the ease by which a material can be magnetised. So, a steep slope indicates that a material can be easily magnetised whilst a shallow slope shows that it is more difficult to magnetise the material.

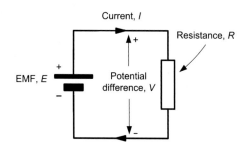

(a) An electric circuit

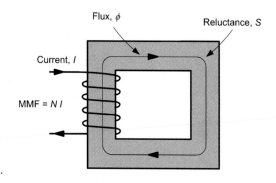

(b) A magnetic circuit

FIGURE 4.11
Comparison of electric and magnetic circuits

> **Key point**
> Flux density is found by dividing the total flux present by the area over which the flux acts.

TABLE 4.2 Comparison of electric and magnetic circuits	
Electric circuit figure 4.11(a)	**Magnetic circuit figure 4.11(b)**
Electromotive force, e.m.f. = V	Magnetomotive force, m.m.f. = $N \times I$
Resistance = R	Reluctance = S
Current = I	Flux = Φ
e.m.f. = current × resistance	m.m.f. = flux × reluctance
$V = I \times R$	$NI = S\Phi$

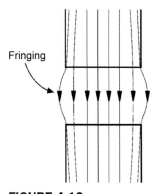

FIGURE 4.12
Fringing of the magnetic flux at an air gap in a magnetic circuit

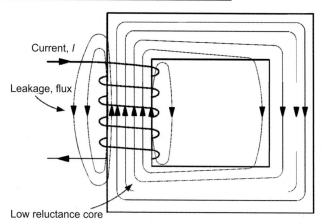

FIGURE 4.13
Leakage flux in a magnetic circuit

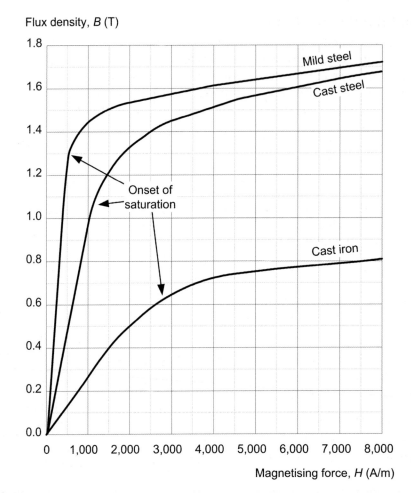

FIGURE 4.14
B-H curves for three ferromagnetic materials

If you look carefully at these curves you will notice that they all begin to flatten off when a certain value of flux density is reached. This is because no more flux can be supported by the material and this condition is known as *magnetic saturation*. Note also how the slope of the curve begins to fall rapidly beyond saturation as the material can no longer be easily magnetised. The point at which saturation occurs is important because it dictates the acceptable working range for a particular magnetic material when it is used as part of a magnetic circuit.

ACTIVITY 4.1

The following data was obtained during measurements made on a steel core:

Magnetising force, H (A/m)	0	500	1,000	1,500	2,500	3,000	5,000	8,000
Flux density, B (T)	0	0.6	1.1	1.3	1.43	1.51	1.55	1.67

Plot a graph showing how flux density, B, varies with magnetising force, H. Label your graph clearly and use it to answer the following questions:

1. What value of flux density corresponds to a magnetising force of 750 A/m?
2. What value of magnetising force corresponds to a flux density of 1.2 T?
3. At what value of magnetising force does saturation begin?

Force Acting on a Conductor

If we place a current carrying conductor in a magnetic field, the conductor has a force exerted on it. Consider the arrangement shown in Figure 4.15, in which a current carrying conductor is placed between the N—S poles of two permanent magnets. The direction of the current passing through it is into the page going away from us. Then by the right-hand screw rule, the direction of the magnetic field, created by the current in the conductor, is clockwise, as shown. We also know that the flux lines from the permanent magnet exit at a North pole and enter at a South pole, in other words, they travel from North to South, as indicated by the direction arrows. The net effect of the coming together of these two magnetic force fields is that at position A, they both travel in the same direction and reinforce one another. While at position B, they travel in the opposite direction and tend to cancel one another. So with a stronger force field at position A and a weaker force at position B the conductor is forced upwards out of the magnetic field.

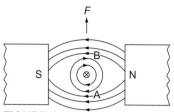

FIGURE 4.15
A current-carrying conductor in a magnetic field

If the direction of the current was reversed, i.e. if it was to travel towards us out of the page, then the direction of the magnetic field in the current carrying conductor would be reversed and therefore so would the direction of motion of the conductor.

A convenient way of establishing the direction of motion of the current carrying conductor is to use *Fleming's left-hand (motor) rule*. This rule is illustrated in Figure 4.16, where the left hand is extended with the thumb, first finger and second finger pointing at right angles to one another. From the figure it can be seen that the first finger represents the magnetic field, the second finger represents the direction of the current in the conductor and the thumb represents the motion of the conductor, due to the forces acting on it.

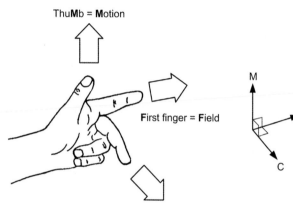

FIGURE 4.16
Fleming's left-hand rule

The amount of force acting on the conductor depends on the current flowing in the conductor, the length of the conductor in the field, and the strength of the magnetic flux (expressed in terms of its *flux density*). The size of the force will be given by the expression:

$$F = BIl$$

where F is the force in Newton (N), B is the flux density in Tesla (T), I is the current (A) and l is the length (m).

EXAMPLE 4.11

In Figure 4.17, a straight current carrying conductor lies at right angles to a magnetic field of flux density 1.2 T such that 250 mm of its length lies within the field. If the current passing through the conductor is 15 A, determine the force on the conductor and the direction of its motion.

In order to find the magnitude of the force we use the relationship $F = BIl$, hence:

$$F = BIl = 1.2 \times 15 \times 250 \times 10^{-3} = 4.5\,N$$

Now the direction of motion is easily found using Fleming's left-hand rule, where we know that the first finger points in the direction of the magnetic field N—S, the second finger points inwards into the page in the direction of the current, which leaves your thumb pointing downwards in the direction of motion.

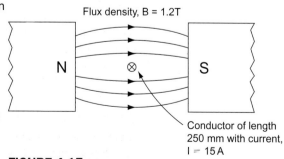

FIGURE 4.17
See Example 4.11

113

Test your knowledge 4.12
Determine the force exerted on a conductor of length 0.5 m and carrying a current of 15 A when it is suspended in a magnetic field having a flux density of 0.25 T.

Test your knowledge 4.13
Determine the flux density required to produce a force of 0.45 N on a conductor having an effective length of 2.5 m carrying a current of 6 A.

Test your knowledge 4.14
Determine the e.m.f. produced across the ends of a conductor having a length of 2 m moving at 1.75 m/s perpendicular to a magnetic field having a flux density of 0.35 T.

Generators

When a conductor is moved through a magnetic field, an e.m.f. will be induced across its ends. An induced e.m.f. will also be generated if the conductor remains stationary whilst the field moves. In either case, cutting at right angles through the lines of magnetic flux (see Fig. 4.18) results in a generated e.m.f. which will have a magnitude given by:

$$E = Blv$$

where B is the magnetic flux density (in Tesla), l is the length of the conductor (in m), and v is the velocity of the field (in m/s).

If the field is cut at an angle, 6, (rather than at right angles) the generated e.m.f. will be given by:

$$E = Blv \sin\theta$$

where θ is the angle between the direction of motion of the conductor and the magnetic field lines.

EXAMPLE 4.12

A conductor of length 20 cm moves at 0.5 m/s through a uniform perpendicular field of 0.6 T. Determine the e.m.f. generated.

Since the field is perpendicular to the conductor, the angle is 90° ('perpendicular' means the same as 'at right angles') we can use the basic equation:

$$E = Blv$$

where $B = 0.6$ T, $l = 20$ cm $= 0.02$ m, and $v = 0.5$ m/s. Thus:

$$E = Blv = 0.6 \times 0.02 \times 0.5 = 0.006\,V = 6\,mV$$

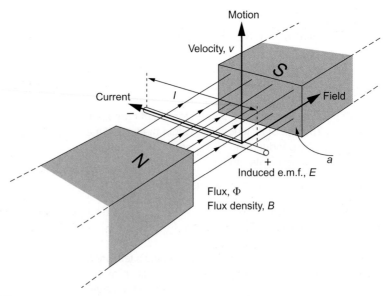

FIGURE 4.18
Generating an e.m.f. by moving a conductor through a magnetic field

Being able to generate a voltage by moving a conductor through a magnetic field is extremely useful as it provides us with an easy way of generating electricity. Unfortunately, moving a wire at a constant linear velocity through a uniform magnetic field presents us with a practical problem simply because the mechanical power that can be derived from an engine is available in rotary, not linear, form!

The solution to this problem is that of using the rotary power available from the engine (via a suitable gearbox and transmission) to rotate a conductor shaped into the form of loop as shown in Figure 4.19. The loop is made to rotate inside a permanent magnetic field with opposite poles (N and S) on either side of the loop.

There now remains the problem of making contact with the loop as it rotates inside the magnetic field but this can be overcome by means of a pair of carbon *brushes* and copper *slip rings.* The brushes are spring loaded and held against the rotating slip rings so that, at any time, there is a path for current to flow from the loop to the load to which it is connected.

The opposite sides of the loop consist of conductors that move through the field. At 0° (with the loop vertical as shown in Figure 4.21) the opposite sides of the loop will be moving in the same direction as the lines of flux. At that instant, the angle, θ, at which the field is cut is 0° and since the sine of 0° is 0 the generated voltage (from $E = Blv \sin \theta$) will consequently also be zero.

If the loop has rotated to a position which is 90° from that shown at the start, the two conductors will effectively be moving at right angles to the field. At that instant, the generated e.m.f. will take a maximum value (since the sine of 90° is 1).

At 180° from the starting position the generated e.m.f. will have fallen back to zero since, once again, the conductors are moving along the flux lines (but in the direction opposite to that at 0°).

At 270° the conductors will once again be moving in a direction which is perpendicular to the flux lines (but in the direction opposite to that at 90°). At this point, a maximum generated e.m.f. will once again be produced. It is, however, important to note that the e.m.f. generated at this instant will be of opposite polarity to that which was generated at 90°. The reason for this is simply that the relative direction of motion (between the conductors and flux lines) has effectively been reversed.

In practice, the single loop shown in Figures 4.19 to 4.21 would comprise a coil of wire wound on a non-magnetic former. This coil of wire effectively increases the length of the conductor within the magnetic field and the generated e.m.f. will then be directly proportional to the number of turns on the coil.

Motors

A simple motor consists of a very similar arrangement to that of the generator that we've just met. A loop of wire that's free to rotate is placed inside a permanent magnetic field (see Fig. 4.22).

Test your knowledge 4.15
Explain how Fleming's left-hand rule can be used to determine the direction of motion of a conductor suspended in a magnetic field. Illustrate your answer with a sketch.

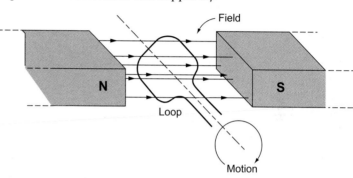

FIGURE 4.19
A rotating loop within a magnetic field

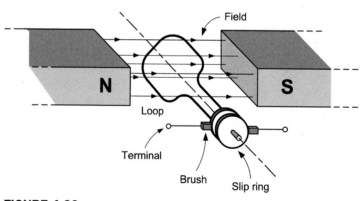

FIGURE 4.20
Using brushes to make contact with the loop

Key point
In a simple generator, a loop of wire rotates inside the magnetic field produced by two opposite magnetic poles. Contact is made to the loop as it rotates by means of slip rings and brushes.

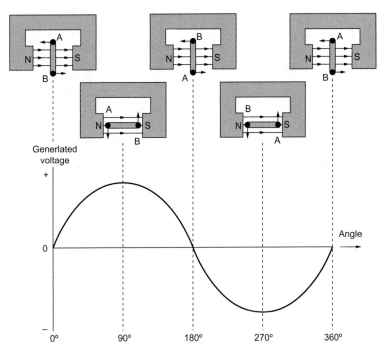

FIGURE 4.21
Voltage generated by the rotating loop

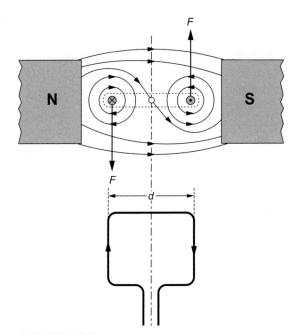

FIGURE 4.22
Torque acting on a current-carrying loop suspended in a magnetic field

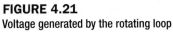

Test your knowledge 4.16
Explain, with the aid of a sketch, how contact can be made with a loop of wire rotating inside a magnetic field.

When a current is applied to the loop of wire, two equal and opposite forces are set up which act on the conductor in the directions indicated. The direction of the forces acting on each arm of the conductor can be established by again using the right-hand grip rule and Fleming's left-hand rule. Now because the conductors are equidistant from their pivot point and the forces acting on them are *equal and opposite,* then they form a *couple.* The *moment* of this couple is equal to the magnitude of a single force multiplied by the distance between them and this moment is known as *torque, T.* Now,

$$T = Fd$$

where T is the torque (in Newton-metres, Nm), F is the force (N) and d is the distance (m).

Test your knowledge 4.17
Explain the purpose and operation of each of the following parts of a d.c. motor:

a. armature
b. solenoid
c. commutator
d. field winding.

We already know that the magnitude of the force F is given by $F = BIl$, therefore the torque produced by the current carrying conductor can be written:

$$T = BIld$$

where T is the torque (Nm), B is the flux-density (T), I is the current (A), l is the length of conductor in the magnetic field (m), and d is the distance (m).

The torque produces a *turning moment* such that the coil or loop rotates within the magnetic field. This rotation continues for as long as a current is applied. A more practical form of d.c. motor consists of a rectangular coil of wire (instead of a single turn loop of wire) mounted on a former and free to rotate about a shaft in a permanent magnetic field, as shown in Figure 4.22. In real motors, this rotating coil is know as the *armature* and it consists of many hundreds of turns of conducting wire. This arrangement is needed in order to maximise the force imposed on the conductor by introducing the *longest possible* conductor into the magnetic field.

The relationship $F = BIl$ tells us that the force used to provide the torque in a motor is directly proportional to the size of the magnetic flux, B. Instead of using a permanent magnet to produce this flux, in a real motor, an electromagnet is used. Here an electromagnetic field is set up using the *solenoid* principle (Fig. 4.23). A long length of conductor is wound into a coil consisting of many turns and a current passed through it. This arrangement constitutes a *field winding* and each of the turns in the field winding assists each of the other turns in order to produce a strong magnetic field, as shown in Figure 4.23. This field may be intensified by inserting a ferromagnetic core inside the coil. Once the current is applied to the conducting coil, the core is magnetised and the result is a strong magnetic field like that supplied by a conventional permanent magnet.

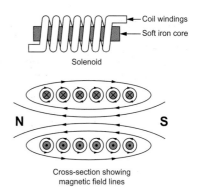

FIGURE 4.23
A solenoid

Returning to the simple motor illustrated in Figure 4.22, we know that when current is supplied to the armature *(rotor)* a torque is produced. In order to produce continuous rotary motion, this torque (turning moment) must always act in the same direction.

Therefore, the current in each of the armature conductors must be reversed as the conductor passes between the North and South magnetic field poles. The *commutator* (see Fig. 4.24) acts like a rotating switch, reversing the current in each armature conductor at the appropriate time to achieve this continuous rotary motion. Without the presence of a commutator in a d.c. motor, only a half turn of movement is possible. Figure 4.25 shows how the commutator acts as a rotary switch that periodically reverses the direction of current flow ensuring that the motor torque continues to turn the rotor in the same direction. Fleming's left-hand rule can be used to the direction of rotation at any instant.

> **Test your knowledge 4.18**
> Show that the torque produced by a d.c. motor is proportional to the product of the current, I, and flux density, B.

> **Key point**
> The torque produced by a d.c. motor is directly proportional to the product of the current flowing in the rotating armature winding.

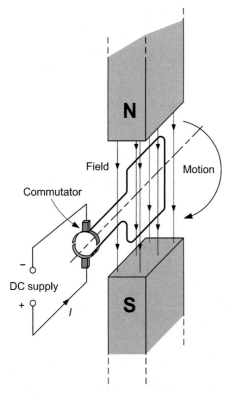

FIGURE 4.24
Simple electric motor with a commutator

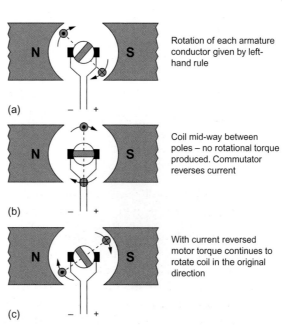

(a) Rotation of each armature conductor given by left-hand rule

(b) Coil mid-way between poles – no rotational torque produced. Commutator reverses current

(c) With current reversed motor torque continues to rotate coil in the original direction

FIGURE 4.25
Action of a commutator

Moving Coil Meter

One other application of the motor principle may be used in simple analogue measuring instruments. Some meters, including multimeters used to measure current, voltage and resistance, operate on the principle of a coil rotating in a magnetic field. The basic construction is shown in Figure 4.28, where the current, I, passes through a pivoted coil and the resultant motor force (the *deflecting torque*) is directly proportional to the current flowing in the coil windings which of course is the current being measured. The magnetic flux is concentrated within the coil by a solid cylindrical ferromagnetic core, in exactly the same manner as the flux is concentrated within a solenoid.

Transformers

Transformers provide us with a means of coupling a.c. power from one circuit to another without a direct connection between the two. Transformers consist of two coil windings; a *primary* (input) and a *secondary* (output), wound on a common magnetic core. A particular advantage of transformers is that voltage may be *stepped-up* (secondary voltage *greater* than primary voltage) or *stepped-down* (secondary voltage *less* than primary voltage). Since no increase in power is possible (like resistors, capacitors and inductors, transformers are *passive* components) an increase in secondary voltage can only be achieved at the expense of a

FIGURE 4.26
A relay which operates on the solenoid principle. Note the coil (left) and contact sets (right)

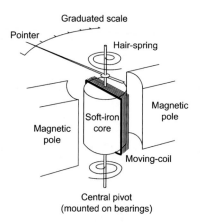

FIGURE 4.27
A moving coil loudspeaker. A strong permanent magnet (centre) provides the field

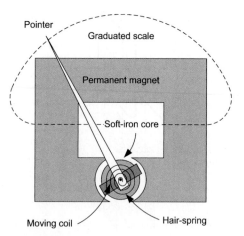

FIGURE 4.28
The moving coil meter

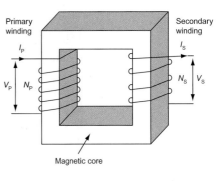

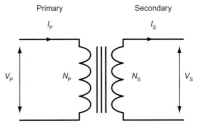

FIGURE 4.29
A selection of transformers

FIGURE 4.30
The transformer principle

corresponding reduction in secondary current, and vice versa (in fact, the secondary power will be very slightly less than the primary power due to losses within the transformer).

The principle of the transformer is illustrated in Figure 4.30. The primary and secondary windings are wound on a common low-reluctance magnetic core consisting of a number of steel laminations. All of the alternating flux generated by the primary winding is therefore coupled into the secondary winding (very little flux escapes due to leakage). A sinusoidal current flowing in the primary winding produces a sinusoidal flux within the transformer core.

At any instant the flux, Φ, in the transformer core is given by the equation:

$$\Phi = \Phi_{max} \sin(2\pi\, ft)$$

where Φ_{max} is the maximum value of flux (in Wb), f is the frequency of the applied current, and t is the time in seconds.

The r.m.s. value of the primary voltage (V_P) is given by:

$$V_P = 4.44 f N_P \Phi_{max}$$

Similarly, the r.m.s. value of the secondary voltage (V_S) is given by:

$$V_S = 4.44 f N_S \Phi_{max}$$

From these two relationships (and since the same magnetic flux appears in both the primary and secondary windings) we can infer that:

$$\frac{V_P}{V_S} = \frac{N_P}{N_S}$$

Furthermore, assuming that no power is lost in the transformer (i.e. as long as the primary and secondary powers are the same) we can conclude that:

$$\frac{I_P}{I_S} = \frac{N_S}{N_P}$$

119

The ratio of primary turns to secondary turns (N_P/N_S) is known as the *turns ratio*. The turns-per-volt rating can be quite useful when it comes to designing transformers with multiple secondary windings.

EXAMPLE 4.13

A transformer has 2,000 primary turns and 120 secondary turns. If the primary is connected to a 220 V a.c. mains supply, determine the secondary voltage.

Since we $V_P/V_S = N_P/N_S$ can make V_S the subject as follows:

$$V_S = \frac{V_P N_S}{N_P} = \frac{220 \times 120}{2000} = 13.2\,V$$

EXAMPLE 4.14

A transformer has 1,200 primary turns and is designed to operate with a 110 V a.c. supply. If the transformer is required to produce an output of 10 V, determine the number of secondary turns required.

Since $V_P/V_S = N_P/N_S$ we can make N_S the subject as follows:

$$N_S = \frac{N_P V_S}{V_P} = \frac{1200 \times 10}{110} = 109$$

Series and Parallel Circuits

When more than one resistor or 'load' is present in an a.c. or d.c. circuit, the loads may be connected in series, Figure 4.31(a), or in parallel, Figure 4.31(b), or a combination of both methods.

The equivalent resistance, R_T, of two resistors connected in series (Fig. 4.32) is given by:

$$R_T = R_1 + R_2$$

The equivalent resistance, R_T, of two resistors connected in parallel (Fig. 4.33) is given by:

$$\frac{1}{R_T} = \frac{1}{R_1} + \frac{1}{R_2}$$

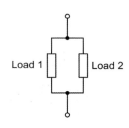

(a) Series connected

(b) Parallel connected

FIGURE 4.31
Series and parallel
circuits

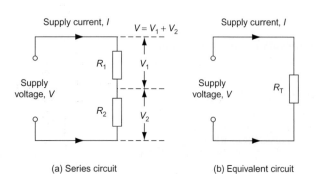

(a) Series circuit

(b) Equivalent circuit

FIGURE 4.32
Resistor connected in series

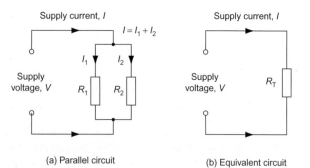

(a) Parallel circuit

(b) Equivalent circuit

FIGURE 4.33
Resistors connected in parallel

Note that this expression can be re-arranged to give:

$$R_T = \frac{R_1 \times R_2}{R_1 + R_2} = \frac{product}{sum}$$

Test your knowledge 4.19
A transformer has 480 primary turns and 36 secondary turns. Determine the secondary voltage if the primary is connected to a 220 V a.c. supply.

EXAMPLE 4.15

Two 15 Ω resistors are connected (a) in series and (b) in parallel. Determine the equivalent resistance of each arrangement.

(a) In the series case, the equivalent resistance will be given by:

$$R_T = R_1 + R_2 = 15 + 15 = 30\,\Omega$$

(b) In the parallel case, the equivalent resistance will be given by:

$$R_T = \frac{R_1 \times R_2}{R_1 + R_2} = \frac{15 \times 15}{15 + 15} = \frac{225}{30} = 7.5\,\Omega$$

ACTIVITY 4.2

1. Determine the equivalent resistance of each of the circuits shown in Figure 4.34. Hint: For circuit (c) solve the parallel branch first.
2. If each of the circuits shown in Figure 4.34 is connected to a 110 V supply, determine:
   a. the current supplied to the circuit
   b. the voltage dropped across each resistor, and
   c. the power dissipated in each resistor.

4.2 MECHANICAL SCIENCE

Force, mass, weight and density are important in many engineering applications and it's quite likely that you already have some idea of what these terms mean. However, do you understand the difference between mass and weight? And, how would you measure a force? This section is designed to help you get up to speed with these important concepts.

Force

A force is a push or pull exerted by one object on another. If the object remains in *equilibrium* (i.e. if it doesn't move or change in any way) then, for each force acting on the object there is another equal and opposite force that acts against it. The force that is applied to an object (or a *body*) is often called an *action* whilst the opposing force is referred to as a *reaction*. As long as the object (or *body*) doesn't move or change in any way, action and reaction will be equal and opposite.

It also follows that, if the forces of action and reaction acting on an object are not equal and opposite, the object will move or change in some way. You can test this theory out very easily by finding a wall and pushing against it. If the wall doesn't move (hopefully it won't!) then you will experience a force pushing back. If you increase the force that you apply to the wall (the *action*) the force pushing back (the *reaction*) will also increase by the same amount.

Now try the same experiment by pushing against a door that is partially open. There will still be some force exerted back by the door but this will be much less than the force that you apply. Because of this imbalance of forces (action being greater than reaction) the door will move and will swing open. This simple experiment leads us to the following conclusions:

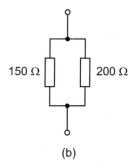

(a)

(b)

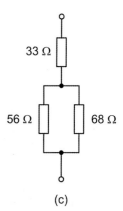

(c)

FIGURE 4.34
See Activity 4.2

121

- When a body is at rest (or in *equilibrium*) the *action* and *reaction* forces acting on it will be equal and opposite.
- If the action and reaction forces acting on a body are not equal and opposite a change (in this case *motion*) will be produced.

Mass and Weight

> **Key point**
>
> Mass is defined as the quantity of matter in a body. The mass of a body remains the same regardless of where it is.

The mass of a body is defined as the *quantity of matter* in the body. It's important to be aware that the mass of a body remains the same regardless of where the body is. So, for example, a mass of 50 kg will be the same on the surface of the Earth as it will be in outer space (where there is 'zero gravity').

The weight of a body is determined by its mass and the gravitational force acting on the body. So, if there is no gravitational force (for example, in outer space) then a body will have no weight! However, in most practical cases we are concerned with what things weigh on the surface of the Earth in which case the relationship between mass and weight is given by:

> **Key point**
>
> The weight of a body is the product of its mass and the gravitational forces acting on it. On the surface of the earth, g is a constant equal to 9.81 m/s^2.

$$W = mg$$

where W is the weight in Newton (N), m is the weight in kg, and g is the gravitational acceleration (in m/s^2). On the surface of the Earth, g is a constant equal to 9.81 m/s^2.

The weight of a body decreases as the body is moved away from the centre of the Earth. Weight obeys the *inverse square law*. This simply means that weight is inversely proportional to the square of the distance from the centre of the Earth. In other words:

$$W \propto \frac{1}{d^2}$$

where W is the weight (in N) and d is the distance from the centre of the Earth (in m).

Finally, it is essential to remember that mass and weight are not the same thing! The mass of an object remains the same wherever it is whereas, the weight of a body is determined by the product of its mass and the gravitational force that is acting on it.

EXAMPLE 4.16

A light alloy beam has a mass of 17.5 kg. Determine the weight of the beam.

Here we will assume that the beam is being used at the Earth's surface. In which case:

$$W = mg = 17.5 \times 9.81 = 171.68 \text{N}$$

EXAMPLE 4.17

A lunar lander weighing 8.25 kN on Earth, is to be used on a mission to explore the surface of the Moon. Given that the gravitational acceleration on the Moon is one sixth (0.16) of that on Earth, determine the weight of the lander on the surface of the Moon.

The weight of the lander will be reduced in direct proportion to the reduction in gravitational acceleration. Hence the lander will weigh 0.16 × 8.25 kN = 1.38 kN on the surface of the Moon.

Density

The density of a body is defined as the mass per unit volume. In other words, the density of an object is found by dividing its mass by its volume. The density of a particular material is a fundamental property of that material. Expressing this as a formula gives:

$$\rho = \frac{m}{V}$$

TABLE 4.3 Density of various materials		
Material	Density (kg/m³)	Relative density
Aluminium	2,700	2.7
Brass	8,500	8.5
Cast iron	7,350	7.35
Concrete	2,400	2.4
Copper	8,960	8.96
Glass	2,600	2.6
Mild steel	7,850	7.85
Wood (oak)	690	0.69

where ρ is the density in kg/m³, m is the mass in kg, and V is the volume in m³.

We sometimes express the density of an object relative to that of pure water (at 4°C). The density of water under these conditions is 1,000 kg/m³. The densities (and relative densities) of various engineering materials are shown in Table 4.3.

EXAMPLE 4.18

Determine the mass of an aluminium block which has the following dimensions; 50 mm × 110 mm × 275 mm.

The total volume of the aluminium block will be given by:

$$V = (50 \times 10^{-3}) \times (110 \times 10^{-3}) \times (275 \times 10^{-3}) = 1.5125 \times 10^{-3}\,m^3$$

Re-arranging the formula for density to make m the subject gives:

$$m = \rho \times V$$

From the table, the value of r for aluminium is 2,700 kg/m³ hence the mass of the block will be given by:

$$m = 2,700 \times (1.5125 \times 10^{-3}) = 4,083.75 \times 10^{-3} = 4.08\,kg$$

Force Diagrams

Every force has three important properties that are used to describe it. These properties are:

- size (or *magnitude*) (see Fig. 4.35a)
- direction (see Fig. 4.35b)
- point of application (see Fig. 4.35c).

Engineers frequently use diagrams to show the effect of forces and also to help solve problems involving a number of forces acting at the same time. In order to specify the direction of a force we use a set of references axes (see Fig. 4.36a). The horizontal direction is generally referred to as the *x*-axis whilst the vertical direction is generally known as the *y*-axis. Note, however, that reference axes are something that we have introduced for our own convenience and thus we need not be constrained to any particular orientation.

> **Test your knowledge 4.20**
> Determine the weight of a solid copper bar having a length of 0.2 m and a cross-sectional area of 4 cm².

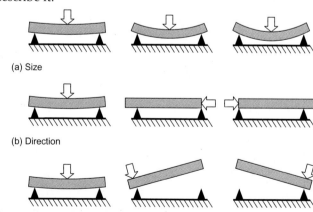

(a) Size

(b) Direction

(c) Point of application

FIGURE 4.35
Properties of a force

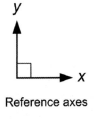

Reference axes

(a)

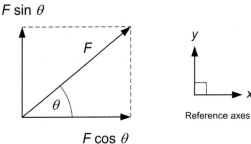

$F \sin \theta$

$F \cos \theta$

FIGURE 4.37
Resolving a force into two components acting at right angles

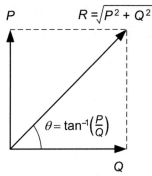

$R = \sqrt{P^2 + Q^2}$

$\theta = \tan^{-1}\left(\frac{P}{Q}\right)$

Q

FIGURE 4.38
Determining the resultant of two forces acting at right angles

Length of arrow indicates magnitude

(b)

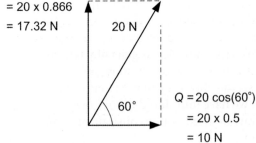

Direction, θ (relative to the x-axis)

(c)

FIGURE 4.36
Diagrammatic representation of a force

The size (or *magnitude*) of force is indicated by its length (see Fig. 4.36b) whilst its direction (usually specified relative to the x-axis) is indicated by the angle, θ, as shown in Fig. 4.36c.

A single force, F, can be resolved into two components acting at right angles, $F \sin \theta$ and $F \cos \theta$ (as shown in Fig. 4.37). Note that we have included the reference axes in this diagram although these are usually not shown in force diagrams.

Just as we can resolve a single force into two components acting at right angles we can also find the one single force (the *resultant*, R) of two forces acting at right angles, as shown in Figure 4.38.

EXAMPLE 4.19

Determine the horizontal and vertical components of a 20 N force acting at a direction of 60° to the horizontal.

The force diagram is shown in Figure 4.39. The 20 N force has been shown acting at 60° to the x-axis.

The horizontal component of the force, Q, will be given by:

$$Q = 20 \times \cos 60° = 20 \times 0.5 = 10N$$

Similarly, the vertical component of the force, P, will be given by:

$$P = 20 \times \sin 60° = 20 \times 0.866 = 17.32N$$

Hence a 20 N force acting at 60° to the horizontal can be replaced by two forces of 10 N and 17.32 N acting at right angles.

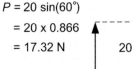

$P = 20 \sin(60°)$

$= 20 \times 0.866$

$= 17.32$ N

20 N

60°

$Q = 20 \cos(60°)$

$= 20 \times 0.5$

$= 10$ N

FIGURE 4.39
See Example 4.19

EXAMPLE 4.20

Determine the resultant of two forces, 3 N and 4 N acting at right angles to one another.

The force diagram is shown in Figure 4.40. The 4 N force has been shown acting along the *x*-axis.

The magnitude of the resultant can be calculated from:

$$R = \sqrt{3^2 + 4^2} = \sqrt{25} = 5 \text{ N}$$

The angle between the resultant and the *x*-axis can be found from:

$$\theta = \tan^{-1}\left(\frac{3}{4}\right) = \tan^{-1}(0.75) = 36.7°$$

Hence the resultant is 5 N acting at 36.7° to the 4 N force, as shown in Figure 4.40.

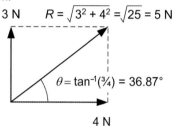

3 N $R = \sqrt{3^2 + 4^2} = \sqrt{25} = 5$ N

$\theta = \tan^{-1}(\frac{3}{4}) = 36.87°$

4 N

FIGURE 4.40
See Example 4.20

Parallelogram of Forces

So far, we have considered the case when two forces are at right angles to one another. When this is not the case we can determine the resultant force by constructing a *parallelogram of forces*, as shown in Figure 4.41.

In Figure 4.41, the resultant, *R*, of the two forces *P* and *Q* is determined by constructing the diagonal, *R*. The magnitude and direction of *R* (relative to one of the other two forces) can be found by scale drawing or by calculation as follows:

1. Resolve each of the known forces (*P* and *Q*) into their horizontal and vertical components (see Example 4.19)
2. Find the total horizontal and total vertical components (taking into account their direction)
3. Determine the magnitude and direction of the total horizontal and total vertical force components (see Example 4.20).

This method may seem a little long-winded but it will usually produce a more precise answer than can be achieved by means of scale drawing.

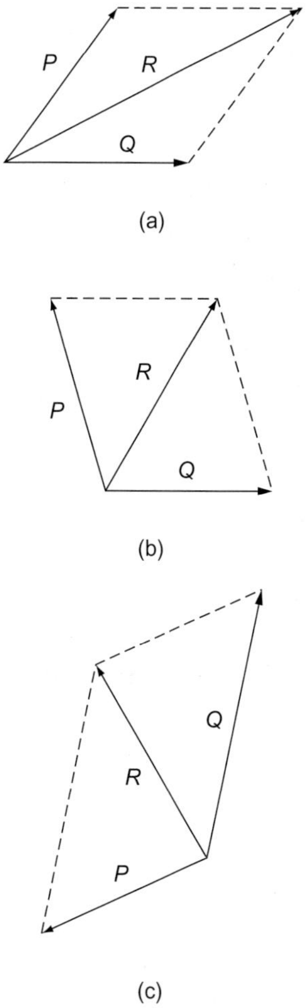

EXAMPLE 4.21

Figure 4.42 shows the arrangement of a small crane with a jib, BC, and a steel cable, AC. Using a graphical method, determine the forces acting in the cable and in the jib when the crane carries a load of 20 kN.

The force diagram (i.e. *triangle of forces*) is shown in Figure 4.43 where *W* is the weight of the load, *T* is the thrust in the jib, and *P* is the tension in the cable.

FIGURE 4.41
Examples of the parallelogram of forces

Test your knowledge 4.21
A force, *F*, of 20 N acts at 27° to the horizontal. Find the horizontal and vertical components of this force.

Test your knowledge 4.22
A force, *F*, acts at 75° to the horizontal. If the horizontal component of this force is 1.5 kN, determine the value of *F*.

Test your knowledge 4.23
A force, *F*, acts at 15° to the horizontal. If the vertical component of this force is 75 N, determine the value of *F*.

Test your knowledge 4.24
Forces of 5 N and 12 N act at right angles to one another. Determine the magnitude and direction of the resultant relative to the 5 N force.

(a)

(b)

(c)

125

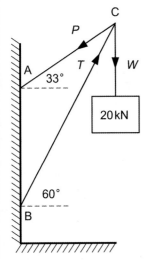

FIGURE 4.42
See Example 4.21

The graphical method (see Fig. 4.44) requires some squared paper, a protractor, a rule, and a drawing pencil. The drawing is constructed to a convenient scale (in this case 1 cm = 2 kN) and the steps are as follows:

1. First draw a line to represent the load, *W* (20 kN). This line must be 10 cm in length and it must be aligned vertically.
2. Construct an angle of 60° at the base of the vertical line and then draw a straight line to represent the force in the jib, *T*. Project this line towards the top edge of the drawing paper.
3. Construct an angle of 33° at the top of the base of the vertical line and then draw a straight line to represent the force in the steel cable, *P*. Project this line towards the top edge of the drawing paper until it meets the line that represents the force in the jib.
4. Locate the point of intersection between the two lines and then measure their lengths. Convert these lengths (using the scaling factor) to force in Newton.
5. The length of the line representing *P* is 11 cm. This indicates that the force in the cable is 22 N.
6. The length of the line representing *T* is 18.5 cm. This indicates that the force in the jib is 37 N.

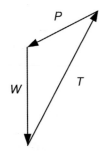

FIGURE 4.43
Triangle of forces,
see Example 4.21

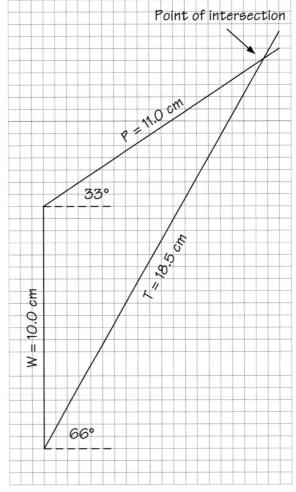

FIGURE 4.44
Scale drawing, see Example 4.21

EXAMPLE 4.22

Repeat Example 4.21 using a non-graphical method.

The non-graphical method is based on resolving the forces in the vertical and horizontal directions. The resultant of the forces in both the horizontal and vertical directions must be zero because the system of forces is in *static equilibrium* (i.e. it isn't in motion). Firstly, resolving the forces into their vertical components gives:

$$20 = T\cos 30° - P\cos 57° \text{ (downward force = upward force)}$$

from which:

$$20 = 0.866T - 0.545T \quad \text{(i)}$$

Secondly, resolving the forces into their horizontal components gives:

$$P\cos 33° = T\cos 60° \text{ (leftward force = rightward force)}$$

from which:

$$0 = 0.839T - 0.5T \quad \text{(ii)}$$

Equations (i) and (ii) need to be solved simultaneously, as follows:

Multiplying (ii) by 1.732 (or 2 × 0.866) gives:

$$0 = 1.453P - 0.866T \quad \text{(iii)}$$

Adding (i) and (iii) gives:

$$20 + 0 = 1.453P - 0.545P \text{ (the terms in } T \text{ are eliminated)}$$

hence:

$$20 = 0.908P \text{ from which } P = 22.06\text{N}$$

Substituting for P in equation (iii) gives:

$$0 = (0.839 × 22.06) - 0.5T$$

thus:

$$0.5T = 18.48 \text{ from which } T = 36.9\text{N}$$

It is useful to compare these calculated values with those found earlier using the graphical method. Which method do you feel is more accurate and why is it more accurate?

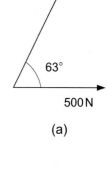

(a)

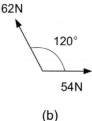

(b)

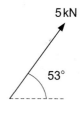

(c)

FIGURE 4.45
See Activity 4.3

ACTIVITY 4.3

1. Determine the resultant of the two forces shown in Figures 4.45(a) and 4.45(b). Show all working.
2. Determine the horizontal and vertical components of the force shown in Figure 4.45(c). Show all working.

ACTIVITY 4.4

Use a graphical method to determine the forces shown in Figure 4.46.

ACTIVITY 4.5

Use a non-graphical method to determine the forces shown in Figure 4.46. Compare your answers with those that you obtained in Activity 4.4. Give reasons for any significant differences in your answers.

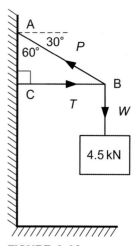

FIGURE 4.46
See Activities 4.4 and 4.5

Pressure

Pressure (or *stress*) is exerted whenever a force is applied to an object such as a floor, wall or the inside surfaces of a container. Pressure is defined as the ratio of force (or *load*) applied perpendicular (i.e. at right angles) to the surface, to the area over which the force (or load) acts. Thus:

$$P = \frac{F}{A}$$

where P is the pressure in Pascal (Pa), F is the force (in N), and A is the area (in m^2).

Test your knowledge 4.25
A mild steel girder has a mass of 560 kg and is supported by two concrete pillars each having a cross-sectional area of 0.044 m^2. Assuming that the load is distributed evenly, determine the pressure exerted on each of the two pillars.

EXAMPLE 4.23

A lathe weighs 425 kN. Determine the pressure exerted on the workshop floor if the load is distributed over a surface area of 0.875 m^2.

Now

$$P = \frac{F}{A} = \frac{425 \times 10^3}{0.875} = 485.7\,\text{kPa}$$

Pressure also exists in several other forms, including *hydrostatic pressure* (the pressure exerted in a stationary fluid), *atmospheric pressure* (the pressure exerted by the atmosphere) and *dynamic pressure* (pressure due to the movement of a fluid). The following laws apply to pressure in a fluid:

1. Pressure at a given depth in a fluid is equal in all directions.
2. Pressure at a given depth in a fluid is independent of the shape of the container in which it is held.
3. Pressure acts at right angles to the surfaces of the container in which it is held.
4. When pressure is applied to a fluid it is transmitted equally in all directions.

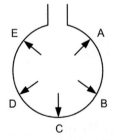

(a) Pressure at a given depth is equal in all directions

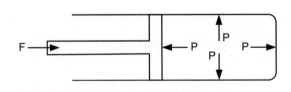

(b) Pressure is independent of the shape of the containing vessel at a given depth

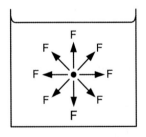

(c) Pressure acts at right angles to the walls of the containing vessel

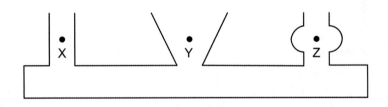

(d) Pressure transmitted through a fluid is equal in all directions

FIGURE 4.47
Fluid pressure laws

Hydrostatic Pressure

Pressure at a point in a fluid can be determined by considering the weight force of the fluid above the point. If the density of the fluid is known, then we can express the weight of the liquid in terms of its density and volume (since density is equal to mass divided by volume). The mass of a liquid is given by:

$$m = \rho \times A \times h \, \text{kg}$$

where m is the mass of the liquid (in kg), ρ is the density of the liquid (in Pa), A is the cross-sectional area, and h is the height.

Since the weight of fluid is equal to the mass multiplied by the acceleration due to gravity, g, the weight will be given by:

$$W = \rho \times A \times g \times h \, \text{Newton}$$

To determine the hydrostatic pressure, P, we simply divide the weight force, W, by the area, A, to arrive at:

$$P = \rho \times g \times h \, \text{Pa}$$

Note that we have ignored the atmospheric pressure above the liquid in arriving at this result. For this reason it would be more correct to refer to P as the *gauge pressure* (i.e. the pressure that a gauge would read since this, too, would be subject to atmospheric pressure).

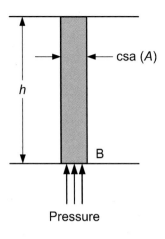

Pressure at B = $\rho g h$

FIGURE 4.48
Hydrostatic pressure

EXAMPLE 4.24

Find the head, h, of mercury in a column when a pressure of 101.32 kPa is applied. Take the density, ρ, of mercury as 13,600 kg/m³.

Rearranging the formula, $P = \rho g h$, to make h the subject gives:

$$h = \frac{P}{\rho g} = \frac{101.32 \times 10^3}{13.6 \times 10^3 \times 9.81} = 0.76 \, \text{m}$$

Moments

A moment is simply a force that produces a *turning effect*. The magnitude of a moment depends on the product of the force applied and the perpendicular distance from the pivot or axis to the line of action of the force. Hence:

$$M = F \times d$$

where M is the moment in Newton-metres, F is the force in Newton and d is the distance in metres. Note that if a load is expressed as a mass (in kg) it will be necessary to determine the force in Newton by multiplying the mass by 9.81.

When a force is in *equilibrium* the total clockwise moment (CW) will be equal to the total anticlockwise moment (ACW).

EXAMPLE 4.25

Figure 4.49 shows a beam of negligible mass having an overall length of 7 m. The beam is supported by a pivot point which is 3 m from the left-hand end and 4 m from the right-hand end. On the right of the pivot, forces of 10 m and 5 m are applied at distance of 1 m and 4 m respectively. On the left of the pivot an unknown force, F, is applied 3 m from the pivot. The direction of action of each of these forces has been shown on the diagram. Determine the value of the unknown force, F, in order to preserve equilibrium.

Key point
When pressure is applied to a fluid it is transmitted equally in all directions.

Since the system is in equilibrium we can infer that:

Clockwise (CW) moment = anti-clockwise (ACW) moment

The CW moment is: $(10 \times 1) + (5 \times 4) = 30\,Nm$

The ACW moment is simply: $F \times 3 = 3F\,Nm$.

Equating the CW and ACW moments gives:

$3F = 30\,Nm$

Thus, $F = 10\,N$.

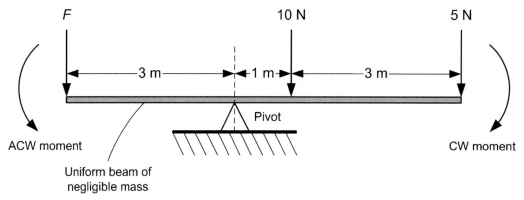

FIGURE 4.49
See Example 4.25

Key point
When a mechanical system is in equilibrium, clockwise and anti-clockwise moments are equal.

Key point
The moment of a force is the product of the force and the perpendicular distance from the point of reference.

FIGURE 4.50
A crane jib is an example of a beam. Note the adjustable counterweight made from concrete blocks

Test your knowledge 4.26
Find the unknown force, F, in Figure 4.52 required to preserve equilibrium.

EXAMPLE 4.26

The diagram shown in Figure 4.51 shows a spanner exerting a force on a nut. Determine the turning effect on the nut.

In this example it's important to note that the 50 N force applied to the spanner is not applied at right-angles but at an angle of 60°. In order to calculate the moment we need to find the effective perpendicular distance, *s*, at which the force is applied. This is found from the right-angled triangle in which the hypotenuse (i.e. the spanner) has a length of 200 mm and the two angles are 60° and 30°. From this:

$$s = 200 \times \sin(60°) = 200 \times 0.866 = 173.2 \text{mm}$$

The moment applied to the nut will thus be:

$$M = F \times d = 50 \times 0.1732 = 8.66 \text{Nm}.$$

Force and Acceleration

In the previous section we briefly introduced as a push or pull exerted by one object on another. We also explained how weight is a force that results from gravity. In fact, gravitational force is something that we all experience all of the time and it's what keeps us firmly in place on the surface of the Earth—without it we would simply drift off into space!

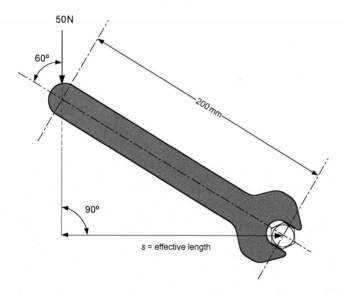

FIGURE 4.51
See Example 4.26

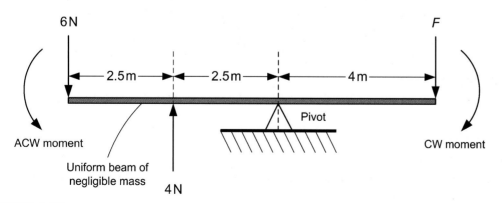

FIGURE 4.52
See Test your knowledge 4.26

When forces of action and reaction acting on a body are not equal and opposite (i.e. when the body is not actually in a state of equilibrium) the body will begin to move. In fact, it will experience an acceleration.

The relationship between the mass of the body, the force applied, and the acceleration that is produced is very important. If we want something to move fast we need to know how much force to apply! It should also be fairly obvious that, the larger the mass of the body the more force we would need to apply in order to make it move. In fact, the unit of force (the Newton) is usually defined in terms of this important relationship. The definition is as follows:

A force of one Newton is that which would produce an acceleration of one m/s^2 in a mass of one kg. The relationship is as follows:

$$F = m \times a$$

where F is the force (in N), m is the mass (in kg), and a is the acceleration (in m/s^2).

EXAMPLE 4.27

A rocket produces a thrust of 9.75 kN. If the rocket has a mass of 2.4 kg, determine the acceleration produced.

Re-arranging $F = m \times a$ to make a the subject gives:

$$a = \frac{F}{m} = \frac{9.75 \times 10^3}{42.4} = 230 \, \text{m/s}^2$$

Velocity and Acceleration

In normal conversation, we use 'speed' to describe how fast something is moving. However, when something is moving an important consideration, apart from its 'speed', is the direction in which it is travelling. When we take into account the direction in which an object or a body is moving we use a more precise word, *velocity*. Velocity therefore means 'speed in a given direction'. For example, an object moving directly from point A to point B might have a velocity of 10 m/s. The same object moving at the same speed but from B to A would have a velocity of −10/s. Notice that we have *arbitrarily* defined a positive velocity as speed in the direction A to B.

Velocity is defined as the ratio of distance travelled to the time taken. Hence:

$$v = \frac{s}{t}$$

where v is velocity (in m/s), s is the distance (in m) and t is the time (in s).

EXAMPLE 4.28

A cruise missile travels in a straight line at a constant speed of 295 m/s. How long will it take to reach a target which is 650 km away?

Rearranging $v = s/t$ the formula to make t the subject gives:

$$t = \frac{s}{v}$$

From which, $t = \dfrac{650}{295} = 2.03$ hours

Test your knowledge 4.27
An object having a mass of 125 kg is to have an acceleration of 15 m/s. What force is required to do this?

Test your knowledge 4.28
A force of 75 N is applied to an object which has a mass of 17.5 kg. Determine the acceleration produced.

Key point
When something moves it needs a force to start it moving. Once it is moving it will continue to move until a force is applied to stop it moving.

Acceleration and Deceleration

If an object is moving at a constant velocity (i.e. its velocity is neither increasing nor decreasing) then its acceleration is zero. If the velocity of the object is increasing the object is undergoing *acceleration*. Conversely, if the velocity of the object is decreasing the object is undergoing *deceleration* (or *retardation*). Figure 4.53 shows this in a graphical form.

Acceleration is defined as the ratio of change in velocity to the time. Hence:

$$a = \frac{v - u}{t}$$

where *a* is acceleration (in m/s²), *u* is the velocity (in m/s) at the beginning of the time interval, *v* is the velocity (in m/s) at the end of the time interval, and *t* is the time (in s).

Note that, when *v* is greater than *u*, *a* will take a *positive* sign (acceleration). Conversely, when *u* is greater than *v*, *a* will have a *negative* sign (deceleration). Figure 4.54 shows this relationship.

Test your knowledge 4.29
A Formula 1 race car is driven at a constant speed on a test track having a length of exactly one mile. Determine the speed at which the car is driven if it completes the test mile in 18.5 s.

EXAMPLE 4.29

An aircraft on a taxiway accelerates from 18 m/s to 27 m/s in 10 seconds. What acceleration does the aircraft experience?

In this example, *u* = 18 m/s, *v* = 27 m/s and *t* = 10 s. Substituting these values into the equations for acceleration gives:

$$a = \frac{v - u}{t} = \frac{27 - 18}{10} = \frac{9}{10} = 0.9 \text{ m/s}^2$$

Key point
Velocity means 'speed in a given direction'. Thus, unlike speed, velocity takes a positive or negative sign according to whether the motion is away from or towards a particular reference point.

EXAMPLE 4.30

The braking system fitted to a high speed train is able to produce a constant deceleration 7.5 m/s². How long will it take for the train to come to rest if it is travelling at 120 m/s?

In this example, *u* = 120 m/s, *v* = 0 m/s (because the train *comes to rest*) and *a* = −5.5 m/s².

First we need to rearrange the equation for acceleration in order to make *t* the subject:

$$t = \frac{v - u}{a} = \frac{0 - 120}{-5.5} = \frac{-120}{-5.5} = 21.82 \text{ s}$$

Key point
When the velocity of a body is increasing it is said to be undergoing acceleration. When the velocity of a body is decreasing it is said to be undergoing deceleration.

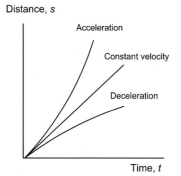

FIGURE 4.53
Distance-time graphs

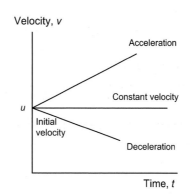

FIGURE 4.54
Velocity-time graphs

We can re-arrange the relationship that we met earlier to make the final velocity, v, the subject as follows:

$$a = \frac{v - u}{t}$$

Multiplying both sides by t gives:

$$a \times t = \frac{v - u}{t} \times t \quad \text{or} \quad at = v - u$$

Adding u to both sides gives:

$$at + u = v \quad \text{or} \quad v = u + at$$

EXAMPLE 4.31

An overhead crane is travelling along a track at a velocity of 0.8 m/s. If the crane is then given a constant acceleration of 0.15 m/s², determine the velocity of the crane after 5 seconds.

Using $v = u + at$ we have:

$$v = u + at = 0.8 + (0.15 \times 5) = 0.8 + 0.75 = 1.55 \text{ m/s}$$

Mechanical Power and Work

Mechanical power is defined as the rate of doing work. Hence:

$$Power = \frac{work\ done}{time\ taken}$$

Note that the work done represents a change in energy and it is measured in Joule (J) whilst the time taken is measured in seconds. If energy is used at the rate of 1 Joule per second this corresponds to a power of 1 W.

Mechanical work done is also measured in terms of the product of the force exerted, F, and the distance moved, d. Hence:

$$P = \frac{F \times d}{t}$$

where P is the power (in W), F is the force (in N), d is the distance in m, and t is the time (in seconds).

EXAMPLE 4.32

A mechanical hoist raises a load of 30 kg through a vertical height of 15 m in a time of 30 s. Determine the work done and power.

Firstly we need to determine the force exerted by a mass of 30 kg. We can find this from:

$$W = mg = 30 \times 9.81 = 294.3 \text{ N}$$

Now we can find the work done (WD):

$$WD = 294.3 \times 15 = 4.414 \text{ kJ}$$

Finally, we can calculate the power from:

$$P = \frac{4.414 \times 10^3}{30} = 147.13 \text{ W}$$

Key point

Mechanical power is the rate of doing work (or using energy).

Friction

When one surface is moved over another with which it is in contact, a resistance is set up to the motion. The amount of resistance will depend on the materials concerned as well as the force that holds the two surfaces together. This resistance to movement is known as *friction*. Note that a slightly greater force *(static friction)* is required to start two surfaces moving over one another compared with the force required to keep them moving *(sliding friction)*.

Figure 4.55 shows a block moving over a horizontal surface. Here, friction will occur between the lower surface of the block and the upper surface of the horizontal plane.

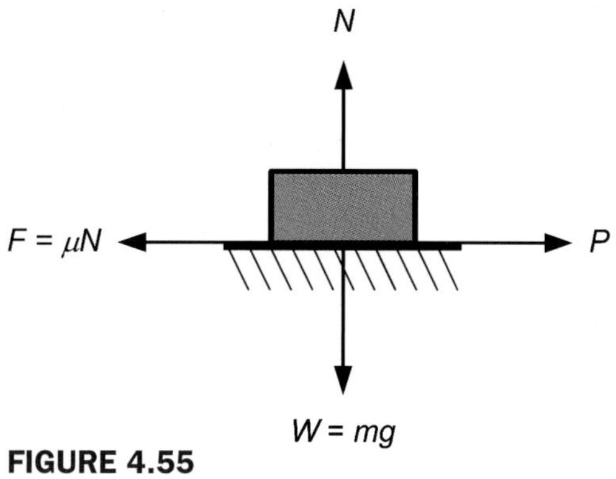

If the system shown in Figure 4.55 is in equilibrium (i.e. just on the point of moving), we can resolve the forces horizontally and vertically in order to obtain the following equations:

Horizontal: $P = F = \mu N$ (where μ is the coefficient of friction)

Vertical: $N = W = mg$ (where g = 9.81 m/s^2)

Combining these two equations gives:

$$F = \mu N = \mu m g$$

FIGURE 4.55
Friction and reaction

Test your knowledge 4.30
A train is travelling at 24 m/s. What deceleration is required in order to bring the train to rest in a time of 30 seconds.

Test your knowledge 4.31
A car starts from rest and accelerates at 2.5 m/s^2 for 11 seconds. What is its final velocity?

Key point
Friction always opposes the motion that produces it.

EXAMPLE 4.33

Calculate the horizontal force required to move a crate having a mass of 100 kg over a floor if the coefficient of friction, μ, between the crate and the floor is 0.35.

The force required to move the crate will be given by:

$$P = F = \mu m g = 0.35 \times 150 \times 9.81 = 515 \text{N}.$$

Key point
Static friction (the force required to start moving) is slightly greater than sliding friction (the force that opposes motion once it has started).

ACTIVITY 4.6

The data in Table 4.4 shows how the distance of a vehicle (relative to a fixed point) varies over a time period of time. Plot a graph showing distance against time and use it to answer the following questions:

1. Is the vehicle accelerating or decelerating? (Give reasons).
2. How far will the vehicle be from the reference point after (a) 4.5 s and (b) 7.5 s.
3. At what time will the vehicle be 45 m from the reference point?
4. What is the velocity of the vehicle?

ACTIVITY 4.7

The data in Table 4.5 shows how the speed of an aircraft varies over a period of time. Plot a graph showing velocity against time and use it to answer the following questions:

1. At what rate is the aircraft decelerating?
2. At what time will the aircraft be travelling at a velocity of 35 m/s?
3. What is the average velocity of the aircraft over the measuring period?
4. If the aircraft continues to decelerate at the same rate, at what time will it be at a standstill?

TABLE 4.4 See Activity 4.6	
Time (s)	Distance (m)
0	0
1	7.7
2	16
3	25
4	33
5	41
6	48
7	55
8	65

TABLE 4.5 See Activity 4.7	
Time (s)	Velocity (m/s)
0	55
1	52
2	50
3	47
4	44
5	41
6	39
7	36
8	33

REVIEW QUESTIONS

1. What current is flowing in a circuit if a charge of 2.5 C is transferred in a time of 20 s?
2. A current of 0.5 A flows in a 15 Ω resistor. Determine the voltage dropped across the resistor and the power that it dissipates.
3. State Ohm's Law.
4. Give three examples of good electrical conductors and explain how they conduct electricity.
5. Determine the equivalent resistance of the circuit shown in Figure 4.56.
6. A flux density of 2.8 mT is developed in free space over an area of 10 cm². Determine the total flux.
7. Sketch a typical B-H graph for a ferromagnetic material. Label the axes clearly and indicate the onset of magnetic saturation.
8. Describe TWO applications of magnetic fields. Illustrate your answer with sketches.
9. Identify the component shown in Figure 4.57 and explain how it works.
10. Determine the e.m.f. produced across the ends of a conductor of length 2 m moving at 2.5 m/s perpendicular to a magnetic field having a flux density of 0.3 mT.
11. Determine the mass and weight of an alloy block having dimensions 100 mm × 100 mm × 250 mm and density 3,300 kg/m³.
12. Determine the horizontal and vertical components of a 50 N force acting at a direction of 40° to the horizontal.
13. Forces of 2 kN and 8 kN act at right angles to one another. What single force could replace these two forces and what must its angle of action be relative to the 8 kN force?
14. A work bench weighs 175 kN. Determine the pressure exerted on the workshop floor if the load is distributed over a surface area of 0.22 m².
15. Explain what is meant by the 'moment' of a force. Illustrate your answer with a diagram.
16. Explain what is meant by the term 'gauge pressure'.

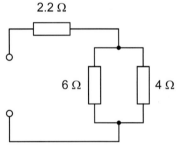

FIGURE 4.56
See Question 5

FIGURE 4.57
See Question 9

17. Calculate the horizontal force required to move a load having a mass of 50 kg over a surface if the coefficient of friction, μ, between the load and the surface is 0.4.

18. Explain the difference between 'static friction' and 'sliding friction'.

19. A race car accelerates from rest at a constant rate of 2.5 m/s^2. What velocity will it have after 12 s?

20. A lift raises a load of 150 kg through a vertical height of 7.5 m in 22 s. Determine the mechanical power.

21. A missile accelerates at a uniform rate of 12.5 m/s^2. If the missile has an initial velocity of 75 m/s, determine the time taken for the missile to reach a speed of 150 m/s.

Electronic Devices and Communication Applications

SUMMARY

The unit will provide you with a fascinating world of electronics. Before we look at the signals that convey information in electronic circuits we will introduce you to some of the units that are used when measuring electrical quantities, such as current, voltage and frequency. You will learn about the difference between analogue and digital signals and how to recognise signals from the shape of their waveforms.

Being able to 'read' and interpret a circuit diagram or 'schematic' is an essential skill required of every electronic technician and engineer. Many different parts and devices are used in electronic circuits and it is important that you should be able to recognise them, from both the symbols that we use to represent them in theoretical circuit diagrams and from their physical appearance.

Having introduced you to a wide range of electronic components we show how these are put together to realise a variety of simple analogue and digital circuits, including those used for communications and data transmission. Finally, you will put your newly acquired knowledge to good use when you build and test some real electronic circuits.

5.1 SIGNALS AND UNITS OF MEASUREMENT

In all forms of communication signals are used to convey *information*. The signals that we use in everyday life can take many forms including flashing lights, shouting, waving our hands, shaking our heads and others forms of "body language". In fact, life would be very difficult without signals – think about driving a car or motorbike in heavy traffic! In this section we shall be looking at how signals are used in electronics, how they can be converted from one form to another, and how they are measured.

Signals

In electronics, signals can take many forms including changes in voltage levels, pulses of current, and sequences of binary coded digits or *bits*. Signals that vary continuously in level are referred to as *analogue* signals whilst those that use discrete (i.e. fixed) levels are referred to as *digital* signals. Figure 5.1 shows some typical analogue and digital signals. Notice how

BTEC First Engineering: Mandatory and Selected Optional Units for BTEC Firsts in Engineering. DOI: 10.1016/B978-1-85617-685-9.00005-6

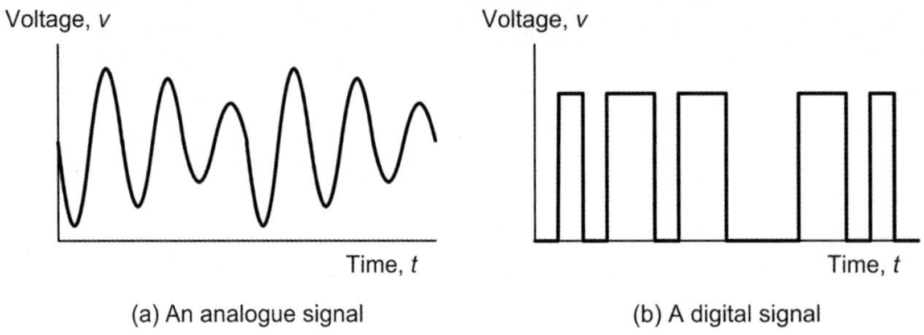

FIGURE 5.1
Typical analogue and
digital signals

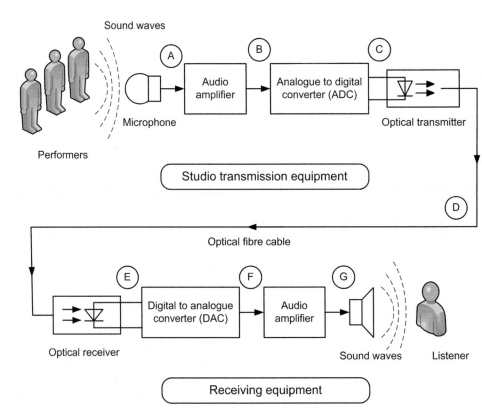

(a) An analogue signal

(b) A digital signal

FIGURE 5.2
A transmission system that uses analogue and digital signals

the digital signal exists only as a series of discrete voltage levels whilst the analogue signal
varies continuously from one voltage level to another.

Signals can also be quite easily converted from one form to another. For example, the signal
from the stage microphone at live radio broadcast will be an analogue signal at the point
at which the original sound is produced (i.e. on stage). After appropriate processing (which
might involve amplification and/or removal of noise and other unwanted sounds) it might
then be converted to a digital signal for radio transmission and then converted back to an
analogue signal before being amplified and sent to the loudspeaker at the point of reception.

SIGNAL CONVERSION

A device that converts an analogue signal to digital format is called an *analogue to digital
converter* (*ADC*) whilst one that converts a digital signal to analogue is referred to as a *digital
to analogue converter* (*DAC*). Figure 5.2 shows a typical transmission system that uses both
analogue and digital signals.

**Test your
knowledge 5.1**
Explain briefly what is
meant by each of the
following terms:

a. analogue
b. digital
c. ADC
d. DAC.

ACTIVITY 5.1

a. Explain the operation of the transmission system shown in Fig. 5.2 by making brief reference to each of the components in the system (i.e. microphone, analogue to digital converter, etc.).
b. At which points (A, B, C etc.) do the signals exist in digital form and at which points do they exist in analogue form?
c. What form do the signals have when present in the optical fibre cable?
d. Can you suggest any advantages and/or disadvantages of the transmission system?

ACTIVITY 5.2

Conventional audio compact disks (CD) are used to store signals that represent sounds. The recorded information comprises digital data derived from sampling the left and right audio channels taken at a rate of 44,100 samples per second. Use library or Internet resources to find out more about the optical technology used for compact disks. Present your findings in the form of a short word-processed article (of not more than 750 words) suitable for publication in a student newspaper.

ACTIVITY 5.3

Signals exist in a wide variety of different forms. Investigate the operation of a domestic weather station that uses remote sensors for sensing temperature, pressure, wind speed, wind direction and rainfall. How are these signals derived, what form do they take, and how are they transmitted from a remote location? Present your work in the form of an illustrated A3 poster suitable for display in the classroom.

Units of Measurement

You will find that a number of units of measurement are commonly used in electronics so let's get started by introducing some of them. In fact, it's important to get to know these units and also to be able to recognise their abbreviations and symbols before you actually need to use them! Later we explain how these units work in much greater detail but for now we simply list them so that at least you can begin to get to know something about them, see Table 5.1.

TABLE 5.1 Some electrical quantities and units of measurement

Parameter	Unit	Abbreviation	Notes
Electric potential	Volt	V	A potential of 1 Volt appears between two points when a current of 1 A flows in a circuit having a resistance of 1 Ω. Note that electric potential is also sometimes referred to as electromotive force (e.m.f.) or potential difference (p.d.) – see page 103
Electric current	Ampere	A	A current of 1 A flows in an electrical conductor when electric charge is being transported at the rate of 1 Coulomb per second
Electrical resistance	Ohm	Ω	An electric circuit has a resistance of 1 Ω when a p.d. (see above) of 1 V is dropped across it when a current of 1 A is flowing in it
Frequency	Hertz	Hz	A signal has a frequency of 1 Hz if one complete cycle of the signal occurs in a time interval of 1 s
Bit rate	Bits per second	bps	A signal has a bit rate of 1 bit per second if one complete binary digit is transmitted in a time interval of 1 s

Key point

Frequency and bit rate are very similar. They both indicate the speed at which a signal is transmitted but bit rate is for digital signals whilst frequency is used with analogue signals.

TABLE 5.2 Some common multiples and sub-multiples

Multiple	Prefix	Abbreviation	Example
×1,000,000	mega	M	92.4 MHz (92.4 million Hertz)
×1,000	kilo	k	4 kbs (4,000 bits per second)
×1	none	none	220 Ω (220 Ohms)
×0.001	milli	m	45 mV (0.045 Volts)
×0.000,001	micro	μ	33 μA (0.000033 Amps)

Key point

In order to avoid some potential confusion between the symbols and abbreviations that we use for units, the former are normally displayed in italic font. For example, V is used as both the abbreviation for voltage and for its unit symbol (the Volt). When used as a symbol in a formula is shown in italic, *V*, and when used as shorthand for Volts it is shown in normal (non-italic) font, V.

Multiples and Sub-Multiples

Unfortunately, because the numbers can be very large or very small, many of the electronic units can be cumbersome for everyday use. For example, the voltage present at the antenna of a mobile phone could be as little as one millionth of a volt, or 0.00001 V. Conversely, the resistance seen at this input of an audio amplifier stage could be as high as one million ohms, or 1,000,000 Ω.

To make life a lot easier we use a standard range of multiples and sub-multiples. These use a prefix letter in order to add a multiplier to the quoted value, as shown in Table 5.2.

CONVERTING TO/FROM MULTIPLES AND SUB-MULTIPLES

Converting to and from multiples and sub-multiples is actually quite easy, as the following examples show:

EXAMPLE 5.1

Convert 10,140 Hz to kHz. To do this you just need to move the decimal point **three** places to the **left**. This is the same as dividing by 1,000 (because there are 1,000 Hz in 1 kHz). Moving the decimal point three places to the left tells us that 10,140 Hz = 10.140 kHz

EXAMPLE 5.2

Convert 1,500,000 Ω to MΩ. To do this you need to move the decimal point **six** places to the **left**. This is the same as dividing by 1,000,000 (because there are 1,000,000 Ω in 1 MΩ. Moving the decimal point three places to the left tells us that 1,500,000 Ω = 1.5 MΩ.

EXAMPLE 5.3

Convert 0.525 V to mV. To do this you need to move the decimal point **three** places to the **right**. This is the same as multiplying by 1,000 (because there are 1,000 mV in 1 V). Moving the decimal point three places to the right tells us that 0.525 V = 525 mV.

Test your knowledge 5.2

State the units for:

a. electric current
b. electric potential
c. resistance
d. frequency
e. bit rate.

EXAMPLE 5.4

Convert 28,000 kbps to Mbps. To do this you need to move the decimal point **three** places to the **left**. This is the same as dividing by 1,000 (because there are 1,000 kbps in 1 Mbps). Moving the decimal point three places to the left tells us that 28,000 kbps = 28 Mbps.

Waveforms and Waveform Measurement

A graph showing the variation of voltage or current present in a circuit is known as a *waveform*. Waveforms show us how voltage or current signals vary with time. There are many common types of waveform encountered in electronic circuits including sine (or sinusoidal), square, triangle, ramp or sawtooth (which may be either positive or negative going), and pulse.

Complex waveforms, like speech and music, usually comprise many different signal components at different frequencies. *Pulse waveforms* are often categorized as either

Test your knowledge 5.3

Explain what is meant by each of the following abbreviations:

a. mV
b. μA
c. kHz
d. Mbps.

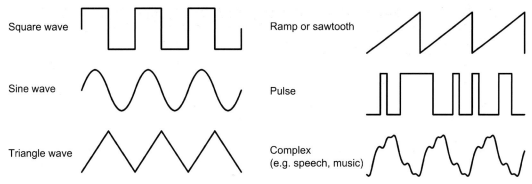

FIGURE 5.3
Some common waveforms

repetitive or *non-repetitive* (the former comprises a pattern of pulses that repeats regularly while the latter comprises pulses which each constitute a unique event). Some common waveforms are shown in Fig. 5.3.

FREQUENCY

The frequency of a repetitive waveform is the number of cycles of the waveform which occur in unit time (i.e. one second). Frequency is expressed in Hertz, Hz, and a frequency of 1 Hz is equivalent to one cycle per second. Hence, if a voltage has a frequency of 400 Hz, 400 cycles of it will occur in every second.

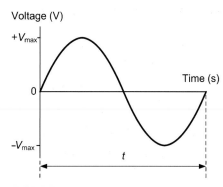

FIGURE 5.4
One cycle of a sine wave voltage showing its periodic time

PERIODIC TIME

The periodic time (or period) of a waveform is the time taken for one complete cycle of the wave (see Fig. 5.4). The relationship between periodic time and frequency is thus:

$$t = 1/f \text{ or } f = 1/t$$

where t is the periodic time (in s) and f is the frequency (in Hz).

EXAMPLE 5.5

A waveform has a frequency of 400 Hz. What is the periodic time of the waveform?

Here we must use the relationship $t = 1/f$ where $f = 400$ Hz.

Hence, $t = 1/400 = 0.0025$ s (or 2.5 ms)

EXAMPLE 5.6

A waveform has a periodic time of 40 ms. What is its frequency?

Here we must use the relationship $f = 1/t$ where $t = 40$ ms or 0.04 s.

Hence, $f = 1/0.04 = 25$ Hz.

AMPLITUDE

The amplitude (or *peak value*) of a waveform is a measure of the extent of its voltage or current excursion from the resting value (usually zero). The *peak-to-peak* value for a wave

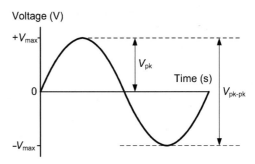

FIGURE 5.5
One cycle of a sine wave voltage showing its peak and peak-peak values

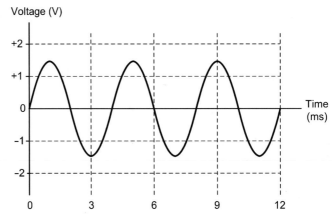

FIGURE 5.6
See Test your knowledge 5.6

which is symmetrical about its resting value is twice its peak value (see Fig. 5.5). These units are usually more convenient to use when taking measurements from a waveform display.

PULSE WAVEFORMS

When describing rectangular and pulse waveforms we use a different set of parameters. These include:

On time, t_{on}

This is the time for which the pulse is present at its maximum amplitude. This is sometimes referred to as the 'mark time'. Note that, when a pulse is not perfectly rectangular (i.e. when it takes some time to change from one level to the other) we define the off time as the time for which the pulse amplitude remains above 50% of its maximum value.

Off time, t_{off}

This is the time for which the pulse is not present (i.e. zero voltage or current). This is sometimes referred to as the 'space time'. Note that, when a pulse is not perfectly rectangular (and takes some time to change from one level to another) we define the off time as the time for which the pulse amplitude falls below 50% of its maximum value.

Pulse period, t

This is the time for one complete cycle of a repetitive pulse waveform. The periodic time is thus equal to the sum of the on and off times (but once again note that this is only valid if the pulse train is repetitive and is meaningless if the pulses occur at random intervals). When a pulse train is not perfectly rectangular the pulse period is measured at the 50% amplitude points.

Pulse repetition frequency, *prf*

The pulse repetition frequency (prf) is the reciprocal of the pulse period. Hence:

$$prf = 1/t = 1/(t_{on} + t_{off})$$

Mark to space ratio

The mark to space ratio of a pulse wave is simply the ratio of the on to off times. Hence:

$$Mark\ to\ space\ ratio = t_{on} : t_{off}$$

Note that, for a perfect square wave the mark to space ratio will be 1:1 because $t_{on} = t_{off}$.

Duty cycle

The duty cycle of a pulse wave is the ratio of the on time to the on plus off time (and usually expressed as a percentage). Hence:

$$Duty\ cycle = t_{on}/(t_{on} + t_{off}) \times 100\% = t_{on}/t \times 100\%$$

For a perfect square wave the duty cycle mark to space ratio will be 50%.

Test your knowledge 5.7

Fig. 5.8 shows a waveform diagram.

a. What type of waveform is shown?
b. What is the amplitude of the waveform?
c. What is the period of the waveform?
d. What is the repetition frequency of the waveform?
e. What is the mark-to-space ratio of the waveform?

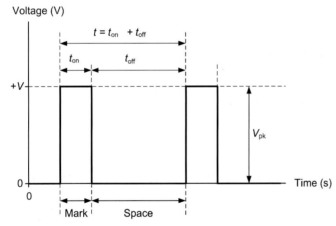

FIGURE 5.7
A rectangular pulse waveform showing 'on' and 'off' times

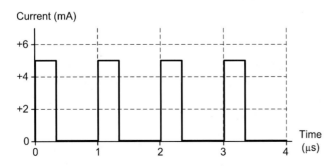

FIGURE 5.8
See Test your knowledge 5.7

145

ACTIVITY 5.4

Waveforms are usually displayed using an instrument called an *oscilloscope*. You will learn more about this instrument in Unit 19 (see Chapter 8, *Electronic circuit construction and testing*). Oscilloscopes can be stand-alone test instruments (see Fig. 5.9) or they can be virtual instruments that use a PC's in-built signal processing capabilities (e.g. the analogue to digital converter in a PC sound card).

Fig. 5.10 shows a typical virtual instrument display obtained by using a soundcard oscilloscope program. The program receives its data from the computer's sound card with a sampling rate of 44.1 kHz and a resolution of 16-bis. The data source can be selected by the PC's own sound card controls (e.g. microphone, line input or wave). The frequency range of the instrument depends on the performance of the computer's sound card, but is typically over the range 20 Hz to 20 kHz. The oscilloscope also contains a simple signal generator producing sine, square, triangle and sawtooth waveforms in the frequency range from 0 to 20 kHz. These signals are available at the speaker output of the sound card.

Use Fig. 5.10 to answer the following questions:

a. What type of waveform is shown?
b. What total time interval is displayed on the screen: (Hint: look at the horizontal scale)
c. What settings are used for the vertical and horizontal scales on the oscilloscope display?
d. What is the greatest positive voltage present in the waveform sample?
e. What is the greatest negative voltage present in the waveform sample?
f. What is the overall peak-peak voltage of the waveform?

Download a copy of the Soundcard Oscilloscope software and investigate the operation of the program using some typical signals applied to the microphone or auxiliary inputs of a PC. The software is available from Christian Zeitnitz's website: http://www.zeitnitz.de/Christian/scope_en

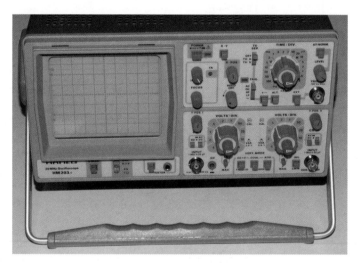

FIGURE 5.9
A typical bench oscilloscope

5.2 ELECTRONIC COMPONENTS AND DEVICES

To be able to understand a circuit diagram you first need to be familiar with the symbols that are used to represent the components and devices. To be able to build or maintain an electronic circuit you also need to know the physical appearance of a wide range of electronic components. These include, cells, batteries, resistors, capacitors, inductors, transformers, diodes and transistors as well as connectors, switches and integrated circuits.

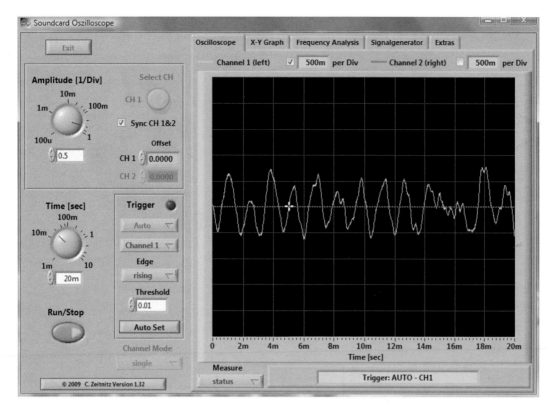

FIGURE 5.10
See Activity 5.4

Cells, Batteries, Power Supplies and Connectors

Cells and batteries provide the power for a wide range of portable and hand-held electronic equipment. There are two basic types of cell, primary and secondary. Primary cells produce electrical energy at the expense of the chemicals from which they are made and once these chemicals are used up, no more electricity can be obtained from the cell. An example of a primary cell is an ordinary 1.5 V AA alkaline battery. In secondary cells, the chemical action is reversible. This means that the chemical energy is converted into electrical energy when the cell is discharged whereas electrical energy is converted into chemical energy when the cell is being charged. An example of a secondary cell is a 1.2 V AA nickel cadmium (NiCd) battery.

FIGURE 5.11
Some typical cells and batteries used in electronic equipment

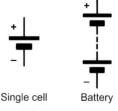

Single cell   Battery

FIGURE 5.12
Symbols for cells and batteries

147

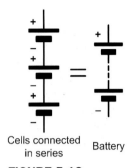

FIGURE 5.13
Series arrangement of cells

Cells connected in series   Battery

FIGURE 5.14
A typical power supply

Key point
We refer to the output voltage produced by a battery or a power voltage as an electromotive force (e.m.f.). Electromotive force is measured in Volts, V. In contrast, we refer to the voltage drop across an electronic component (such as a resistor or capacitor) as a potential difference (p.d.). Potential difference is also measured in Volts (V). The best way to distinguish between e.m.f. and p.d. is to remember that e.m.f. is the 'cause' and p.d. is the 'effect'.

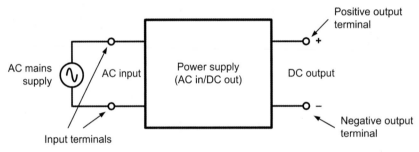

FIGURE 5.15
A simplified block schematic representation of the power supply in Fig. 5.14

In order to produce a battery, individual cells are usually connected in series with one another, as shown in Fig. 5.13. The voltage produced by a battery with n cells will be n times the voltage of one individual cell (assuming that all of the cells are identical). Furthermore, each cell in the battery will supply the same current. Series connected cells are often used to form batteries. For example, the popular PP3, PP6 and PP9 batteries are made from six "layered" 1.5 V primary alkaline cells which are effectively connected in series. A 12 V car battery, on the other hand, uses six 2 V lead-acid secondary cells connected in series.

Where an electronic circuit derives its power from an AC mains supply, we sometimes show the supply as a box with two terminals (one marked positive and one marked negative). Treating the power supply as a separate unit helps keep the circuit simple. If the power supply fails we can simply replace the entire unit in much the same way as we would replace a set of exhausted batteries. A typical power supply which has an AC mains input and DC output is shown in Fig. 5.14. Fig. 5.15 how we can represent the power supply using a simple *block schematic diagram*. Note that we have not shown any switches, fuses or indicators in this diagram!

The input and output of the power supply in Fig. 5.15 will be connected using appropriately designed and rated connectors. Depending on the particular application, electronic circuits use a wide range of different types of connector.

Test your knowledge 5.8
A portable TV uses a battery which has six 2V cells connected in series. What e.m.f. does this battery supply?

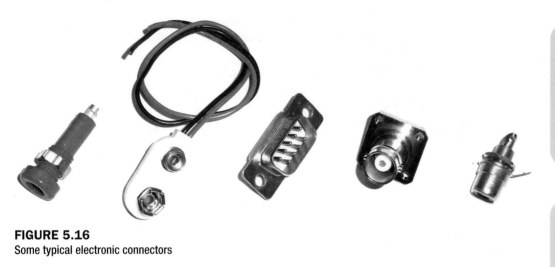

FIGURE 5.16
Some typical electronic connectors

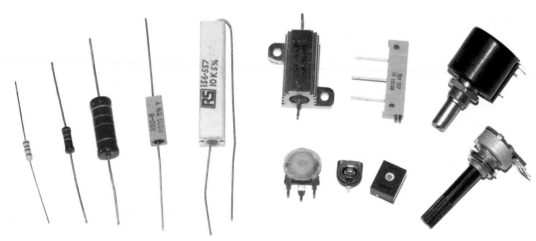

FIGURE 5.17
Various types of resistor, including fixed, preset and variable types

Resistors

Resistance can be thought of as an opposition to the flow of electric current. The amount of current that will flow in a circuit when a given electromotive force (**e.m.f.**) is applied to it is inversely proportional to its resistance. In other words, the larger the resistance, the greater the opposition to current flow when an e.m.f. is applied.

Various types of fixed, preset and variable resistor are found in electronic circuits, including carbon film, metal film, and wire-wound types, see Fig. 5.17. Resistors are used for determining the voltages and currents in circuits, as 'loads' to consume power, and in preset and variable form for making adjustments (for example, volume and tone controls). Typical circuit symbols for various types of resistor are shown in Fig. 5.18. More resistors are shown in Figs. 8.14 to 8.17 on pages 262.

The terms *potentiometer* and variable resistor are often used interchangeably. However, strictly speaking, preset and variable resistors have only two terminals whilst potentiometers (either preset or rotary types) have three terminals. Note also that a preset or variable potentiometer can be used as a variable resistor by simply ignoring one of its end terminals, or by connecting its moving contact to one of its outer terminals.

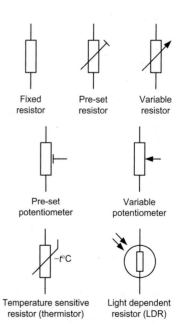

Fixed resistor   Pre-set resistor   Variable resistor

Pre-set potentiometer   Variable potentiometer

Temperature sensitive resistor (thermistor)   Light dependent resistor (LDR)

FIGURE 5.18
Symbols used for resistors

149

Test your knowledge 5.11

An electronic circuit uses resistors having the following values: $68\,\Omega$, $1.2\,k\Omega$, $27\,\Omega$, and $3.3\,M\Omega$. Assuming that a tolerance of $\pm10\%$ is required, from which series of preferred values should these resistors be taken?

TABLE 5.3 The E6, E12 and E24 series of preferred values

Series of preferred values	Values available
E6	1.0, 1.5, 2.2, 3.3, 4.7, 6.8
E12	1.0, 1.2, 1.5, 1.8, 2.2, 2.7, 3.3, 3.9, 4.7, 5.6, 6.8, 8.2
E24	1.0, 1.1, 1.2, 1.3, 1.5, 1.6, 1.8, 2.0, 2.2, 2.4, 2.7, 3.0, 3.3, 3.6, 3.9, 4.3, 4.7, 5.1, 5.6, 6.2, 6.8, 7.5, 8.2, 9.1

The specifications for a resistor usually include the value of resistance (expressed in Ω, $k\Omega$ or $M\Omega$), the accuracy or *tolerance* of the marked value (quoted as the maximum permissible percentage deviation from the marked value), and the power rating (which must be equal to, or greater than, the maximum expected power dissipation).

Fixed resistors are available in several series of preferred values, see Table 5.3. The number of values provided with each series (i.e. 6, 12 and 24) is determined by the tolerance involved. In order to cover the full range of resistance values using resistors having a $\pm20\%$ tolerance it will be necessary to provide six basic values (known as the E6 series). More values will be required in the series that offers a tolerance of $\pm10\%$ and consequently the E12 series provides twelve basic values. The E24 series for resistors of $\pm5\%$ tolerance provides 24 basic values and, as with the E6 and E12 series, decade multiples (i.e., $\times1$, $\times10$, $\times100$, $\times1\,k$, $\times10\,k$, $\times100\,k$ and $\times1\,M$) of the basic series. Carbon and metal oxide resistors are normally marked with colour codes that indicate their value and tolerance. Resistor markings are described and explained on page 264.

Capacitors

Capacitors store energy in the form of an electric field. When a potential difference is applied to two conducting plates an electric charge will appear on the plates and an electric field will appear between the plates. The field can be concentrated intensified by placing an insulating material (such as polyester film, mica or a ceramic material) between the plates. This material acts as a dielectric and its electrical properties help to increase the *capacitance* of the component.

Capacitors provide us with a means of storing and conserving electric charge. They are widely used in power supplies where they act as "reservoirs" for charge and also in many timing and wave shaping circuits. Capacitors will pass alternating currents but they will "block" direct current (once charged). They are thus used for coupling signals (which are AC) in and out of amplifier stages.

The specifications for a capacitor usually include the value of capacitance (expressed in μF, nF or pF), the accuracy or tolerance of the marked value (quoted as the maximum permissible percentage deviation from the marked value), the voltage rating (which must be equal to, or greater than, the maximum expected voltage applied to the capacitor). Capacitors are usually available with values in the E6 series (see Table 5.3).

Large value capacitors often use a chemical dielectric material and they require the application of a DC *polarising voltage* in order to work properly. This voltage must be applied with the correct polarity (invariably this is clearly marked on the case of the capacitor) with a positive ($+$) sign or negative ($-$) sign or a coloured stripe or other marking. Failure to observe the correct polarity can result in over-heating, leakage, and even a risk of explosion! Capacitor markings are described and explained on page 266.

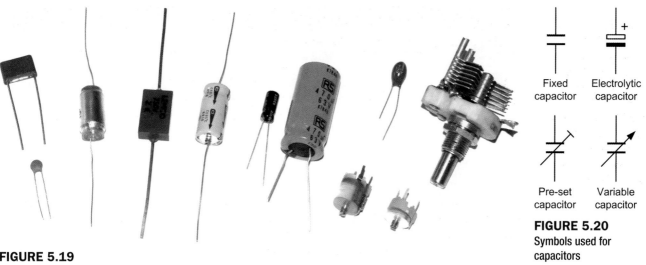

FIGURE 5.19
Various types of capacitor, including fixed, preset and variable types

Fixed capacitor

Electrolytic capacitor

Pre-set capacitor

Variable capacitor

FIGURE 5.20
Symbols used for capacitors

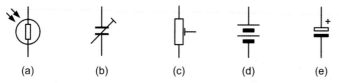

(a)   (b)   (c)   (d)   (e)

FIGURE 5.21
See Test your knowledge 5.12

Test your knowledge 5.12
Identify each of the five electronic component symbols shown in Fig. 5.21.

Inductors and Chokes

Inductors store energy in the form of a magnetic field. When current flows in a conductor a magnetic field appears in the space that surrounds it. The field can be concentrated intensified by winding the conductor into a coil and intensified by placing a magnetic material (such as steel, iron or ferrite) inside the coil. This material acts as *core* and its magnetic properties help to increase the *inductance* of the component. Larger inductors with steel cores are sometimes referred to as *chokes*.

Inductor specifications normally include the value of inductance (expressed in H, mH, μH, or nH), the current rating (i.e. the maximum current which can be continuously applied to the inductor under a given set of conditions), and the accuracy or tolerance (quoted as the maximum permissible percentage deviation from the marked value).

Fixed inductors are usually available with values in the E6 series (see Table 5.3). Variable inductors used in high-frequency applications often have ferrite dust cores that can be adjusted in order to obtain a precise value of inductance.

Transformers

Transformers are another useful type of electronic component. Transformers consist of two (or more) coil windings wound on a common magnetic core (usually steel or ferrite). The input winding is referred to as the primary whilst the output winding is called the secondary. An alternating current flowing in the primary winding will produce an alternating magnetic flux within the transformer core. This will link with the secondary winding and induce an alternating voltage in it.

FIGURE 5.22
Various types of inductor

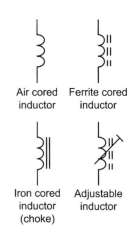

FIGURE 5.23
Symbols used for
inductors

FIGURE 5.24
Various types of transformer

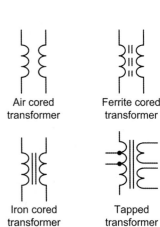

FIGURE 5.25
Symbols used for transformers

Transformers provide us with a means of coupling AC power from one circuit to another without a direct connection between the two. A further advantage of transformers is that voltage may be stepped-up (secondary voltage *greater* than primary voltage) or stepped-down (secondary voltage *less* than primary voltage).

The specifications for a transformer usually include the rated primary and secondary voltages and currents, the required power rating, usually expressed in Volt-Amperes, VA).

Diodes

Semiconductors form the basis of many important electronic components such as diodes, transistors and integrated circuits. Semiconductor devices are made from materials that are neither conductors nor insulators. During manufacture they can be given properties that make them ideal for a wide range of applications in electronic circuits.

Test your knowledge 5.13
Identify each of the five electronic component symbols shown in Fig. 5.26.

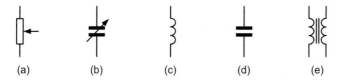

FIGURE 5.26
See Test your knowledge 5.13

(a)　　(b)　　(c)　　(d)　　(e)

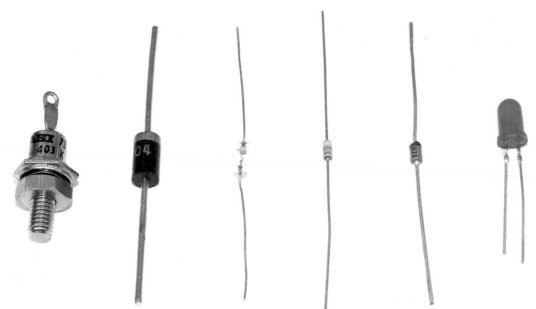

FIGURE 5.27
Various types of diode

Diodes conduct current in one direction but not the other. They can thus act as a simple 'one way street' turning 'on' when the current is applied in one direction and 'off' when it is in the opposite direction. Because of this, diodes are frequently used as a means of converting alternating current (AC) to direct current (DC).

Diodes are often divided into signal or rectifier types according to their use. Signal types operate with low current and low voltage whilst rectifier types used in power supplies operate with much higher voltages and currents.

ZENER DIODES

Zener diodes are a special type of diode that conducts heavily at a particular voltage in the reverse direction (known as the *Zener voltage*). This property makes the Zener diode ideal for use as a *voltage reference* in power supply and other circuits. Zener diodes are available with various voltage ratings in the E6 and E12 series (see page 150).

LIGHT EMITTING DIODES

Light emitting diodes (LED) produce light when a small current is applied to them and they are ideal for use as general-purpose *indicators*. Compared with conventional filament lamps they operate from significantly smaller voltages and currents. They are also very much more reliable than filament lamps. Most LED will provide a reasonable level of light output when a forward current of as little as 10 mA to 20 mA is supplied.

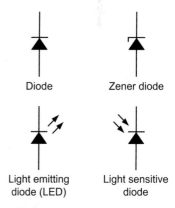

Diode　　Zener diode

Light emitting
diode (LED)

Light sensitive
diode

FIGURE 5.28
Symbols used for diodes

ACTIVITY 5.5

Use library or internet resources to obtain data on each of the following diodes:

a. 1N4148
b. BZX85 C9V1

For each diode:

1. State the name of the manufacturer and/or supplier.
2. State the type of diode (e.g. general purpose silicon)
3. Draw the circuit schematic symbol for the diode and identify the individual diode connections by labelling the terminals (e.g. anode, cathode)
4. State the manufacturer's maximum ratings for the transistor in terms of voltage, current and/or power.

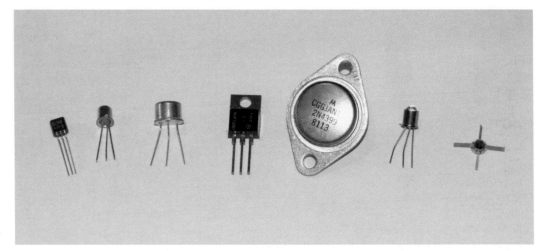

FIGURE 5.29
Various types of transistor

NPN bipolar junction transistor (BJT)

PNP bipolar junction transistor (BJT)

N-channel junction gate field effect transistor (JFET)

P-channel junction gate field effect transistor (JFET)

NPN light sensitive transistor

NPN Darlington transistor

FIGURE 5.30
Symbols used for transistors

Transistors

Transistors fall into two main categories: *bipolar junction transistors* (BJT) and *unipolar field-effect transistors* (FET). BJT comprise NPN or PNP junctions of N-type and P-type semiconductor material (usually silicon). The junctions are extremely small and they are produced in a single slice of silicon by diffusing impurities through a photographically reduced mask. Field effect transistors (FET) comprise a channel of P-type or N-type material surrounded by material of the opposite polarity.

Transistors are often classified according to their field of application (e.g. small-signal, general purpose, power, switching, high-frequency, etc.).

Switches

Switches provide us with a means of interrupting the current in a circuit. An obvious application for a switch is that of connecting or disconnecting the supply to a circuit. Switches come in many shapes and forms according to the application concerned. They include push button, rocker, toggle, rotary, microswitches and also specialised switches that detect tilt, pressure, and the presence of a nearby magnetic field. Fig. 5.33 shows some common types of switch whilst Fig. 5.34 shows a selection of circuit symbols used for switches.

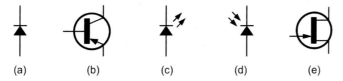

FIGURE 5.31
See Test your knowledge 5.14

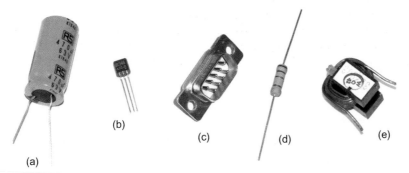

FIGURE 5.32
See Test your knowledge 5.15

ACTIVITY 5.6

Use library or internet resources to obtain data on each of the following transistors:

a. BC548
b. 2N3819

For each transistor:

1. State the name of the manufacturer and/or supplier.
2. State the type of transistor (e.g. PNP silicon BJT)
3. Draw the circuit schematic symbol for the transistor and identify the individual transistor connections by labelling the terminals (e.g. collector, base, emitter)
4. State the manufacturer's maximum ratings for the transistor in terms of voltage, current and/or power.

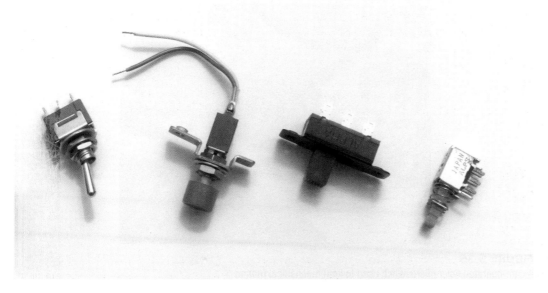

FIGURE 5.33
Some common types of switch

155

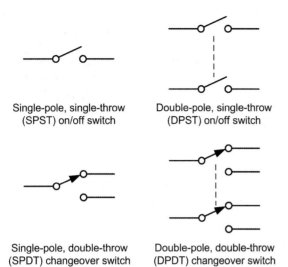

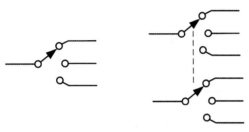

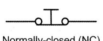

Single-pole, single-throw (SPST) on/off switch

Double-pole, single-throw (DPST) on/off switch

Single-pole, double-throw (SPDT) changeover switch

Double-pole, double-throw (DPDT) changeover switch

One-pole, three-way (1P 3W) (multipole selector switch)

Two-pole, three-way (2P 3W) (multipole selector switch)

Normally-open (NO) push-button switch

Normally-closed (NC) push-button switch

FIGURE 5.34
Symbols used for switches

The most basic form of switch is the *single-pole single-throw* (*SPST*) switch. This switch has a simple on/off action when respectively closed and opened (see Fig. 5.33). The *double-pole single-throw* (*DPST*) switch is similar to the (SPST) switch with its on/off action but is capable of switching two circuits independently (see Fig. 5.33). A further switch type has a changeover action and is available in both *single-pole double-throw* (*SPDT*) and *double-pole double-throw* (*DPDT*) variants (see Fig. 5.33).

Some switches provide momentary operation, other switches 'toggle' and remain in whatever state they are set to until operated again. The contacts of a momentary action switch will break or make only for the duration for which they are operated. Switches of this type are usually push button types but there are also lever-operated *microswitches* which are used in mechanisms such as those that detect a door or window opening (see Fig. 5.35).

When the contacts of a momentary action switch are normally open and they close, completing the circuit when operated, they are referred to as *normally open* (*NO*) contacts. Conversely, when the contacts of a momentary action switch are normally closed and they open, breaking the circuit when operated, they are referred to as *normally closed* (*NC*) contacts.

Sensors and Transducers

Sensors and transducers are important components in electronic instrumentation and control systems. Transducers are devices that convert energy in the form of sound, light, heat, etc., into an equivalent electrical signal, or vice versa. A sensor is a transducer that is used to generate an input signal to

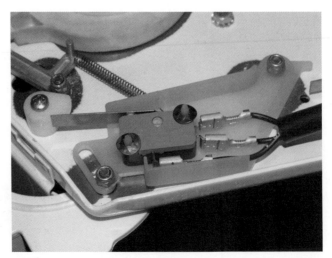

FIGURE 5.35
A cam-operated lever microswitch used to limit mechanical motion

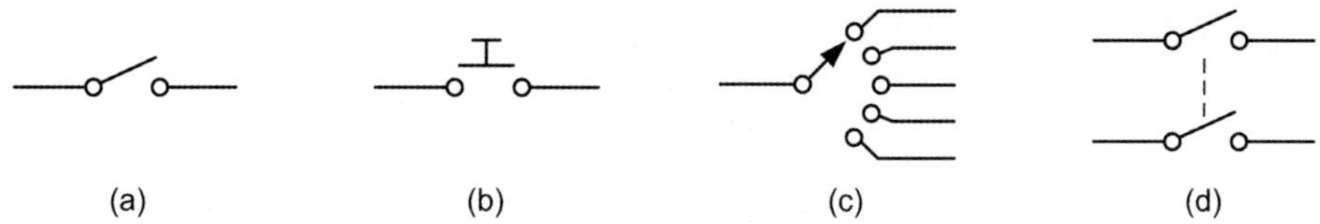

FIGURE 5.36
See Test your knowledge 5.16

Test your knowledge 5.16
Identify each of the four electronic component symbols shown in Fig. 5.36.

a control or measurement system. Before we go further, let's consider a couple of examples that you will already be familiar with. A loudspeaker is a device that converts low-frequency electric current into sound. A thermocouple, on the other hand, is a transducer that converts a temperature difference into a voltage.

It's important to appreciate that transducers are used both as *inputs* to and *outputs* from electronic systems. From the two previous examples, it should be obvious that a loudspeaker is an output transducer designed for use in conjunction with an audio system whereas a thermocouple is an input transducer which can be used in a temperature control system.

SENSORS AND INPUT TRANSDUCERS

The signal produced by a sensor is an electrical analogy of a physical quantity, such as angular position, distance, velocity, acceleration, temperature, pressure, light level, etc. The signals returned from a sensor, together with control inputs from the operator (where appropriate) will subsequently be used to determine the output from the system. The choice of sensor is governed by a number of factors including accuracy, resolution, cost, electrical specification and physical size.

Sensors can be categorised as either active or passive. An *active sensor* generates a current or voltage output. A *passive transducer* requires a source of current or voltage and it modifies this in some way (e.g. by virtue of a change in the sensor's resistance). The result may still be a voltage or current but it is not generated by the sensor on its own.

Sensors can also be classed as either digital or analogue. The output of a *digital sensor* can exist in only two discrete states, either 'on' or 'off', 'low' or 'high', 'logic 1' or 'logic 0', etc. The output of an *analogue sensor* can take any one of an infinite number of voltage or current levels. It is thus said to be continuously variable. A variety of common types of sensor are summarised in Table 5.4.

INDICATORS AND OUTPUT TRANSDUCERS

Output transducers convert electrical signals into physical quantities such as heat, light, sound and motion. Typical examples of output transducers include lamps, light emitting diodes (LED), buzzers, loudspeakers, actuators, motors, solenoids and heaters. Fig. 5.28 on page 153 shows the symbols that are used to represent several of these components in circuit schematics.

RELAYS

Relays are switching devices that are operated from low-voltage and low-current but which are capable of switching much higher voltages and currents. The simplest form of relay comprises a coil wound on a soft-iron core and a set of spring-loaded contacts which are driven by a moving armature. When a current flows in the coil a magnetic field is produced and the armature moves, closing or opening the relay contacts. More complex forms of relay use solid state devices. They tend to be more reliable and operate much faster than simple electromechanical types.

157

TABLE 5.4 Some common sensors and input transducers

Quantity sensed	Transducer name	Notes
Angular position	Resistive rotary position sensor (potentiometer)	Rotary track potentiometer with linear law produces analogue voltage proportional to angular position
Angular velocity (rotary speed)	Tachogenerator	Small DC generator with linear output characteristic. Analogue output voltage proportional to shaft speed
Light level	Light dependent resistor (LDR)	An analogue output voltage results from a change of resistance within a cadmium sulphide (CdS) sensing element
Linear position	Resistive linear position sensor	Linear track potentiometer with produces an analogue voltage proportional to linear position
Liquid level	Float switch	Simple switch that operates when a particular liquid level is reached (often 'full' or 'empty')
Liquid flow	Flow rate sensor	Small turbine comprising vanes driven by moving fluid. Turbine interrupts infrared light beam and generates digital pulses. Pulse repetition frequency is proportional to flow rate
Pressure	Strain gauge	Force exerted on a resistive element causes a change in resistance. Multiple transducers are used to generate an analogue output voltage which is proportional to the pressure applied
Sound	Microphone	Changes in sound pressure impact on a diaphragm connected to a coil that moves inside a magnetic field. When the coil moves a small voltage is generated
Temperature	Thermocouple	Junction of wires made from dissimilar metals. A small voltage is generated when a difference of temperature exists between the measuring and reference junctions

FIGURE 5.37
An analogue linear position sensor based on a linear resistive potentiometer

FIGURE 5.38
A float switch liquid level sensor

FIGURE 5.39
A turbine flow-rate sensor with digital output

ACTIVITY 5.7

Use library or internet resources to obtain a data sheet for an AD590 semiconductor temperature sensor. Use the data sheet to answer the following questions:

1. What output current would be produced when the sensor is used at a temperature of −15°C?
2. What temperature is being sensed when the output current is 0.32 mA?
3. Sketch and label the pin connections for an AD590 sensor.
4. Suggest TWO applications for an AD590 sensor.

Test your knowledge 5.17

Identify sensors and transducers for use in each of the following applications:

a. Sensing the temperature in a gas boiler
b. Sensing the light level in an automatic night light
c. Measuring the flow of fuel supplied to a car engine
d. Measuring the strain in a bridge support
e. Generating a warning signal when a water tank is empty
f. Alerting parents when a baby is crying.

TABLE 5.5 Some common indicators and output transducers

Output generated	Transducer name	Notes
Rotary motion	Motor	Current flowing in a coil suspended in a magnetic field produces rotation of an output shaft (electrical energy is converted into angular kinetic energy)
Linear motion	Linear actuator	Motor arrangement (see above) drives linear track via a toothed gear wheel (electrical energy is converted into linear kinetic energy)
Sound	Loudspeaker	Current flowing in a coil suspended in a magnetic field produces motion. The coil is attached to a diaphragm that exerts pressure on a mass of air and produces sound vibrations
Heat	Heating element	Current flowing in a resistor produces heat (electrical energy is converted into heat energy)
Light (white)	Lamp or light bulb	Current flowing in a resistive element placed in an evacuated glass envelope produces white light (electrical energy is converted into light and heat)
Light (red or infrared)	Light emitting diode (LED)	Current flowing in a semiconductor junction diode produces light in the infrared spectrum (electrical energy is converted into visible and invisible light)

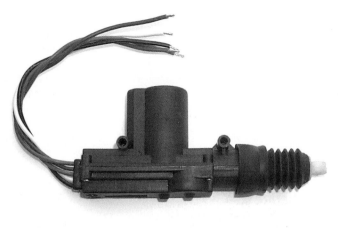

FIGURE 5.40
A linear actuator

FIGURE 5.41
A stepper motor

FIGURE 5.42
Some common types of relay

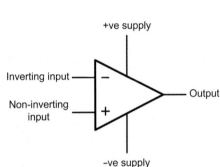

FIGURE 5.43
Symbol for an operational amplifier

Operational Amplifiers

Operational amplifiers are analogue integrated circuits designed for linear amplification that offer near-ideal characteristics (very high voltage gain and input resistance coupled with low output resistance and wide bandwidth). Operational amps can be thought of as universal gain blocks to which external components are added in order to define their function within a circuit. For example, by adding just two or three resistors, we can produce an amplifier having a precisely defined *gain*.

The symbol for an operational amplifier is shown in Fig. 5.43. There are several important things to note about this. The device has two inputs and one output and no common connection. Furthermore, we often do not show the supply connections – it is often clearer to leave them out of the circuit altogether!

In Fig. 5.43 one of the inputs is marked '−' and the other is marked '+'. These polarity markings have nothing to do with the supply connections – they indicate the overall phase shift between each input and the output. The '+' sign indicates zero phase shift whilst the '−' sign indicates 180° phase shift. Since 180° phase shift produces an inverted (turned 'upside down') waveform, the '−' input is often referred to as the *inverting input*. Similarly, the '+' input is known as the *non-inverting input*.

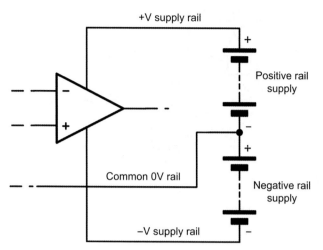

FIGURE 5.44

Supply connections for an operational amplifier

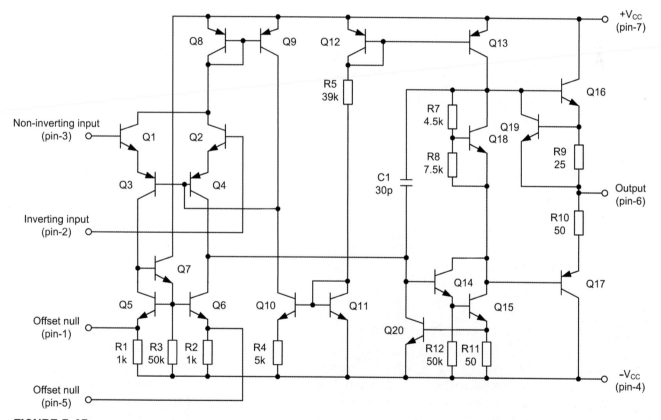

FIGURE 5.45

Internal circuit of a 741 operational amplifier

Most (but not all) operational amplifiers require a symmetrical supply (typically $\pm 5\,V$ to $\pm 15\,V$). This allows the output voltage to swing both positive (above $0\,V$) and negative (below $0\,V$). Fig. 5.44 shows how the supply connections would appear if we decided to include them. Note that we usually have two separate supplies; a positive supply and an equal, but opposite, negative supply. The common connection to these two supplies (i.e. the $0\,V$ rail) acts as the common rail in our circuit. The input and output voltages are usually measured relative to this rail.

Operational amplifiers are supplied as *integrated circuits*. These are 'chips' of semiconductor material that contain a large number of individual devices, all fabricated on the same small

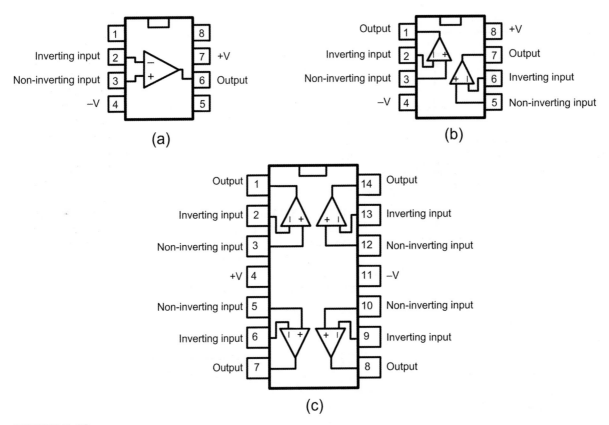

(a)

(b)

(c)

FIGURE 5.46
Typical operational amplifier integrated circuit packages. (a) Single operational amplifier type TL071. (b) Dual operational amplifier type TL072. (c) Quad operational amplifier type TL074

FIGURE 5.47
An LM324 quad operational amplifier used both as an amplifier and as a comparator

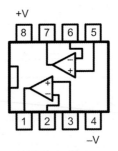

FIGURE 5.48
See Test your knowledge 5.19

slice of semiconductor material. Operational amplifiers can be packaged singly, as dual operational amplifiers with two independent operational amplifiers, or as quads with four independent operational amplifiers all operating from the same supply. Fig. 5.46 shows typical *dual-in-line* (*DIL*) packages used for single, dual and quad operational amplifiers.

ACTIVITY 5.8

Use library or internet resources to obtain data on each of the following operational amplifiers:

a. 741

b. LM324

For each operational amplifier:

1. State the name of the manufacturer and/or supplier.
2. State the number of individual operational amplifiers supplied within a single package
3. Draw and label the pin connections for the operational amplifier
4. State the manufacturer's recommended range of supply voltages.

> **Key point**
> Operational amplifiers are analogue integrated circuits that provide very high voltage gain and very high input resistance. Operational amplifiers are often thought of as universal gain blocks to which a few external discrete components are added in order to define their function within a circuit.

Operational amplifiers can be used for amplification; applying gain and increasing signal voltages. They can also be used as *comparators* where they compare one voltage with another and provide a high or low output voltage depending on whether the voltage at the inverting input is respectively less than or greater than the voltage present at the non-inverting input.

Logic Gates

We all make decisions based on logic. The decisions that we make are invariably conditional (i.e. they depend on a particular set of circumstances) and they usually take the form:

If (condition) then (action).

For example:

If cold then put on the heating

or

If warm then open the window.

We also make compound decisions that are based on more than one set of circumstances. They take the form:

If (condition 1) or (condition 2) then (action).

or

If (condition 1) and (condition 2) then (action).

For example:

If hungry or thirsty then go to the cafe

or

If night and sleepy then go to bed.

Another possibility is that we might want to make a decision based on the absence of a condition rather than its presence.

For example:

If (condition 1) or (condition 2) and not (condition 3) then (action).

An example of this might be:

If hungry or thirsty and not raining then walk to the shops.

The important words in the above conditional statements are 'and', 'or' and 'not'. These words describe logical operations and we will next look at ways in which electronic logic circuits can emulate these functions so that they can make decisions automatically.

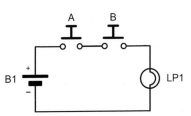

FIGURE 5.49
Switch and lamp AND logic

Switch, A	Switch, B	Lamp, LP1
Open	Open	Off
Open	Closed	Off
Closed	Open	Off
Closed	Closed	On

FIGURE 5.50
Possible states for the circuit of Fig. 5.48

A	B	Y
0	0	0
0	1	0
1	0	0
1	1	1

FIGURE 5.51
Truth table for the AND switch and lamp logic

FIGURE 5.52
Switch and lamp OR logic

Switch, A	Switch, B	Lamp, LP1
Open	Open	Off
Open	Closed	On
Closed	Open	On
Closed	Closed	On

FIGURE 5.53
Possible states for the circuit of Fig. 5.52

A	B	Y
0	0	0
0	1	1
1	0	1
1	1	1

FIGURE 5.54
Truth table for the OR switch and lamp logic

> **Key point**
> Logic circuits involve signals that can only exist in one of two, mutually exclusive, states. These two states are usually denoted by 1 and 0, 'on' or 'off', 'closed' and 'open', 'high' and 'low', etc.

Consider the simple circuit shown in Fig. 5.49. In this circuit a battery, B1, is connected to a lamp, LP1, via two switches, A and B. It should be obvious that the lamp will only operate when both of the switches are closed (i.e. both *A AND B* are closed). Let's look at the operation of the circuit in a little more detail. Since there are two switches (A and B) and there are two possible states for each switch (open or closed), there is a total of four possible conditions for the circuit. We can summarize these conditions in Fig. 5.50.

Note that the two states (i.e. open or closed) are mutually exclusive and that the switches cannot exist in any other state than completely open or completely closed. Because of this, we can represent the state of the switches using the *binary digits*, 0 and 1, where an open switch is represented by 0 and a closed switch by a 1. Furthermore, if we assume that 'no light' is represented by a 0 and 'light on' is represented by a 1, we can rewrite Fig. 5.50 in the form of a *Truth Table* as shown in Fig. 5.51, where the output, Y, is the result of the combination of the inputs, A and B.

Fig. 5.52 shows another circuit with two switches. This circuit differs from that shown in Fig. 5.49 by virtue of the fact that the two switches are connected in *parallel* rather than in *series*. In this case the lamp will operate when either of the two switches is closed (in other words, when *A OR B* is closed). As before, there is a total of four possible conditions for the circuit. We can summarize these conditions in Fig. 5.53. Once again, adopting the convention that an open switch can be represented by 0 and a closed switch by 1, we can rewrite the truth table in terms of the binary states as shown in Fig. 5.54.

Logic Functions

Logic gates are circuits designed to produce the basic logic functions such as AND and OR. These circuits are designed to be interconnected into larger, more complex, logic circuit arrangements. Each type of logic gate has its own symbol and in Fig. 5.55 we have shown both the British Standard (BS) symbol together with the more universally accepted American Standard (MIL/ANSI) symbol. Note that, while inverters and buffers each have only one input, exclusive-OR gates have two inputs and the other basic gates (e.g. AND, OR, NAND and NOR) are commonly available with up to eight inputs.

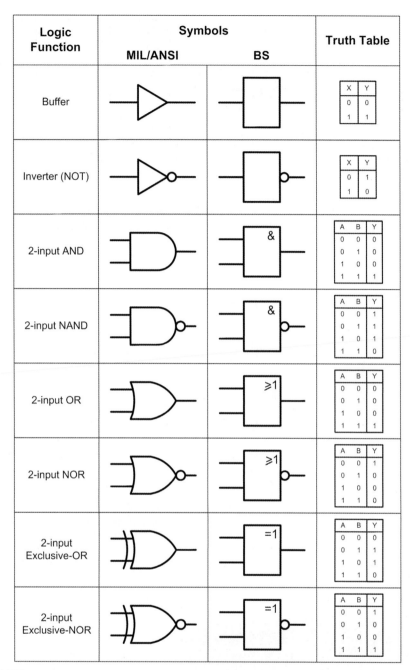

FIGURE 5.55
Logic gate symbols and truth tables

Some of the logic gates shown in Fig. 5.55 have inverted outputs. These gates are the NOT, NAND, NOR and Exclusive-NOR and the small circle at the output of the gate (see Fig. 5.56a) indicates this inversion. It is important to note that the output of an inverted gate (e.g. NOR) is identical to that of the same (i.e. non-inverted) function with its output connected to an inverter (or NOT gate) as shown in Fig. 5.56b.

Logic gates, like operational amplifiers are supplied as *integrated circuits* with many components fabricated on the same small slice of silicon. Logic gates can be packaged singly or as dual, triple or quadruple independent logic devices, all operating from the same supply. Fig. 5.57 shows some typical logic gate packages.

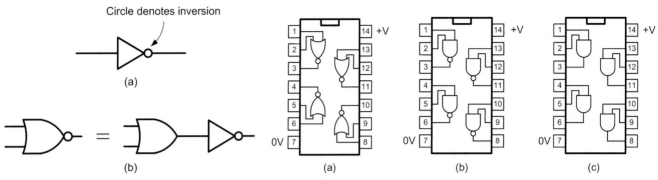

FIGURE 5.56
Logic gates with inverted outputs

FIGURE 5.57
Some typical logic gate package. (a) Quad two-input NOR gate type 4001. (b) Quad two-input NAND gate type 7400. (c) Quad two-input AND gate type 7408

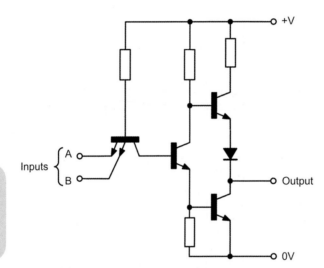

Test your knowledge 5.20
Identify each of the logic gate symbols shown in Fig. 5.60.

FIGURE 5.58
Internal circuitry of a typical two-input NAND gate

FIGURE 5.59
A dual four-input AND gate

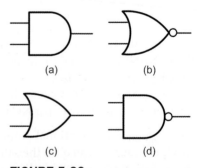

FIGURE 5.60
See Test your knowledge 5.20

ACTIVITY 5.9

Use library or internet resources to obtain data on each of the following integrated circuit logic devices:

a. 4011
b. 7432

For each logic device:

1. State the name of the manufacturer and/or supplier.
2. State the logical function and number of individual logic gates supplied within a single package
3. Draw and label the pin connections for the logic device
4. State the manufacturer's recommended range of supply voltages

Monostable, Bistable, and Astable Devices

The basic logic gates, such as two-input NAND and two-input NOR devices, can be combined with other components to produce a variety of more complex logic devices. These include monostables, bistables and astables.

MONOSTABLE, BISTABLE, AND ASTABLE DEVICES

A *monostable* is a logic device that has only one stable output state. The output of a monostable is initially at logic 0 (low) until an appropriate level change occurs at its *trigger input*. Depending on the individual monostable device, this level change can be from logic 0 (low) to logic 1 (high). This is known as a *positive edge trigger*. Alternatively, triggering can require a change from logic 1 (high) to logic 0 (low). This is referred to as a *negative edge trigger*. In either case, following the receipt of a valid trigger pulse the output of the monostable changes state to logic 1. Then, after a time interval determined by some external timing components (see page 169), the output reverts to logic 0. The device then awaits the arrival of the next trigger pulse. A typical application for a monostable device is in stretching a pulse of very short duration.

By contrast, the output of a *bistable* has two stables states (logic 0 or logic 1) and, once set the output of the device will remain at a particular logic level for an indefinite period until reset. A bistable thus constitutes a simple form of *memory circuit* because it will remain in its latched state (known as *set* or *reset*) until commanded to change its state (or until the supply is disconnected). In other words, it remembers the state that it is placed in. Various forms of bistable are available including R-S, D-type and J-K types.

The simplest form of bistable is the *R-S bistable*. This device has two inputs, SET and RESET, and complementary outputs, Q and Q̄. A logic 1 applied to the SET input will cause the Q output to become (or remain at) logic 1 whilst a logic 1 applied to the RESET input will cause the Q output to become (or remain at) logic 0. In either case, the bistable will remain in its SET or RESET state until an input is applied in such a sense as to change the state.

R-S bistables can be easily implemented using cross-coupled NAND or NOR gates, as shown in Fig. 5.61. Unfortunately, these arrangements can be unreliable as the output state is indeterminate when SET and RESET are simultaneously at logic 1 (the output can't be both SET and RESET at the same time!).

The *D-type bistable* has two principal inputs; D (standing variously for data or delay) and CLOCK (CK). The data input (logic 0 or logic 1) is clocked into the bistable such that the output state only changes when the clock changes state.

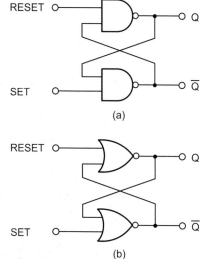

FIGURE 5.61
R-S bistables can be built from (a) cross-coupled NAND and (b) cross coupled NOR gates

Operation is thus said to be *synchronous*. Additional subsidiary inputs (which are invariably active low) are provided so that the output can be directly set or reset. These inputs are usually labelled PRESET (PR) and CLEAR (CLR). D-type bistables are used both as *latches* (a simple form of memory) and as *binary dividers*.

J-K bistables are the most sophisticated and flexible of the bistable types and they can be configured in various ways including binary dividers, shift registers, and latches. J-K bistables have two clocked inputs (J and K), two direct inputs (PRESET and CLEAR), a CLOCK (CK) input, and outputs (Q and Q̄). As with R-S bistables, the two outputs are complementary (i.e. when one is 0 the other is 1, and vice versa). Similarly, the PRESET and CLEAR inputs are invariably both active low (i.e. a 0 on the PRESET input will set the Q output to 1 whereas a 0 on the CLEAR input will set the Q output to 0).

Astable devices have outputs that alternate continuously between the two logic states, logic 0 (low) and logic 1 (high). Astable devices are frequently used in *oscillators* or *clocks* where they provide a continuous train of square or rectangular pulses.

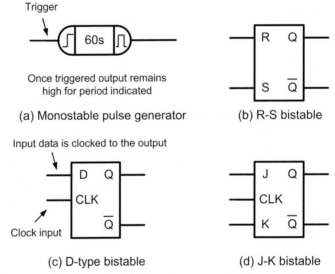

(a) Monostable pulse generator

(b) R-S bistable

(c) D-type bistable

(d) J-K bistable

FIGURE 5.62
Symbols for monostables, R-S, D-type and J-K bistables

FIGURE 5.63
A dual D-type bistable logic device used in a typical logic circuit

Timers

The timer is without doubt one of the most versatile integrated circuit chips ever produced. Not only is it a neat mixture of analogue and digital circuitry but its applications are virtually limitless in the world of timing and digital pulse generation. The device also makes an excellent case study for students because it combines a number of basic circuit concepts and techniques, both analogue and digital.

To begin to understand how timer circuits operate, it is worth spending a few moments studying the internal circuitry of the most commonly available timer, 555 timer. This chip contains two operational amplifiers together with an R-S bistable element (see page 167). In addition, an inverting buffer (see page 165) is incorporated so that an appreciable current can be delivered to a load. The main features of the device are shown in Fig. 5.64 and briefly explained in Table 5.6.

Test your knowledge 5.21

Explain the difference between monostable, bistable and astable logic circuits.

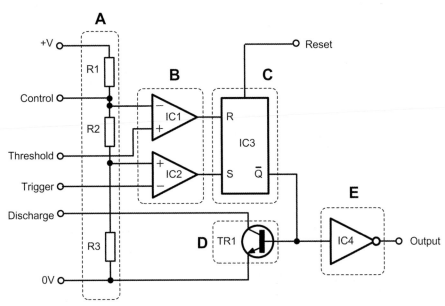

FIGURE 5.64
Internal arrangement of a 555 timer

TABLE 5.6	Internal features of the 555 timer
Feature	**Function**
A	A potential divider (see page 173) comprising R1, R2 and R3 connected in series. Since all three resistors have the same values the input voltage (+V) will be divided into thirds, i.e. one third of +V will appear at the junction of R2 and R3 whilst two thirds of +V will appear at the junction of R1 and R2.
B	Two operational amplifiers connected as comparators (see page 163). The operational amplifiers are used to examine the voltages at the threshold and trigger inputs and compare these with the fixed voltages from the potential divider (two thirds and one third of +V respectively).
C	An R–S bistable stage (see page 167). This stage can be either set or reset depending upon the output from the comparator stage. An external reset input is also provided.
D	An open-collector transistor switch. This stage is used to discharge an external capacitor by effectively shorting it out whenever the base of the transistor is driven positive.
E	An inverting power amplifier. This stage is capable of supplying enough current to drive a small relay, buzzer or other low-resistance load connected to the output.

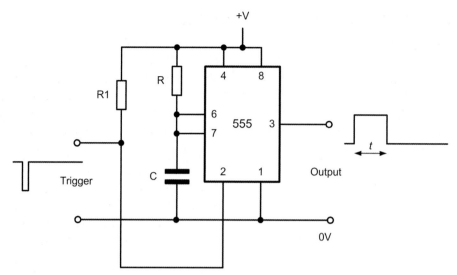

FIGURE 5.65
A 555 timer used in monostable mode

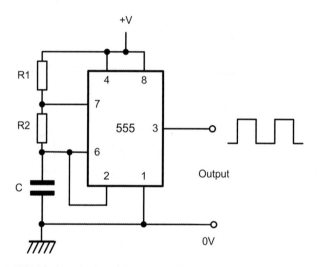

FIGURE 5.66
A 555 timer used in astable mode

ACTIVITY 5.10

Use library or internet resources to obtain data on each of the following integrated circuit timers:

a. 555
b. 556

For each logic device:

1. State the name of the manufacturer and/or supplier
2. Draw and label the pin connections for the logic device
3. State the manufacturer's recommended range of supply voltages.

Key point
Timers are integrated circuits that contain both analogue and digital circuitry. They can be configured for both monostable and astable operation and are used in a wide variety of timing applications.

The 555 timer can be used in both monostable and astable modes. Fig. 5.65 shows a monostable circuit where C and R are the timing components whilst Fig. 5.66 shows an astable oscillator where C, R1 and R2 determine the output pulse repetition frequency and duty cycle (see page 145).

5.3 ANALOGUE AND DIGITAL CIRCUITS

Circuit Diagrams

Before you can make sense of the circuits that you will meet later in this unit it's important to be able to read and understand simple electronic circuit diagrams. Circuit diagrams use standard symbols (like those that we met in the last section) and conventions to represent the components and wiring used in an electronic circuit. Visually, they usually bear very little relationship to the physical layout of a circuit but, instead, they provide us with a theoretical view of the circuit.

As a general rule, the input to a circuit should be shown on the left of a circuit diagram and the output shown on the right. The *supply* (usually the most positive voltage) is normally shown at the top of the diagram and the *common, 0 V,* or *ground* connection is normally shown at the bottom. This rule is not always obeyed, particularly for complex diagrams where many signals and supply voltages may be present. Fig. 5.67 shows some commonly used conventions and some additional symbols that you need to be able to recognise.

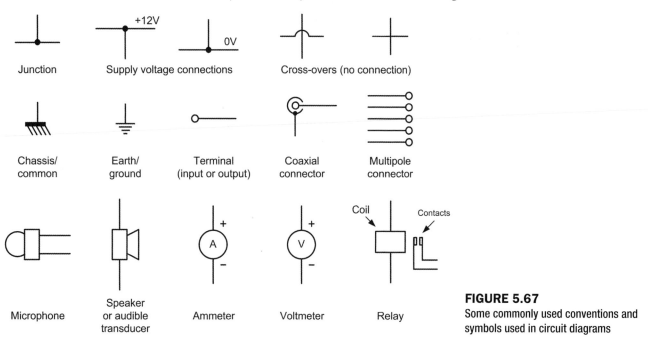

FIGURE 5.67
Some commonly used conventions and symbols used in circuit diagrams

---

ACTIVITY 5.11

Fig. 5.68 shows the circuit diagram of a simple radio receiver. The radio is powered from a battery and it uses two transistors and two diodes. Look at this diagram carefully and then see if you can answer the following questions:

1. Which component is used for the tuning control?
2. What component is used for the volume control?
3. What type of switch is used to switch the power on and off?
4. Which component couples the loudspeaker to the circuit?
5. What is the battery voltage?
6. Which component provides power indication?
7. What type of component is:
   a. C3
   b. L1
   c. TR1
   d. TR2.
8. Which component links the output of TR1 to the input of TR2?
9. Which component supplies current to the LED?
10. Which component links the collector to the base of TR2?

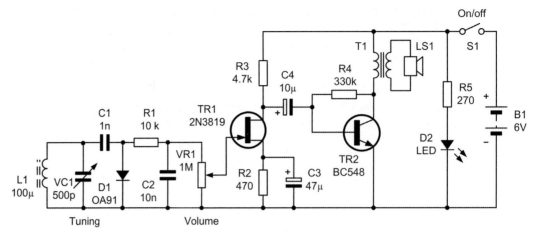

FIGURE 5.68
See Activity 5.11

Passive Circuits

Electronic circuits are sometimes referred to as passive or active depending upon whether or not they contain devices such as transistors, logic gates, operational amplifiers and other integrated circuits. Simple passive circuits can comprise of nothing more than combinations of resistors, capacitors and/or inductors.

SERIES AND PARALLEL CIRCUITS

Before we look at some more complex circuits it's important to understand what we mean by series and parallel circuits. A series circuit consists of two or more components connected one after another. In a series circuit, the same current flows in all of the components whilst the applied voltage is divided between them. In a parallel circuit, the same voltage appears across all of the components whilst the current is divided between them.

Fig. 5.69(a) shows two resistors, R1 and R2, connected in series whilst Fig. 5.69(b) shows two resistors, R1 and R2, connected in parallel. In each case, the equivalent resistance of the circuit (i.e. the one single resistor that could replace R1 and R2) is shown as resistor R.

In the series circuit shown in Fig. 5.69(a), the same current flows in each of the resistors and the value of R (the effective resistance of the series circuit) is given by the sum of the two individual resistances, R1 and R2. Hence, for the series case:

$$R = R1 + R2$$

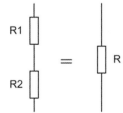

(a) Two resistors in series

In the parallel circuit shown in Fig. 5.69(b), the same voltage appears across each of the resistors and the reciprocal of the value of R (i.e. 1/R) is given by the sum of the reciprocals of the other two resistances, 1/R1 and 1/R2. Hence, for the parallel case:

$$\frac{1}{R} = \frac{1}{R1} + \frac{1}{R2}$$

This formula can be re-arranged so that:

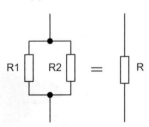

(b) Two resistors in parallel

FIGURE 5.69
Two resistors connected (a) in series and (b) in parallel

$$R = \frac{R1 \times R2}{R1 + R2}$$

This can be more conveniently remembered as 'product over sum'.

Fig. 5.70(a) shows two capacitors, C1 and C2, connected in series whilst Fig. 5.69(b) shows two capacitors, C1 and C2, connected in parallel. In each case, the equivalent resistance of the circuit (i.e. the one single capacitor that could replace C1 and C2) is shown as C.

In the series circuit shown in Fig. 5.69(a), the same charge appears in each of the capacitors and the value of the reciprocal of the effective capacitance of the series circuit (i.e. 1/C) is given by the sum of the reciprocals of the two individual capacitances, 1/C1 and 1/C2. Hence, for the series case:

$$\frac{1}{C} = \frac{1}{C1} + \frac{1}{C2}$$

This formula can be re-arranged so that:

$$C = \frac{C1 \times C2}{C1 + C2}$$

In the parallel circuit shown in Fig. 5.70(b), the same voltage appears across each of the capacitors and the value of C (the effective capacitance of the series circuit) is given by the sum of the two capacitances, C1 and C2. Hence, for the parallel case:

$$C = C1 + C2$$

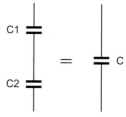

(a) Two capacitors in series

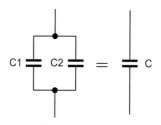

(b) Two capacitors in parallel

FIGURE 5.70
Two capacitors connected
(a) in series and (b) in parallel

ACTIVITY 5.12

1. Calculate the effective resistance of each of the circuits shown in Fig. 5.71. Hint: Solve circuit (c) in two stages, starting with the series branch.
2. Calculate the effective capacitance of each of the circuits shown in Fig. 5.72. Hint: Solve circuit (c) in two stages, starting with the parallel branch.

POTENTIAL DIVIDERS

The potential divider (see Fig. 5.73) is an extremely useful circuit since, by selecting appropriate values for the two resistors, R1 and R2, it allows you to obtain a fraction of the input voltage, V_{IN}. Note that the circuit works equally well with AC and DC signals. The value of output voltage, V_{OUT}, produced by the voltage divider is given by the relationship:

$$V_{OUT} = V_{IN} \times \frac{R1}{R1 + R2}$$

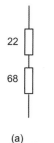

(a)

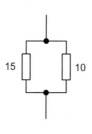

(b)

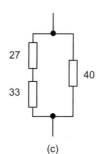

(c)

FIGURE 5.71
See Activity 5.12

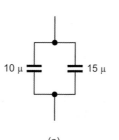

(a)

(b)

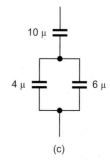

(c)

FIGURE 5.72
See Activity 5.12

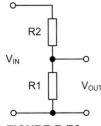

FIGURE 5.73
A potential divider

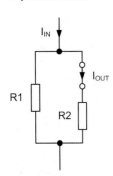

FIGURE 5.74
A current divider

As an example, suppose that we need to produce a voltage of precisely 5 V from a 15 V DC supply. We would need to make the value of R2 twice that of R1. Values of 1000 Ω for R1 and 2000 Ω for R2 would be suitable.

CURRENT DIVIDERS

The current divider (see Fig. 5.74) is another useful circuit since, by selecting appropriate values for the two resistors, R1 and R2, it allows you to obtain a fraction of the input current, I_{IN}. Like the potential divider, this circuit works equally well with AC and DC signals. The value of output current, I_{OUT}, produced by the current divider is given by the relationship:

$$I_{OUT} = I_{IN} \times \frac{R2}{R1 + R2}$$

As an example, suppose that we need to produce a current of precisely 10 mA from a 100 mA supply. We would need to make the value of R2 nine times that of R1. Suitable values would be 10 Ω for R1 and 90 Ω for R2.

C-R CIRCUITS

Earlier we mentioned that a capacitor is a device for storing electric charge. This charge can be stored in a capacitor by connecting it to a battery or power supply via a series resistor. Later, the stored charge can be drained away by connecting a resistor in parallel with the capacitor. After a period of time there will then be no charge remaining in the capacitor. The time that it takes to charge and discharge a capacitor depends on the values of capacitance and resistance and this makes capacitors ideal for use in timing and delay circuits. Because this is so important, it's worth looking at this in a little more detail.

Simple charging and discharging circuits are shown in Figs. 5.75 and 5.77. In the charging arrangement shown in Fig. 5.75 switch S1 is closed and S2 is left open. If the capacitor is initially uncharged current will flow and charge will build up inside the capacitor. As the capacitor becomes charged, the capacitor voltage (V_C) will increase until it eventually becomes close, but never quite equal, to the voltage of the supply (V). At that point (when V_C is approximately equal to V) we say that the capacitor is fully charged. A graph showing how the capacitor voltage (V_C) increases with time is shown in Fig. 5.76. This graph is known as an *exponential growth* curve.

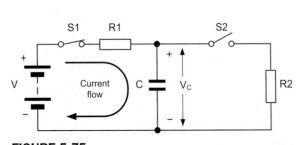

FIGURE 5.75
A C-R charging circuit

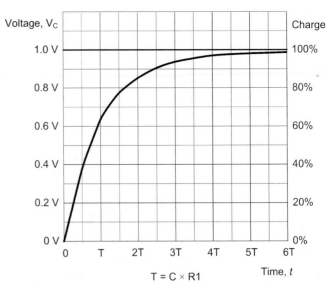

FIGURE 5.76
Graph of capacitor voltage against time for the C-R charging circuit

The speed at which the capacitor becomes charged depends on the *time constant*, T, of the circuit. This is the product of the capacitance, C, and the charging resistance, R. Hence:

$$T = C \times R$$

where C is the value of capacitance (in F), R is the resistance (in Ω), and T is the time constant (in seconds).

For our charging circuit (Fig. 5.75) the time constant is $T = C \times R1$.

You might now be wondering how it takes to *fully* charge the capacitor? The true answer is that the capacitor voltage never quite reaches the supply voltage even if you wait for a *very* long time. However, it does get closer and closer to it and for this reason we say that the capacitor is fully charged after a time interval equal to five times the time constant (5T or $5C \times R$).

In the discharging arrangement shown in Fig. 5.77 switch S1 is open and S2 is closed. If the capacitor is initially fully charged current will flow while the charge inside the capacitor decays away. As the capacitor becomes discharged, the capacitor voltage (V_C) will decrease until it eventually becomes close, but never quite equal, to zero (0V). At that point (when V_C is approximately equal to 0V) we say that the capacitor is fully discharged. A graph showing how the capacitor voltage (V_C) decreases with time is shown in Fig. 5.78. This graph is known as an *exponential decay* curve.

The speed at which the capacitor becomes discharged depends on the *time constant*, T, of the circuit. For our charging circuit (Fig. 5.77) the time constant is $T = C \times R2$.

Once again, you might now be wondering how long it takes to *fully* discharge the capacitor? The true answer is that the capacitor voltage never quite reaches 0V even if you wait for a *very* long time. However, it does get closer and closer to 0V and for this reason we say that the capacitor is fully charged after a time interval equal to five times the time constant (5T or $5C \times R$).

ENERGY STORAGE

A charged capacitor acts as a reservoir for charge and the stored energy can be put to good use some time later. The amount of energy stored in a capacitor depends on the product of

Test your knowledge 5.22

1. A C-R circuit consists of $C = 100\,\mu F$ and $R = 2.7\,M\Omega$. What is the time constant of the circuit? If it is initially uncharged, how long would it take to fully charge the capacitor?
2. A $470\,\mu F$ capacitor is required to become fully charged after an interval of 2 seconds. What value of resistance should be connected between the capacitor and its charging source?

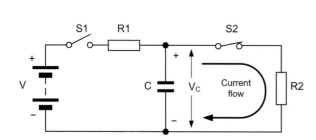

FIGURE 5.77
A C-R discharging circuit

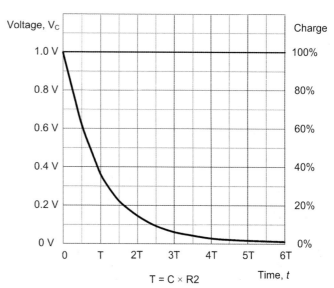

FIGURE 5.78
Graph of capacitor voltage against time for the C-R discharging circuit

175

capacitance (C) and the square of the applied voltage (V). Hence, to store a large amount of energy we need a correspondingly larger value of capacitance for a given value of charging voltage. The following relationship applies:

$$W = \frac{1}{2} C V^2$$

where C is the value of capacitance (in F), V is the capacitor voltage, and W is the stored energy (in Joules).

Analogue circuits

Having spent some time introducing different types of component and basic circuit configurations it's time to bring things together by looking at a variety of useful analogue and digital circuit applications. If you decide to continue with your study of electronics further you will undoubtedly meet these circuits again so we will not waste space by discussing the theoretical aspects of these circuits here. Instead, it is sufficient for you to be able to recognise the individual components and appreciate what the circuits do as a whole.

DIODE APPLICATIONS

The most important characteristic of a diode is that it passes current in one direction only. In effect, a diode acts as a 'one way street', passing current one way but not the other. This feature yields a number of useful applications include protecting circuits from currents that might otherwise flow the wrong way round (for example, due to the inadvertent reverse connection of a battery) and also for converting alternating current (AC) into a direct current (DC). Fig. 5.79 shows some typical diode applications.

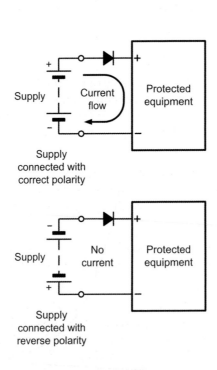

(a) Reverse polarity protection

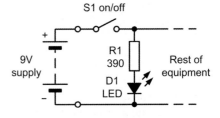

(b) LED power indicator

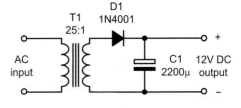

(c) Simple 12V power supply

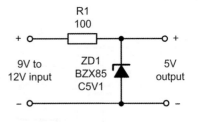

(d) Zener stabilised 5V DC supply

FIGURE 5.79
Some typical diode applications

Fig. 5.79(a) shows how an item of electronic equipment can be protected against the inadvertent reverse connection of its DC supply by simply inserting a diode in the supply lead. The diode conducts when the battery is connected with the correct polarity but fails to conduct when the battery is connected the wrong way round.

Fig. 5.79(b) shows a simple LED indicator, typical of that found in most items of electronic equipment. The value of the series resistor, R1, is selected so that the LED operates with the correct forward current (usually around 10 mA to 20 mA).

In Fig. 5.79(c) the diode is used as a half-wave rectifier. The diode only conducts on positive half-cycles of the AC voltage from the secondary of T1. To maintain the DC output voltage when the diode is not conducting (i.e. during the negative half-cycles of secondary voltage) a large reservoir capacitor, C1, helps by storing the charge deposited in it during the period of time for which the diode is not conducting.

A simple zener diode voltage stabiliser is shown in Fig. 5.79(d). The diode conducts heavily when the voltage across it reaches its rated zener voltage (5.1 V in this case). This *clamps* the voltage and holds it constant regardless of changes in the input voltage.

TRANSISTOR APPLICATIONS

Transistors act as amplifier for current. A small current applied to the input (the *base* connection) gives rise to a much larger current flowing in the output (the *collector* or *emitter* connections). This makes transistors extremely useful for increasing the amplitude of signals (i.e. amplification) and also for switching (for example, operating an LED, lamp or relay when an input signal is applied to the base). Fig. 5.80 shows some typical transistor applications.

A simple single-stage transistor amplifier is shown in Fig. 5.80(a). This circuit provides a modest voltage gain (about 75) over the audio frequency (AF) range. The circuit operates from a supply of 9 V DC at around 1 mA.

A simple transistor intruder alarm is shown in Fig. 5.80(b). In this circuit the transistor operates as a switch controlling the current that flows through the relay, RL1. The transistor will switch 'on' (conducting heavily) whenever the door switch, S1, opens. The current flowing in the relay coil will then cause its contacts to close which in turn operate an external bell or siren. The diode, D1, protects the transistor against the high reverse voltage appearing across the relay coil when the transistor switches 'off' and the magnetic flux suddenly collapses.

Fig. 5.80(c) shows a voltage stabiliser that provides an output voltage of approximately 9 V for an input voltage of between 12 V and 17 V. The output voltage is determined by the zener diode, ZD1. Circuits like this are commonly used in electronic equipment where the supply voltage varies.

An audio amplifier suitable for driving headphones or a small loudspeaker is shown in Fig. 5.80(d). Similar circuits are commonly found in most portable consumer electronic equipment.

Finally, Fig. 5.80(e) shows an *astable oscillator* (also referred to as a *multivibrator*). This circuit produces a 5 V square wave output at a frequency of approximately 1 kHz. This circuit could be used as the basis of a simple digital signal source, test oscillator or clock. The frequency of operation and duty cycle can be easily changed by altering the values of C1, C2, R2 and R3.

OPERATIONAL AMPLIFIER APPLICATIONS

As mentioned earlier, operational amplifiers are extremely versatile and they can be used both as 'gain blocks' and comparators. Fig. 5.81 shows some typical operational amplifier

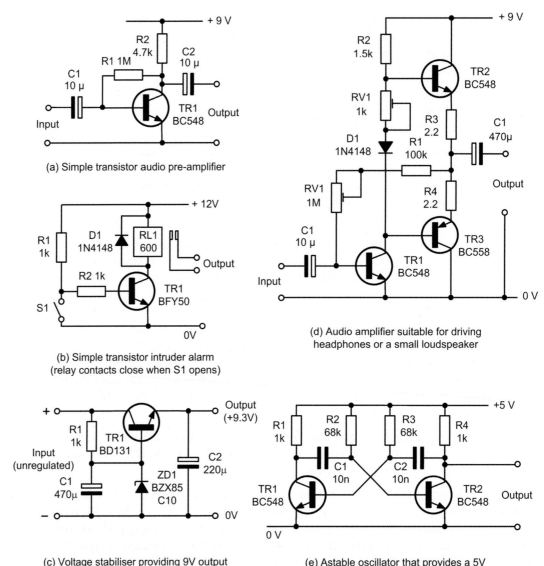

(a) Simple transistor audio pre-amplifier

(b) Simple transistor intruder alarm
(relay contacts close when S1 opens)

(c) Voltage stabiliser providing 9V output
for input voltages between 12V and 17V

(d) Audio amplifier suitable for driving
headphones or a small loudspeaker

(e) Astable oscillator that provides a 5V
square wave output at approximately 1 kHz

FIGURE 5.80
Some typical transistor applications

applications (note that, for clarity, we have not shown the positive and negative supply connections on these circuits (see page 161 for more information)).

Fig. 5.81(a) shows a DC inverting amplifier with a voltage gain of 100. Because of the inverting action, the output of this circuit will go negative when the input goes positive, and vice versa. The voltage gain of the circuit is determined by the ratio of R2 to R1. By contrast, Fig. 5.81(b) shows a DC non-inverting amplifier with a voltage gain of 5. In this case, the output of this circuit will go positive when the input is positive, and negative when the input is negative.

A differential amplifier is shown in Fig. 5.81(c). The output voltage produced by this circuit will be the difference of the two input voltages (i.e. $V2 - V1$). A simple audio amplifier is shown in Fig. 5.81(d). This circuit provides a voltage gain of 10 (once again this is determined by the ratio of R2 to R1) and a frequency response extending from around 70 Hz to 16 kHz. The low frequency response is determined by C1 and R1 whilst the high frequency response is defined by C2 and R2.

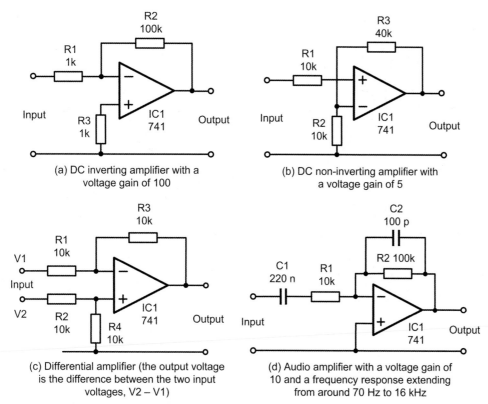

FIGURE 5.81

Some typical operational amplifier applications. Note that, for clarity, we have omitted the positive and negative supply connections (see page 161)

TIMER APPLICATIONS

Timers contain a mixture of analogue and digital circuitry and they rather neatly bridge the divide between digital and analogue circuits. We have already introduced the basic monostable and astable configurations (see Figs. 5.65 and 5.66) but some typical applications of these versatile chips are shown in Fig. 5.82.

Fig. 5.82(a) shows a one minute timer. In this circuit the 555 timer operates in monostable mode with S1 and S2 acting as start and stop switches respectively. The output of the circuit operates a simple buzzer, BZ1. Fig. 5.82(b) shows a 1 kHz test oscillator that produces a 1 V square wave output. This circuit acts as a test signal generator. The output frequency can be varied by changing the values of C and R2.

Digital Circuits

Digital circuits are often divided into two main classes, *combinational logic* and *sequential logic*. In the former case the logical state of the output depends on whatever the current state of the input happens to be. In the latter case, the output depends also on what has happened before. Putting this another way, in sequential logic it is the sequence of events that is important whilst in combinational logic it is only what is happening at the present time that matters.

Some examples of combinational logic circuits are shown in Fig. 5.83(a). The output of each circuit can be explained either by stating its *logical function* (using terms like AND and OR and using brackets to indicate precedence) or by using *truth tables*. Note that, as with all digital logic circuits, the inputs and outputs can only exist in two states, logic 0 and logic 1 (see page 181).

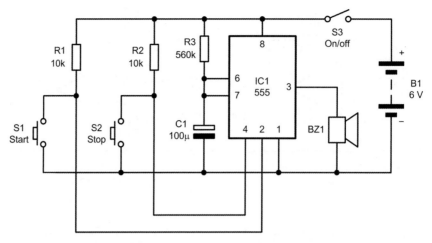

(a) One minute timer with audible output

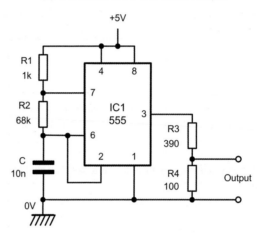

(b) 1kHz test oscillator with 1 V square wave output

FIGURE 5.82
Some typical timer applications

A sequential logic application is shown in Fig. 5.84(a). This circuit is a three stage pulse counter and each stage is based on a single J-K bistable element (see page 182). The current output state of the counter is displayed in binary format using three light emitting diodes, D1, D2 and D3. For example, after two complete input pulses have been counted the output state will be 010 (D1 'off', D2 'on', D3 'off'). Note that D3 is the most significant binary digit (leftmost digit of the binary number) and D1 is the least significant binary digit (rightmost digit of the binary number). The counting sequence can be seen from the waveforms in Fig. 5.84(b) and also from Table 5.6. Note from Fig. 5.84(b) that the outputs (taken from Q1, Q2 and Q3) are successively divided by 2.

Circuit Construction and Testing

As part of this unit you are expected to know about the different methods and techniques used for circuit construction and testing. You will find this information in Section 8.8 (*Circuit construction*) on page 270 and Section 8.9 (*Test equipment and measurements*) on page 279. You are strongly advised to work through both of these sections before carrying out the practical assignments and investigations for this unit.

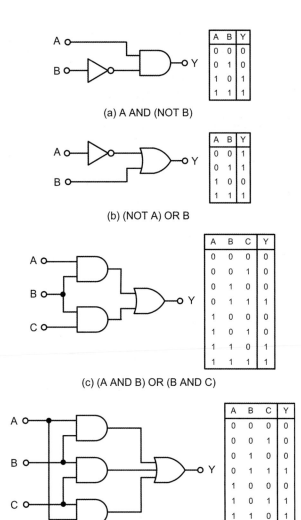

(a) A AND (NOT B)

(b) (NOT A) OR B

(c) (A AND B) OR (B AND C)

(d) Majority vote

FIGURE 5.83
Examples of combinational logic

ACTIVITY 5.12

Investigate the operation of any TWO of the following analogue circuits:

a. A single-stage transistor amplifier using a BC548 or similar device
b. An operational amplifier comparator using a 741 or similar device
c. A 555 timer operating in monostable mode
d. A 555 timer operating in astable mode.

In each case build and test the circuit, draw a labelled circuit diagram and list all of the components used.

ACTIVITY 5.13

Investigate the operation of any TWO of the following digital circuits:

a. A four-input OR gate made from three two-input OR gates
b. A two input OR gate made only from two-input NAND gates
c. An R-S bistable made from two two-input NAND or NOR gates
d. A four-stage binary counter using J-K bistable devices

In each case build and test the circuit, draw a labelled circuit diagram and list all of the components used.

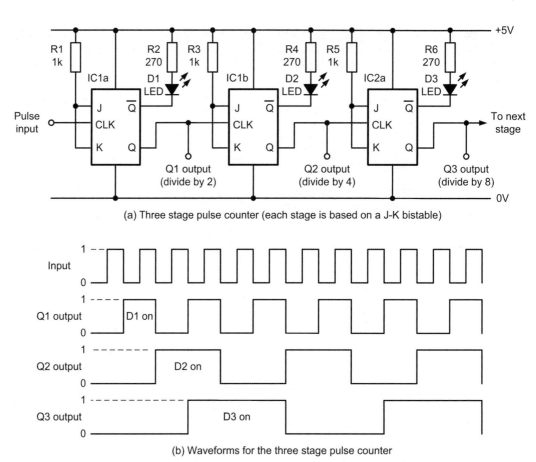

(a) Three stage pulse counter (each stage is based on a J-K bistable)

(b) Waveforms for the three stage pulse counter

FIGURE 5.84
A three stage pulse counter based on J-K bistables

TABLE 5.7 Sequential output states for the three stage pulse counter

Input pulse number	Q3	Q2	Q1	Notes
0	0	0	0	Initial state with all Q outputs at logic 0
1	0	0	1	D1 'on'
2	0	1	0	D2 'on'
3	0	1	1	D1 and D2 'on'
4	1	0	0	D3 'on'
5	1	0	1	D3 and D1 'on'
6	1	1	0	D3 and D2 'on'
7	1	1	1	D3, D2 and D1 'on'
8	0	0	0	Count restarts from zero
9	0	0	1	D1 'on'

5.4 COMMUNICATION SYSTEMS AND DATA TRANSMISSION

In recent years, electronic systems have become increasingly more complex. This, in turn, has led to the need for systems to be interconnected so that they can share information and data. Various different methods of interconnection are used according to the complexity and required performance (notably speed and volume of data). This section provides you with a general introduction to data communications systems and the techniques that we use in order to transmit and receive data. We shall also look at three typical examples of modern

TABLE 5.8 Examples of electronic communication systems

Communication system	Wireless LAN	Optical LAN	Universal serial bus (USB)
Transmitter	Modulator, radio transmitter and antenna	Optical transmitter (photodiode)	Logic gate buffer/line driver
Transmission medium	Air/space	Optical fibre	Twisted pair copper cable
Receiver	Antenna and radio receiver/demodulator	Optical receiver (phototransistor)	Line receiver/logic gate buffer
Protocol/standards	IEEE 802.11 Wireless LAN	ARINC 629 aircraft data bus	USB 2.0 packet transfer protocol
Encoding/modulation	Digital modulation of an analogue UHF radio carrier	Digital modulation of an infrared light beam	Digital data packets of up to 1024 bytes
Example application	Linking several home computers to a broadband hub	Linking multiple avionic systems in the LAN in a modern passenger aircraft	Linking a personal computer to a nearby printer

data communication systems as well as techniques for representing and controlling the data that flows within them.

Data Communication Systems

Even the most complex of electronic communication systems comprise several basic elements. These include:

- a *transmitter* that interfaces with the sending equipment, encodes the outgoing signal information and presents it to the transmission medium
- a channel, link or *transmission medium* through which the signals are passed
- a *receiver* that accepts signals from the transmission medium, decodes them and interfaces with the receiving equipment.

In addition, there is a need for *protocols* that control the flow of information and deal with errors and a system for efficiently *encoding* and/or *modulating* the information that travels within the transmission medium. Three different examples of electronic communication systems are listed in Table 5.8.

SERIAL AND PARALLEL DATA COMMUNICATION

Data communication systems invariably transmit one binary digit (or *bit*) of data after another. This method of data transfer is known as *serial data communication* shown in Fig. 5.85(a) Where more than one bit is sent at a time, for example when eight bits are sent simultaneously, this is referred to as *parallel data communication* shown in Fig. 5.85(b) Data can be easily converted from serial form into parallel form or from parallel form to serial form using conventional digital logic.

SIMPLEX AND DUPLEX DATA COMMUNICATION

In the simple data communication systems shown in Fig. 5.85(c) data flow is in one direction only. A system of this type is said to support *simplex* data communication.

Test your knowledge 5.24

Explain what is meant by the following terms:

1. Serial data
2. Parallel data
3. Simplex communication
4. Duplex communication
5. Transmission medium.

Test your knowledge 5.25

Give THREE examples of different forms of transmission medium and give an example of where each might be used.

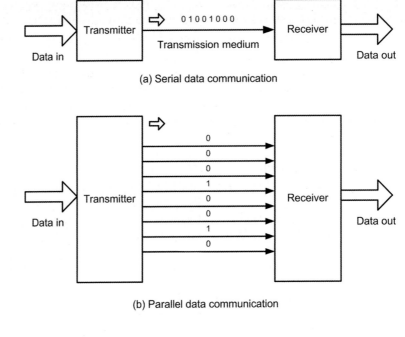

(a) Serial data communication

(b) Parallel data communication

(c) Simplex serial data communication

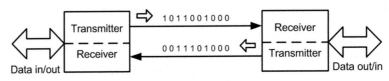

(d) Duplex serial data communication

FIGURE 5.85
Data communication systems

A full two-way (i.e. bi-directional) flow of data requires a transmitter and receiver at *both* ends of the transmission medium. Such an arrangement is referred to as a *duplex* data communication system.

Electronic Communication Systems

In this section we shall take a brief look at the basic techniques and methods used in each of the three examples of electronic communication systems listed earlier in Table 5.7. We shall begin with a method of data communication that you probably use every day at home, your own 'wireless' network!

Data Communication Via Radio: 'Wireless' Networks

Radio signals travel as invisible electromagnetic waves in air or space and are generally understood to occupy a frequency range that extends from a few tens of kilohertz (kHz) to several hundred Gigahertz (GHz). The lowest part of the radio frequency range that is of practical use (below 30 kHz) is only suitable for narrow-band communication. At this frequency, signals propagate as ground waves (following the curvature of the Earth) over very long distances. At the other extreme, the highest frequency range that is of practical importance extends above 30 GHz. At these *microwave* frequencies, considerable bandwidths are available (sufficient to transmit many television channels using point-to-point links or to permit very high definition radar systems) and signals tend to propagate strictly along line-of-sight paths. At other frequencies signals may propagate by various means including reflection from ionised layers in the ionosphere. Most wireless systems operate in the 2.4 GHz band. This is ideal for short-range data communications using low power and relatively small antennas.

MODULATION AND DEMODULATION

In order to convey information using a radio frequency carrier, the signal information must be superimposed or 'modulated' onto the carrier. Modulation is the name given to the process of changing a particular property of the carrier wave in sympathy with the instantaneous voltage (or current) signal. Modulation is achieved by means of a *modulator* that combines the data signal to be transmitted with a radio frequency carrier wave, see Fig. 5.86(a).

Demodulation is the reverse of modulation and is the means by which the signal information is recovered from the modulated carrier. Demodulation is achieved by means of a *demodulator*. The output of a demodulator consists of a reconstructed version of the original signal information present at the input of the modulator stage within the transmitter, see Fig. 5.86(b).

A simple wireless data communication system is shown in Fig. 5.87. This system permits communication in one direction only but in practice we would need to duplicate the system for sending data in the opposite direction (i.e. a duplex system is required).

Data Communication Via Optical Fibres: An Airborne Local Area Network (LAN)

Optical fibres are now widely used as a transmission medium for ground-based long-haul data communications and in *local area networks* (*LAN*). They are also being introduced into the latest commercial aircraft in order to satisfy the need for a wideband network linking an increasing number of highly sophisticated avionic and cabin entertainment systems.

Optical fibres offer some very significant advantages over conventional copper cables. They are lightweight and have a very small physical size. They also offer exceptionally wide bandwidth and very high data rates. In a commercial aircraft, the reduction in weight that results from the use of fibre optical cabling can yield significant fuel savings. Copper cabling is typically more than ten times heavier than optical fibre cabling and on a large, latest generation aircraft the total saving in weight can amount to more than a tonne!

Essentially, an optical fibre consists of a cylindrical silica glass *core* surrounded by further glass *cladding*. The fibre acts as a *channel* (or *waveguide*) along which an electromagnetic wave can pass with very little loss. The more dense medium (the *core*) is surrounded by the less dense medium (the *cladding*) and the light wave will propagate inside the core by means of a series of *total internal reflections*. The construction of a typical optical fibre data cable is shown in Fig. 5.91.

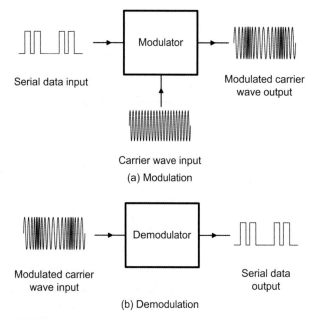

Serial data input → Modulator → Modulated carrier wave output

Carrier wave input

(a) Modulation

Modulated carrier wave input → Demodulator → Serial data output

(b) Demodulation

FIGURE 5.86
Modulation and demodulation

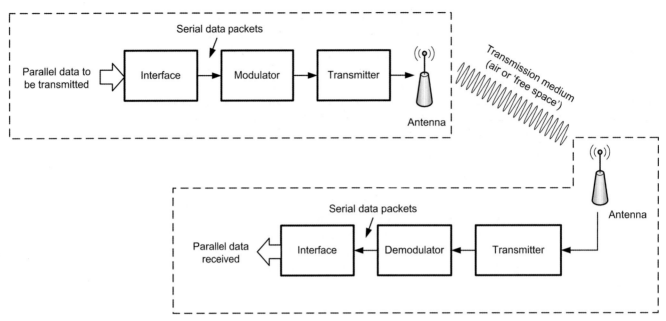

FIGURE 5.87
A simplex wireless data communication system

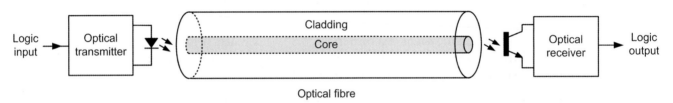

FIGURE 5.88
A simplex optical data fibre data communication system

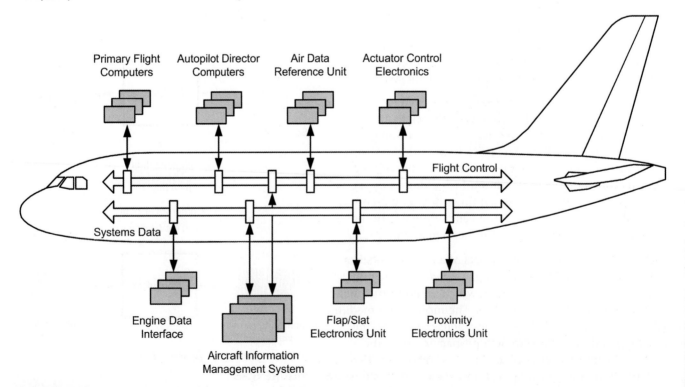

FIGURE 5.89
An aircraft local area network (LAN)

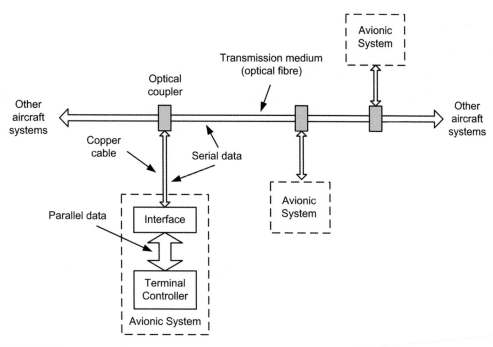

FIGURE 5.90
Data interface devices in the aircraft LAN

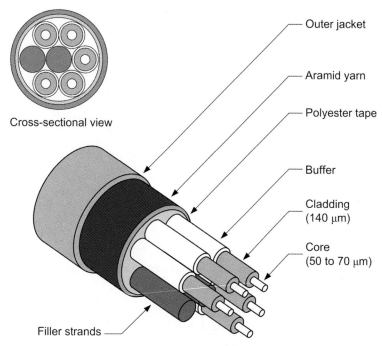

FIGURE 5.91
Typical optical fibre construction

Data Communication Via Cables

Arguably the easiest and lowest cost method for transferring data between two nearby devices is with the aid of an ordinary *copper cable* and one of the best examples of this is the *Universal Serial Bus* (*USB*). The transmission medium comprises a *twisted pair* of insulated

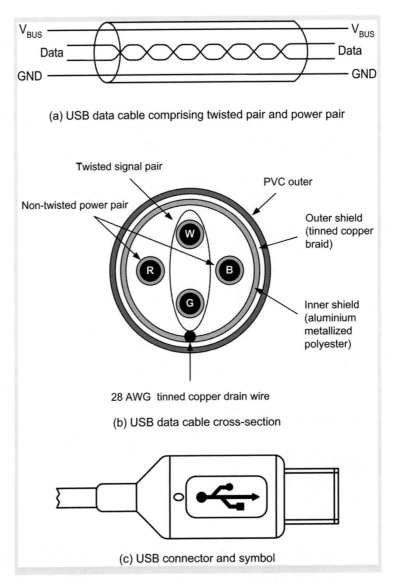

(a) USB data cable comprising twisted pair and power pair

(b) USB data cable cross-section

(c) USB connector and symbol

FIGURE 5.92
USC data cable, connector and symbol

wires together with a power connection of USB devices (such as memory sticks) that don't have their own in-built power supply (see Fig. 5.92).

The USB standard replaces several earlier data transfer standards, including those that are based on the RS-232 and parallel ports found on older computers. USB can be used to connect a wide range of computer peripherals such as keyboards, printers, mice, digital cameras, and external hard drives, to a *host controller* within a PC. A USB host controller can support the connection of up to 127 individual USB devices.

USB devices can be linked together through a series of *hubs*. At the lowest level the *root hub* is built into the host controller. More hubs can then be added in order to extend the capacity of a system. USB *endpoints* are found on each connected device and the channels that link them to the host controller are referred to as *pipes*. Each pipe acts as a logical channel (i.e. a channel through which data transfer can take place) and each USB device can have up to 32 active pipes (16 into the host controller and 16 out of the host controller).

Data Transmission

Finally, we shall consider the ways in which data is represented within data communication systems and how the flow of data is managed and controlled.

DATA REPRESENTATION

In most conventional digital logic systems logic 1 (high) is ideally represented by a voltage of $+5\,V$ whilst logic 0 (low) is represented by $0\,V$. In practice, and in order to allow for some variation in supply voltage levels and component tolerances, a range of voltages is normally used. The logic representation that we use in data communications systems are often quite different from those that we use in conventional digital logic but it is usually quite easy to convert from one form to another. The important thing to remember is that, in a data communication system all of the data must be converted into a binary code using only the two binary digits, 1 and 0. In different data communication systems these binary digits (**bits**) can be represented by voltages, currents, audible tones, light or no light.

NUMBERS AND CODING SYSTEMS

Binary numbers – particularly large ones – are not very convenient. To make numbers easier to handle we often convert binary numbers to *hexadecimal* (base 16). This format is easier for mere humans to comprehend and offers the advantage over denary (base 10) in that it can be converted to and from binary with ease. The first sixteen numbers in binary, denary, and hexadecimal are shown in Table 5.8. Note that a single hexadecimal character (in the range zero to F) is used to represent a group of four binary digits. This group of four bits (or single hex. character) is sometimes called a *nibble*.

A **byte** of data comprises a group of eight binary digits (or *bits*). Thus a byte can be represented by just two hexadecimal (hex.) characters. A group of sixteen bits (a *word*) can be represented by four hex. characters, thirty-two bits (a double word by eight hex. characters, and so on. The value of a byte expressed in binary can be easily converted to hex. by arranging the bits in groups of four and converting each nibble into hexadecimal using the table shown below:

TABLE 5.9 Binary, denary and hexadecimal numbers		
Binary (base 2)	**Denary (base 10)**	**Hexadecimal (base 16)**
0000	1	0
0001	1	1
0010	2	2
0011	3	3
0100	4	4
0101	5	5
0110	6	6
0111	7	7
1000	8	8
1001	9	9
1010	10	A
1011	11	B
1100	12	C
1101	13	D
1110	14	E
1111	15	F

Note that, to avoid confusion about whether a number is hexadecimal or decimal, we sometimes place a $ symbol before a hexadecimal number or add an H to the end of the number. For example, 64 means decimal 'sixty-four'; whereas, $64 or 64H means hexadecimal 'six-four', which is equivalent to decimal 100. Similarly, 7FH means hexadecimal 'seven-F' which is equivalent to decimal 127.

EXAMPLE 5.7

Convert hexadecimal A3 into binary.

From the table above, A = 1010 and 3 = 0101. Thus A3 in hexadecimal is equivalent to 10100101 in binary.

EXAMPLE 5.8

Convert binary 11101000 binary to hexadecimal.

From the table above, 1110 = E and 1000 = 8. Thus 11101000 in binary is equivalent to E8 in hexadecimal.

ASCII CODE

In many data communications applications, such as text messaging, there is a need to transfer messages and information using plain text and punctuation. To do this we normally use a code referred to as *ASCII* (American Standard Code for Information Interchange). This code uses seven or eight bits to represent text characters. Some examples of ASCII code are shown in Table 5.10.

COMMUNICATION PROTOCOLS

The notion of a communication protocol might need a little explaining so imagine for a moment that you are faced with the problem of organizing a discussion between a large number of people sitting around a table who are blindfolded and therefore cannot see one another. In order to ensure that they didn't all speak at once, you would need to establish

TABLE 5.10 Binary, denary and hexadecimal numbers

ASCII character	Hexadecimal (base 16)	Binary (base 2)
Space	20	00100000
A	41	01000001
B	42	01000010
C	43	01000011
D	44	01000100
E	45	01000101
F	46	01000110
G	47	01000111
H	48	01001000
I	49	01001001
J	4A	01001010
K	4B	01001011
L	4C	01001100
M	4D	01001101
N	4E	01001110
O	4F	01001111

some ground rules, including how the delegates would go about indicating that they had something to say and also establishing some priorities as to who should be allowed to speak in the event that several delegates indicate that they wish to speak at the same time. These (and other) considerations would form an agreed protocol between the delegates for conducting the discussion. The debate should proceed without too many problems provided that everybody in the room understands and is willing to accept the protocol that you have established.

In computers and digital systems *communications protocols* are established to enable the efficient exchange of data between multiple devices connected in the same system. These protocols consist of a set of rules and specifications governing, amongst other things, data format and physical connections. An important requirement of a protocol is that it supports the following features, handshaking, flow control and error checking. We shall briefly look at each of these in turn:

HANDSHAKING

Handshaking is a process during which a logical connection is established between two devices. Handshaking takes place before any data is actually transferred and a physical connection is required in order for the handshaking process to take place. During a handshaking exchange, one device will request attention from the other device and wait for a response. When the response is received and acknowledged by the second device, the first device will respond in order to complete the handshake. Following the handshake it is then possible to exchange data between the two devices. As an example, the handshaking exchange of data between a printer and a computer might look something like this if it was written in normal everyday language:

Printer: "Hello, I'm a printer waiting to accept data from you!"

Computer: "I'm a little busy at the moment – just wait a little …"

Then, sometime later …..

Printer: "Hello, I'm a printer waiting to accept data from you!"

Computer: "I'm a computer and I have data for you. Are you ready to accept the data?"

Printer: "Yes, I'm ready to accept data. Please send me your data"

Computer: "Fine – sending data now …."

FLOW CONTROL

Flow control is the process of managing the transfer of data between two devices. It ensures that the data is sent at an appropriate rate and that the receiver is not overloaded by sending data that it can't accept. In the dialogue between the printer and computer that we met earlier you should note that the computer is implementing a form of flow control by telling the printer that it is too busy to send data and is requesting that the printer should wait.

ERROR CHECKING

Error checking is used to ensure that the received data is the same as that which has actually been sent. Errors can be detected by various methods including *automatic repeat request* (*ARQ*) where the transmitter sends the data together with a code which the receiver uses to check for errors and *forward error correction* (*FEC*) where the transmitter encodes the data with an *error correcting code* (*ECC*). It is also possible to combine ARQ with FEC so that minor errors are corrected without retransmission, and major errors are corrected via a request for retransmission.

Test your knowledge 5.28

1. Explain briefly what a communication protocol is and why it is necessary.
2. State THREE features that normally form part of a communication protocol.

ACTIVITY 5.14

Use library or Internet resources to investigate the Bluetooth standard for data communication. Write an article (using not more than 750 words) for your local newspaper explaining what Bluetooth is and how it might be of use to readers. Include relevant diagrams and illustrations in your article.

ACTIVITY 5.15

Interview the Network Manager in your school or college. Ask him or her to explain the 'network architecture' and how data is exchanged between the computers and servers in the network. Use the notes made at the interview to write an article (using not more than 750 words) for your student magazine. Include relevant diagrams and illustrations in your article.

QUESTIONS

1. State the units used for each of the following quantities:
   a. electric potential
   b. electric current
   c. electrical resistance
   d. frequency
   e. bit rate.
2. Express 159 kHz in MHz.
3. Express 0.655 mA in μA.
4. Express 15 mV in V.
5. Sketch an example of each of the following types of waveform:
   a. sine
   b. square
   c. triangle
   d. sawtooth
   e. pulse
   f. complex (e.g. speech or music).
6. A signal has a frequency of 50 kHz. What is the period of the signal when expressed in μs?
7. A repetitive pulse waveform is 'high' for 20 ms and 'low' for 80 ms. What is:
   a. the period of the waveform
   b. the pulse repetition frequency of the waveform
   c. the mark to space ratio of the waveform
   d. the duty cycle of the waveform.

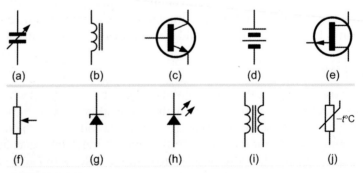

FIGURE 5.93
See Question 8

8. Identify each of the component symbols shown in Fig. 5.93
9. Sketch the symbol used for each of the following components:
   a. a fixed resistor
   b. an electrolytic capacitor
   c. an air-cored inductor#
   d. a light emitting diode
   e. a PNP transistor
   f. an operational amplifier.
10. Identify each of the components shown in Fig. 5.94.
11. Refer to the circuit shown in Fig. 5.95.
    a. State the function of the circuit.
    b. What type of device is IC1?
    c. What type of switch is S1?
    d. What is the supply voltage for the circuit?
    e. Which component is connected in series with C3?
    f. Which component is connected to the non-inverting input of IC1?
    g. What voltage will appear across C5 when S1 is switched 'on'?
12. Calculate the resistance of the series-parallel circuit shown in Fig. 5.96.
13. Calculate the capacitance of the series-parallel circuit shown in Fig. 5.97.
14. Identify each of the logic devices shown in Fig. 5.98.
15. The logic circuit shown in Fig. 5.98 produces an output when either one of the two inputs, A or B, is present but not when both of them are present.
    a. What is the logical function of the circuit?
    b. Sketch the truth table for the circuit.
16. Sketch a labelled circuit of an astable oscillator using two transistors.
17. Explain what is meant by each of the following terms used in data communication:
    a. transmission medium
    b. protocol
    c. handshaking
    d. flow control.

(a)

(b)

(c)

(d)

(e)

(f)

FIGURE 5.94
See Question 10

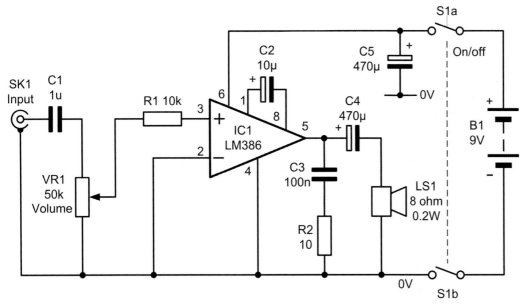

FIGURE 5.95
See Question 11

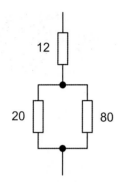

FIGURE 5.96
See Question 12

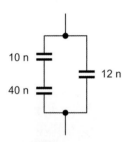

FIGURE 5.97
See Question 13

18. Convert:
   a. binary 10011101 to hexadecimal
   b. hexadecimal 7C to binary.
19. Explain the advantages of an optical fibre LAN when compared with a LAN that uses copper cabling.
20. Explain why error checking is important in a data communication system. State two methods of error checking.

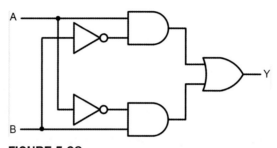

FIGURE 5.98
See Question 14

Selecting Engineering Materials

SUMMARY

Welcome to the fascinating world of engineering materials! All branches of engineering are concerned with the use and processing of materials and so this unit makes an excellent companion to the others that you study as part of your BTEC First award. In fact, the correct choice and processing of materials can be crucial. For example, Civil Engineers need to ensure that the materials from which they build their roads, bridges and dams are not only able to withstand the forces that will be applied to them but also that they will not deteriorate with time. Think about what might happen if the materials chosen were not appropriate for the job. For example, if a suspension bridge was built from a material that is not sufficiently strong it might collapse under a heavy load. Even on a much smaller scale knowledge of materials is vitally important. An Electronic Engineer designing a complex flight control system must ensure that the materials chosen do not prematurely deteriorate in the harsh environment in which the system will operate. Failure of the system could be equally catastrophic, particularly if the failure just happened to occur at a critical point in flight, such as landing or take-off. A study of materials is, therefore, an essential part of every engineer's portfolio and for this reason alone, is worthy of your attention.

6.1 INTRODUCING ENGINEERING MATERIALS

This unit is about the materials used in engineering so let's start by looking at the materials used in a typical engineered product. Figure 6.1 shows the interior of a low-voltage d.c. power supply of the type used to supply power to small items of consumer electronic equipment. The power supply is rated at 12 V, 1 A maximum. For safety reasons, the power supply is totally enclosed and connection to the a.c. mains supply is made by means of cable fitted with a fused 13 A mains plug. Connection to the equipment being powered is made by a second power lead fitted with a d.c. power connector.

The main components used in the low-voltage d.c. power supply have been labelled A to H in Figure 6.1. They are as follows:

A—Enclosure

The enclosure is made from a plastic material. This not only provides protection for the components inside the power supply but it also reduces the risk of damage and electric shock

BTEC First Engineering: Mandatory and Selected Optional Units for BTEC Firsts in Engineering. DOI: 10.1016/B978-1-85617-685-9.00006-8

that might result from contact with 'live' circuitry. The plastic material is chosen because it provides insulation, is inexpensive, and can be moulded into a complex shape (note how the supporting pillars and ventilation slots are integral parts of the case).

B—Transformer core

The transformer core is made from laminated strips of steel that are stamped from thin sheets of metal. Steel is chosen as the material for the transformer core because of its excellent *ferromagnetic* properties (see Unit 4). The core is laminated and has a small gap in it to prevent the passage of *eddy currents* that would otherwise be induced in it (recall, from Unit 4, that steel is a good conductor of electricity as well as an excellent magnetic material).

C—Transformer windings

The coils that make up the transformer windings (not easily seen in Fig. 6.1) are made from enamelled copper wire. Copper is used because it has a low resistance and is non-magnetic. The wire is *drawn* (or pulled out) in a continuous length of the required wire diameter and then given a thin *enamel* surface coating in order to provide insulation. The insulation is needed in order to avoid the risk of shorts between adjacent turns of wire.

D—Power lead

The power lead uses copper wire which has multiple strands (to make it flexible). Once again, copper is chosen because it is an excellent conductor. The stranded copper wire is given a plastic insulating coating.

E—Printed circuit board

The printed circuit board is made from a *composite* material (glass-fibre reinforced plastic). This material is strong and is a good electrical insulator. The material is given a copper laminating coating which is etched in order to provide the required conducting track pattern which links the component leads and pins together.

F—Heat dissipator

The heat dissipator (or 'heat sink') is designed to conduct heat away from an integrated circuit voltage regulator (not visible in Fig. 6.1) and convect it into the surrounding space. The heat dissipator is made from aluminium alloy (which is a good conductor of heat and is also relatively light). The complex finned shape is made from an *extrusion* in which hot alloy is forced under pressure into a shape which has a continuous cross-section.

G—Semiconductors

The power supply uses several different types of semiconductor device. These include four rectifier diodes, a light-emitting diode, and an integrated circuit voltage regulator. These devices are based on semiconductor materials, a class of material that is neither a conductor nor an insulator. The ability of a semiconductor device to carry an electric current depends on the amount and polarity of the supplied voltage. The materials used in the manufacture of the semiconductors in the power supply include silicon and gallium arsenide.

H—Retaining screws

The five pan head retaining screws are made from steel with a black oxide finish. The screws are designed to form their own *threads* when inserted into the pillars in the (much softer) plastic case.

Finally, it's important to note that the materials used for these components have different properties and they require quite different treatments and manufacturing processes. Being able to select appropriate materials and specify the correct treatments and manufacturing processes to be used with them is an important skill that you must develop along the road to becoming an engineer.

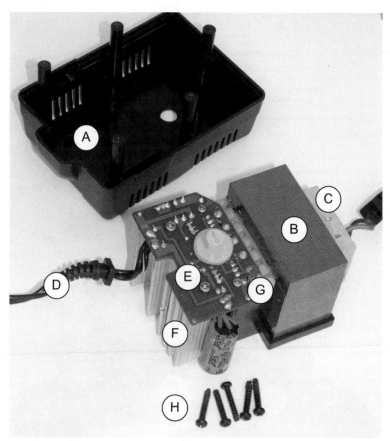

FIGURE 6.1
A low-voltage d.c. power supply showing some of the main components

Test your knowledge 6.1
In Figure 6.1, which components are:

a. moulded
b. stamped
c. extruded
d. etched.

Test your knowledge 6.2
In Figure 6.1, explain why copper is used in the manufacture of the parts labelled C, D, and E.

Test your knowledge 6.3
What property distinguishes semiconductors from other materials?

Test your knowledge 6.4
Give three reasons for using plastic for the power supply enclosure shown in Figure 6.1.

6.2 PROPERTIES OF MATERIALS

By now, you should be beginning to appreciate that a very wide range of materials, parts and components are used in engineering! Figure 6.2 shows the main groups of engineering materials. The Latin name for iron is *ferrum*, so it is not surprising that ferrous metals and alloys are all based on the metal iron. Alloys consist of two or more metals (or metals and non-metals) that have been brought together as compounds or solid solutions to produce a metallic material with special properties. For example an alloy of iron, carbon, nickel and chromium is *stainless steel*. This is a corrosion resistant ferrous alloy. Non-ferrous metals and alloys are the rest of the metallic materials available. Non-metals can be natural, such as rubber, or they can be synthetic such as the plastic compound PVC.

When selecting a material you need to make certain that it has properties that are appropriate for the job it has to do. For example, you should ask yourself the following questions:

- Will it corrode in its working environment?
- Will it weaken or melt in a hot environment?
- Will it break under normal working conditions?
- Can it be easily cast, formed or cut to shape?

You can assess the suitability of different materials for a particular engineering application by comparing their properties. The properties of a material used in engineering are extremely important and careful consideration is usually given to it.

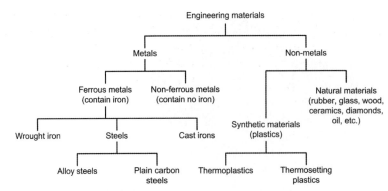

FIGURE 6.2
The main groups of engineering materials

FIGURE 6.3
Off-road vehicle—see
Activity 6.1

Chemical Properties

When engineers look at the chemical properties of a material they are usually concerned
with two things; *corrosion* and *degradation*.

CORROSION

This is caused by the metals and metal alloys being attacked and eaten away by chemical
substances. For example the rusting of ferrous metals and alloys is caused by the action of
atmospheric oxygen in the presence of water. Another example is the attack on aluminium
and some of its alloys by strong alkali solutions. Take care when using *degreasing agents* on
such metals. Copper and copper based alloys are stained and corroded by the active sulphur
and chlorine products found in some heavy duty cutting lubricants.

TABLE 6.1 See Activity 6.1			
Component reference	Component name and/or function	Material(s) used in manufacture of the component	Reason(s) for using this material
A	Frame		
B	Front fairing		
C	Driver's helmet		
D	Tyres		
E	Shock absorber (spring)		
F	Engine block/casing		
G	Wheel rims		

DEGRADATION

Non-metallic materials do not corrode but they can be attacked by chemical substances. Since this weakens or even destroys the material it is referred to as *degradation*. Unless specially compounded, rubber is attacked by prolonged exposure to oil. Synthetic (plastic) materials can be softened by chemical solvents. Exposure to the ultraviolet rays of sunlight can weaken (perish) rubbers and plastics unless they contain compounds that filter out such rays.

Key point
Engineers are usually concerned with corrosion and degradation when they look at the chemical properties of a material.

Electrical Properties

The electrical properties of a material are important when selecting it for an engineering application that requires the material to either allow or prevent the flow of an electric current to some degree. We specify this property by referring to its electrical resistance.

ELECTRICAL RESISTANCE

Materials with a very low resistance to the flow of an electric current are good electrical *conductors*. Materials with a very high resistance to the flow of electric current are good *insulators*. Generally, metals are good conductors and non-metals are good insulators (poor conductors). A notable exception is carbon which conducts electricity despite being a non-metal. The electrical resistance of a metal conductor depends upon:

Key point
Materials that have a low electrical resistance are described as 'good conductors'.

- Its length (the longer it is the greater its resistance)
- Its thickness (the thicker it is the lower its resistance)
- Its temperature (the higher its temperature the greater is its resistance)
- Its resistivity (this is the resistance measured between the opposite faces of a metre cube of the material).

Note that a small number of non-metallic materials, such as silicon, have atomic structures that fall between those of electrical conductors and insulators. These materials are called semiconductors and are used for making solid state devices such as diodes and transistors.

Key point
Materials that have a very high electrical resistance are described as 'insulators'.

Magnetic Properties

All materials respond to strong magnetic fields to some extent. Only the ferromagnetic materials respond sufficiently to be of interest. The more important ferromagnetic materials are the metals iron, nickel and cobalt. Magnetic materials are often divided into two types: soft and hard.

Soft magnetic materials, such as soft iron, can be magnetised by placing them in a magnetic field. They cease to be magnetised as soon as the field is removed.

Hard magnetic materials, such as high carbon steel that has been hardened by cooling it rapidly (*quenching*) from red heat, also become magnetised when placed in a magnetic field.

Key point
The most important ferromagnetic materials are iron, nickel, and cobalt.

Key point
Metals are good conductors of heat whilst non-metals are poor heat conductors.

Test your knowledge 6.6
What is meant by the term *ferromagnetic* material? Give TWO examples of ferromagnetic materials.

Test your knowledge 6.7
Explain the difference between *corrosion* and *degradation*.

Test your knowledge 6.8
Explain the difference between *soft* and *hard* magnetic materials.

Another view
You must be careful when interpreting the strength data quoted for various materials. A material may appear to be strong when subjected to a static load, but will break when subjected to an impact load. Materials also show different strength characteristics when the load is applied quickly than when the load is applied slowly.

Hard magnetic materials retain their magnetism when the field is removed. They become permanent magnets.

Permanent magnets can be made more powerful for a given size by adding cobalt to the steel to make an alloy. Soft magnetic materials can be made more efficient by adding silicon or nickel to the pure iron. Silicon-iron alloys are used for the rotor and stator cores of electric motors and generators. Silicon-iron alloys are also used for the cores of power transformers.

Thermal Properties

Thermal properties are to do with how a material responds to heat and different temperatures. Thermal properties include:

MELTING TEMPERATURE

The melting temperature of a material is the temperature at which a material loses its solid properties. Most plastic materials and all metals become soft and eventually melt. Note that some plastics do not soften when heated, they only become charred and are destroyed. This will be considered later in this unit.

THERMAL CONDUCTIVITY

This is the ease with which materials conduct heat. Metals are good conductors of heat. Non-metals are poor conductors of heat. Therefore non-metals are heat insulators.

EXPANSION

Metals expand appreciably when heated and contract again when cooled. They have high coefficients of linear expansion. Non-metals expand to a lesser extent when heated. They have low coefficients of linear expansion. Again, these properties will be considered in more detail in your science unit.

Mechanical Properties

Mechanical properties are important when engineering materials are to withstand loads and forces imposed on them.

STRENGTH

This is the ability of a material to resist an applied force (load) without fracturing (breaking). It is also the ability of a material not to *yield*. Yielding is when the material 'gives' suddenly under load and changes shape permanently but does not break. This is what happens when metal is bent or folded to shape. The load or force can be applied in various ways as shown in Figure 6.4.

TOUGHNESS

This is the ability of a material to resist impact loads as shown in Figure 6.5. Here, the toughness of a piece of high carbon steel in the soft (annealed) condition is compared with a piece of the same steel after it has been hardened by raising it to red-heat and cooling it quickly (*quenching* it in cold water). The hardened steel shows a greater strength, but it lacks toughness.

A test for toughness, called the *Izod test*, uses a notched specimen that is hit by a heavy pendulum. The test conditions are carefully controlled, and the energy absorbed in bending or breaking the specimen is a measure of the toughness of the material from which it was made.

ELASTICITY

Materials that change shape when subjected to an applied force but spring back to their original size and shape when that force is removed are said to be elastic. They have the property of elasticity.

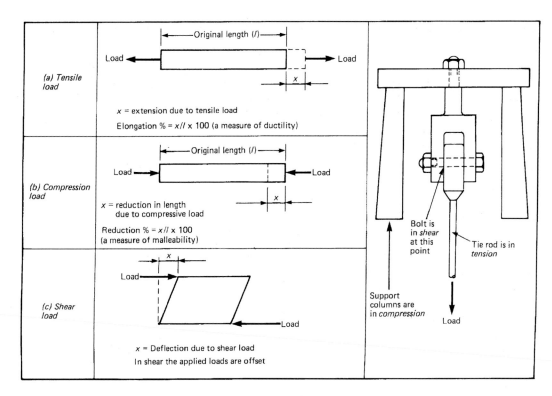

FIGURE 6.4
Different ways in which a load can be applied

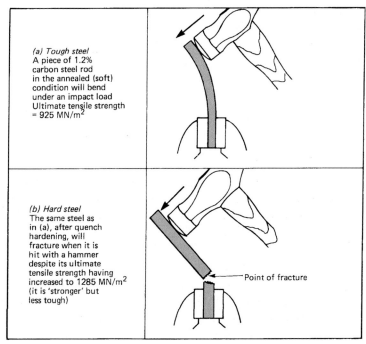

FIGURE 6.5
Impact loads

PLASTICITY

Materials that flow to a new shape when subjected to an applied force and keep that shape when the applied force is removed are said to be plastic. They have the property of plasticity.

DUCTILITY

Materials that can change shape by plastic flow when they are subjected to a pulling (tensile) force are said to be ductile. They have the property of ductility. This is shown in Figure 6.6(a).

Test your knowledge 6.9

Explain what is meant by *quenching* and state what effect the process has on the mechanical properties of steel.

201

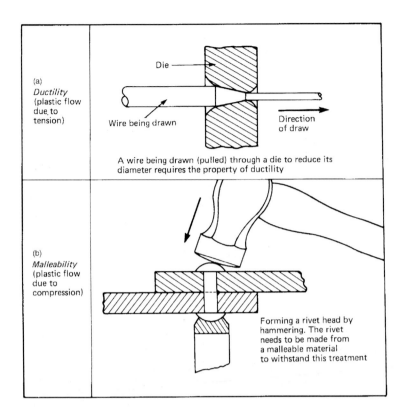

FIGURE 6.6

Two common engineering processes; drawing and riveting. Drawing exploits ductility whilst riveting exploits malleability

MALLEABILITY

Materials that can change shape by plastic flow when they are subjected to a squeezing (compressive) force are said to be malleable. They have the property of malleability. This is shown in Figure 6.6(b).

HARDNESS

Materials that can withstand scratching or indentation by an even harder object are said to be hard. They have the property of hardness. Figure 6.7 shows the effect of pressing a hard steel ball into two pieces of metal with the same force. The ball sinks further into the softer of the two pieces of metal than it does into the harder.

There are various hardness tests available. The *Brinell hardness test* uses the principles set out above. A hardened steel ball is pressed into the specimen by a controlled load. The diameter of the indentation is measured using a special microscope. The hardness number is obtained from the measured diameter by use of conversion tables.

The *Vickers test* is similar but uses a diamond pyramid instead of a hard steel ball. This enables harder materials to be tested. The diamond pyramid leaves a square indentation and the diagonal distance across the square is measured. Again, conversion tables are used to obtain the hardness number from the measured distance.

The *Rockwell test* uses a diamond cone. A minor load is applied and a small indentation is made. A major load is then added and the indentation increases in depth. This increase in depth of the indentation is directly converted into the hardness number and it can be read from a dial on the machine.

RIGIDITY

Materials that resist changing shape under load are said to be rigid. They have the property of rigidity. The opposite of rigidity is flexibility. Rigid materials are usually less strong than flexible materials. For example, cast iron is more rigid than steel but steel is the stronger and

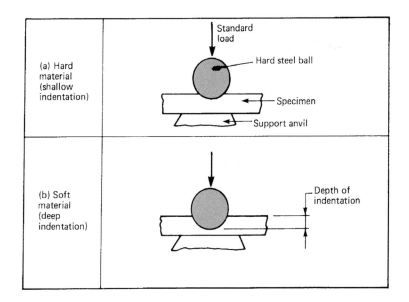

FIGURE 6.7
The effect of pressing a hard steel ball into two materials with different hardness properties

FIGURE 6.8
Casting metal components in a foundry

tougher. However the rigidity of cast iron makes it a useful material for machine frames and beds. If such components were made from a more flexible material the machine would lack accuracy and it would be deflected by the cutting forces.

6.3 CLASSES OF MATERIAL

For most engineering applications the most important criteria for the selection of materials is that they do the job properly (i.e. that they perform according to the specification that has been agreed) and that they do it as cheaply as possible. Whether a material does its job depends on its *properties*, which are a measure of how the material reacts to the various influences to which it is exposed. For example, loads, atmospheric environment, electromotive forces, heat, light, chemicals, and so on.

To aid your understanding of the different types of materials and their properties I have divided them into four classes; metals, polymers, ceramics, and composites. Strictly speaking, composites are not really a separate group since they are made up from the other categories of material. However, because they display unique properties and are a very important engineering group, they are treated separately here.

Test your knowledge 6.12
State the properties required by each of the following engineering products. For each answer, give the reason for your choice.

a. A wire rope sling used for lifting heavy loads.
b. The axle of a motor vehicle.
c. A spring for use in a lock mechanism.

203

Metals

You will be familiar with metals such as aluminium, iron, and copper, in a wide variety of everyday applications: i.e. aluminium saucepans, copper water pipes and iron stoves. Metals can be mixed with other elements (often other metals) to form an *alloy*. Metal alloys are used to provide improved properties because they are often stronger or tougher than the parent pure metal. Other improvements can be made to metal alloys by *heat-treating* them as part of the manufacturing process. Thus steel is an alloy of iron and carbon and small quantities of other elements. If, after alloying, the steel is quickly cooled by quenching in oil or brine, a very hard steel can be produced. Much more will be said later about alloys and the heat-treatment of metals.

Polymers

Polymers are characterised by their ability to be (initially at least) moulded into shape. They are chemical materials and often have long and unattractive chemical names. There is considerable incentive to seek more convenient names and abbreviations for everyday use. Thus you will be familiar with PVC (polyvinyl chloride) and PTFE (polytetrafluoroethylene).

Polymers are made from molecules which join together to form long chains in a process known as *polymerisation*. There are essentially three major types of polymer. *Thermoplastics,* which have the ability to be remoulded and reheated after manufacture. *Thermosetting plastics,* which once manufactured remain in their original moulded form and cannot be re-worked. *Elastomers or rubbers,* which often have very large elastic strains, elastic bands and car tyres are two familiar forms of rubber.

Ceramics

This class of material is again a chemical compound, formed from oxides such as silica (sand), sodium and calcium, as well as silicates such as clay. Glass is an example of a ceramic material, with its main constituent being silica. The oxides and silicates mentioned above have very high melting temperatures and on their own are very difficult to form. They are usually made more manageable by mixing them in powder form, with water, and then hardening them by heating. Ceramics include, brick, earthenware, pots, clay, glasses and refractory (furnace) materials. Ceramics are usually hard and brittle, good electrical and thermal insulators and have good resistance to chemical attack.

Composites

A composite is a material with two or more distinct constituents. These separate constituents act together to give the necessary strength and stiffness to the composite material. The most common example of a composite material today is that of fibre reinforcement of a resin matrix but the term can also be applied to other materials such as metal-skinned honeycomb panels. The property that these materials have in common is that they are light, stiff and strong and, what's more, they can be extremely tough.

Reinforced concrete is another example of a composite material that is invaluable in engineering. The steel and concrete retain their individual identities in the finished structure. However, because they work together, the steel carries the tension loads and the concrete carries the compression loads. Furthermore, although not considered as a separate class of material, some *natural materials* exist in the form of a composite. The best known examples are wood, mollusc shells and bone. Wood is an interesting example of a natural fibre composite; the longitudinal hollow cells of wood are made up of layers of spirally wound cellulose fibres with varying spiral angle, bonded together with lignin during the growing of the tree.

The previous classification of materials is rather crude and many important subdivisions exist within each category. The natural materials, except for those mentioned above under composites, have been deliberately left out. The study of materials such as wool and cotton is

ACTIVITY 6.2

How much do you already know about different materials and their properties? Test your knowledge by trying the exercise set out below. In attempting to tackle this task you might find it helpful to explore the objects that exist within your school, college or home, and ask yourself why they are made from those particular materials. Complete the table by using a grading scheme such as: excellent, good, fair, poor; or high, above average, below average, low, or some other similar scheme.

Material Properties	Class of Material			
	Metals	Polymers	Ceramics	Composites
Density				
Stiffness				
Strength				
Toughness				
Shock resistance				
Hardness				
Thermal conductivity				
Electrical conductivity				
Corrosion resistance				
Melting temperature				

ACTIVITY 6.3

As you progress through Unit 8 you will find it extremely useful to have data on a variety of materials that are commonly used in engineering. Sources of data include reference books and text books as well as several on-line databases accessible via the Internet. You are advised to become familiar with several of these sources. However, in order to provide you with a simplified set of data, I have created the matSdata database specifically for engineering students. The database will provide you with data and datasheets on a wide range of commonly used engineering materials. Furthermore, unlike other databases, matSdata allows you to maintain your own personal copy of the database. This means that you can modify existing records and add your own data as you progress through the course.

You can download a copy of the matSdata database from www.key2study.com/matsdata or from the URL shown in Appendix 5. To get started you should first download the database and then extract the files in order to make your own personal copy in a folder on your hard drive, USB memory stick, or even on a floppy disk.

Next you should run the matSdata program (this is called matSdata.exe) and read the Help file. Finally, it's worth printing out datasheets for each of the following materials that you will be meeting in the next section of this unit:

- Brass (common brass)
- Carbon steel
- Polyvinyl chloride (PVC).

better placed in a course concerned with the textile industry. Here, we will be concentrating on the engineering application of materials and will only mention naturally occurring materials where appropriate.

Finally, what you may have discovered from Activity 6.2 was how difficult it is to make generalisations about the properties of different classes of materials. We shall now continue this theme by examining each class of material in a little more detail but, before we do, I suggest that you take a look at Activity 6.3.

Test your knowledge 6.14

a. Name the THREE main classes of polymer materials
b. Name TWO examples of composite materials.

205

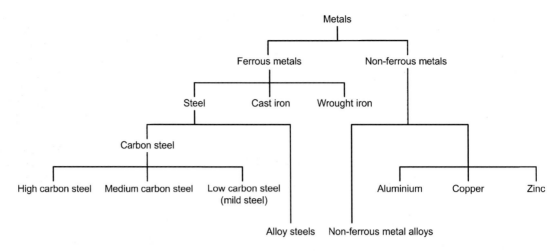

FIGURE 6.9
Classification of metals

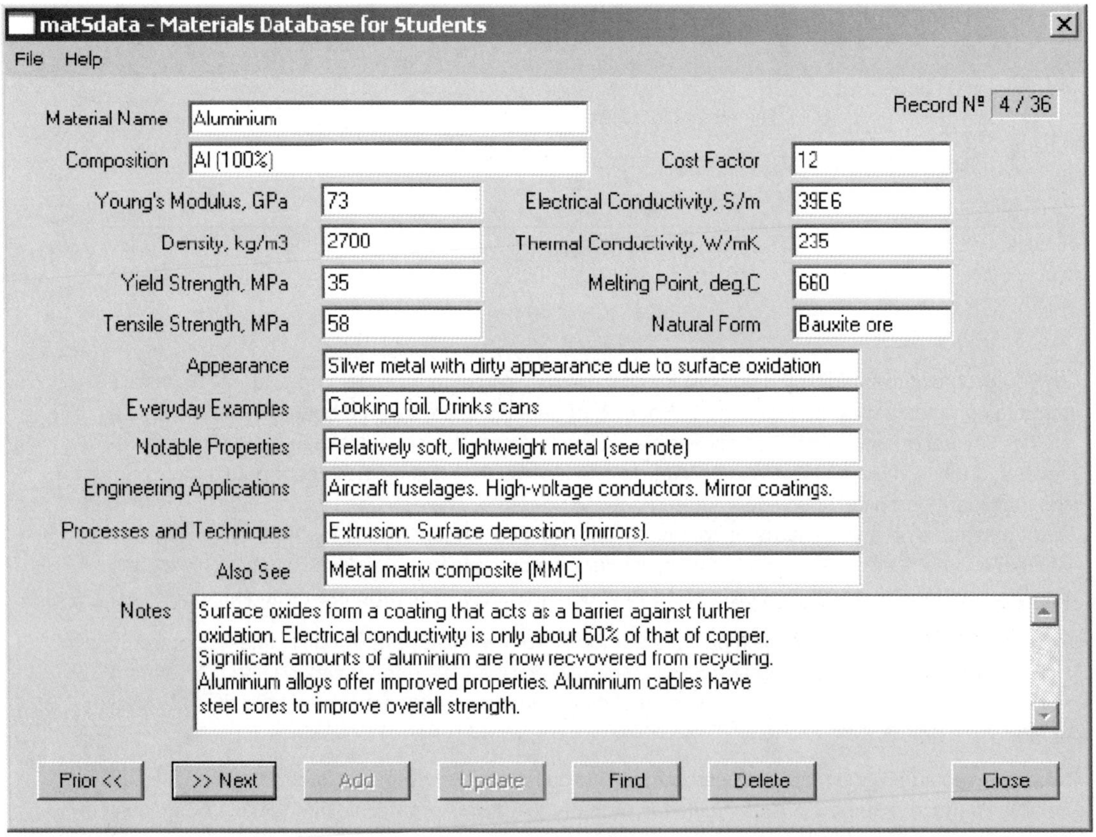

FIGURE 6.10
The matSdata record for aluminium. You can download your own copy of matSdata from www.key2study.com/matsdata

6.4 FERROUS METALS

As previously stated, ferrous metals are based upon the metal iron. For engineering purposes iron is usually associated with various amounts of the non-metal carbon. When the amount of carbon present is less than 1.8% we call the material steel. The figure of 1.8% is the theoretical maximum. In practice there is no advantage in increasing the amount of carbon present above 1.4%. We are only going to consider the plain carbon steels. Alloy steels are beyond the scope of this book. The effects of the carbon content on the properties of plain carbon steels are shown in Figure 6.13.

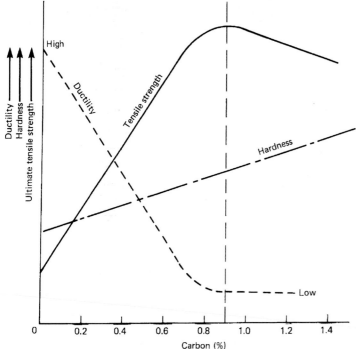

FIGURE 6.11
A sprocket made from medium carbon steel

FIGURE 6.12
High quality adjustable spanners made from high carbon steel

FIGURE 6.13
Effect of carbon content on the properties of plain carbon steels

Cast irons are also ferrous metals. They have substantially more carbon than the plain carbon steels. Grey cast irons usually have a carbon content between 3.2% and 3.5%. Not all this carbon can be taken up by the iron and some is left over as flakes of graphite between the crystals of metal. It is these flakes of graphite that gives cast iron its particular properties and makes it a 'dirty' metal to machine.

Low Carbon Steels

Low carbon steels (also known as *mild steels)* are the cheapest and most widely used group of steels. Although they are the weakest of the steels, nevertheless they are stronger than most of the non-ferrous metals and alloys. They can be hot and cold worked and machined with ease.

Medium Carbon Steels

These are harder, tougher, stronger and more costly than the low carbon steels. They are less ductile than the low carbon steels and cannot be bent or formed to any great extent in the cold condition without risk of cracking. Greater force is required to bend and form them. Medium carbon steels hot forge well but close temperature control is essential. Two carbon ranges are shown. The lower carbon range can only be toughened by heating and quenching (cooling quickly by dipping in water). They cannot be hardened. The higher carbon range can be hardened and tempered by heating and quenching.

High Carbon Steels

These are harder, stronger and more costly than medium carbon steels. They are also less tough. High carbon steels are available as hot rolled bars and forgings. Cold drawn high carbon steel wire (piano wire) is available in a limited range of sizes. Centreless ground high carbon steel rods (silver steel) are available in a wide range of diameters (inch and metric sizes) in lengths of 333 mm, 1 m and 2 m. High carbon steels can only be bent cold to a limited extent before cracking. They are mostly used for making cutting tools.

Another view
The term 'carbon steel' can be a little misleading because all steel contains carbon. Iron, the basic ingredient of steel, has so much carbon in its basic 'pig iron' form that carbon actually has to be *removed* to produce what we call 'carbon steel'. Just remember that carbon steel has carbon as the *primary* additional element and that it has no intentionally mixed alloys included to change the mechanical properties of the steel.

207

TABLE 6.2 Ferrous metals

Name	Group	Carbon content (%)	Typical applications
Dead mild steel (low carbon steel)	Plain carbon steel	0.10–0.15	Sheet for pressing out components such as motor car body panels. General sheet-metal work. Thin wire, rod and drawn tubes.
Mild steel (low carbon steel)	Plain carbon steel	0.15–0.30	General purpose workshop rod, bars and sections. Boiler plate. Rolled steel beams, joists, angles, etc.
Medium carbon steel	Plain carbon steel	0.30–0.50	Crankshafts, forgings, axles, and other stressed components.
		0.50–0.60	Leaf springs, hammer heads, cold chisels, etc.
High carbon steel	Plain carbon steel	0.8–1.0	Coil springs, wood chisels.
		1.0–1.2	Files, drills, taps and dies.
		1.2–1.4	Fine-edge tools (knives, etc.)
Grey cast iron	Cast iron	3.2–3.5	Machine castings.

6.5 NON-FERROUS METALS

Non-ferrous metals (i.e. metals that are *not* based on iron) include metals such as aluminium and zinc as well as alloys such as brass and bronze. We shall start by looking at copper—a material that is widely used in electrical engineering.

Copper

Pure copper is widely used for electrical conductors and switchgear components. It is second only to silver in conductivity but it is much more plentiful and very much less costly. Pure copper is too soft and ductile for most mechanical applications.

For general purpose applications such as roofing, chemical plant, decorative metal work and copper-smithing, tough-pitch copper is used. This contains some copper oxide which makes it stronger, more rigid and less likely to tear when being machined. Because it is not so highly refined, it is less expensive than high conductivity copper.

There are many other grades of copper for special applications. Copper is also the basis of many important alloys such as brass and bronze, and we will be considering these next. The general properties of copper are:

- relatively high strength
- very ductile so that it is usually cold worked. An annealed (softened) copper wire can be stretched to nearly twice its length before it snaps
- corrosion resistant
- second only to silver as a conductor of heat and electricity
- easily joined by soldering and brazing. For welding, a phosphorous deoxidised grade of copper must be used.

Copper is available as cold-drawn rods, wires and tubes. It is also available as cold-rolled sheet, strip and plate. Hot worked copper is available as extruded sections and hot stampings. It can also be cast. Copper powders are used for making sintered components. It is one of the few pure metals of use to the engineer as a structural material.

Brass

Brass is an alloy of copper and zinc. The properties of a brass alloy and the applications for which you can use it depends upon the amount of zinc present. Most brasses are attacked by sea water. The salt water eats away the zinc (known as *dezincification*) and leaves a weak, porous, spongy mass of copper. To prevent this happening, a small amount of tin is added to the alloy. There are two types of brass that can be used at sea or on land near the sea. These are Naval brass and Admiralty brass.

Brass is a difficult metal to cast and brass castings tend to be coarse grained and porous. Brass depends upon hot rolling from cast ingots, followed by cold rolling or drawing to give it its mechanical strength. It can also be hot extruded and plumbing fittings are made by hot stamping. Brass machines to a better finish than copper as it is more rigid and less ductile than that metal. You will find the different types of brass and their properties listed in the matSdata database.

Tin Bronze

As the name implies, the tin bronzes are alloys of copper and tin. These alloys also have to have a deoxidising element present to prevent the tin from oxidising during casting and hot working. If the tin oxidises the metal becomes hard and 'scratchy' and is weakened. The two deoxidising elements commonly used are:

- zinc in the gun-metal alloys.
- phosphorus in the phosphor–bronze alloys.

Unlike the brass alloys, the bronze alloys are usually used as castings. However low-tin content phosphor–bronze alloys can be extensively cold worked. Tin–bronze alloys are extremely resistant to corrosion and wear and are used for high pressure valve bodies and heavy duty bearings.

Aluminium

Aluminium has a density approximately one third that of steel. However it is also very much weaker so its strength/weight ratio is inferior. For stressed components, such as those found in aircraft, aluminium alloys have to be used. These can be as strong as steel and nearly as light as pure aluminium.

High purity aluminium is second only to copper as a conductor of heat and electricity. It is very difficult to join by welding or soldering and aluminium conductors are often terminated by crimping. Despite these difficulties, it is increasingly used for electrical conductors where its light weight and low cost compared with copper is an advantage. Pure aluminium is resistant to normal atmospheric corrosion but it is unsuitable for marine environments. It is available as wire, rod, cold-rolled sheet and extruded sections for heat sinks.

Commercially pure aluminium is not as pure as high purity aluminium and it also contains up to 1% silicon to improve its strength and stiffness. As a result it is not such a good conductor of electricity nor is it so corrosion resistant. It is available as wire, rod, cold-rolled sheet and extruded sections. It is also available as castings and forgings. Being stiffer than high purity aluminium it machines better with less tendency to tear. It forms non-toxic oxides on its surface which makes it suitable for food processing plant and utensils. It is also used for forged and die-cast small machine parts. Because of their range and complexity, the light alloys based upon aluminium are beyond the scope of this unit.

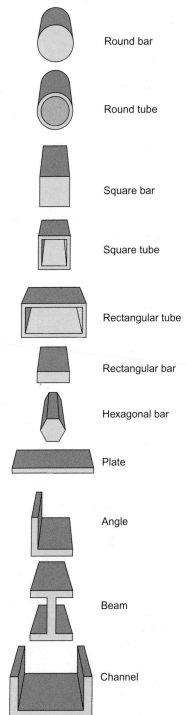

Round bar

Round tube

Square bar

Square tube

Rectangular tube

Rectangular bar

Hexagonal bar

Plate

Angle

Beam

Channel

FIGURE 6.14
Various forms in which ferrous and non-ferrous metals are supplied

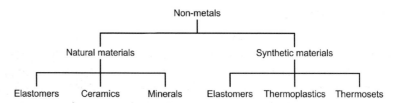

FIGURE 6.15
Classes of non-metals

Test your knowledge 6.18
Use the matSdata database or other sources of information on materials to determine the tensile strength of aluminium, phosphor bronze, carbon steel, copper, lead, and titanium. Rank these materials (from the strongest to the weakest) in order of tensile strength.

6.6 NON-METALS

Non-metals materials can be grouped under the headings shown in Figure 6.15. In addition, wood is also used for making the patterns which, in turn, are used in producing moulds for castings. We are only going to consider some ceramics, thermosets and thermoplastics.

Ceramics

The word ceramic comes from a Greek word meaning potter's clay. Originally, ceramics referred to objects made from potter's clay. Nowadays, ceramic technology has developed a range of materials far beyond the traditional concepts of the potter's art. These include:

- glass products
- abrasive and cutting tool materials
- construction industry materials
- electrical insulators
- cements and plasters for investment moulding
- refractory (heat resistant) lining for furnaces
- refractory coatings for metals.

The four main groups of ceramics are amorphous, crystalline, bonded and cements. Ceramic materials are generally very hard compared with other engineering materials. Crystalline ceramics (such as *silicon carbide*) are used as abrasives and cutting tool materials. The properties of ceramic materials can be summarised as follows:

STRENGTH

Ceramic materials are reasonably strong in compression, but tend to be weak in tension and shear. They are brittle and lack ductility. They also suffer from micro-cracks which occur during the firing process. These lead to fatigue failure. Many ceramics retain their high compressive strength at very high temperatures.

HARDNESS

Most ceramic materials are harder than other engineering materials. Because of this, they are widely used for cutting tool tips and abrasives. They retain their hardness at very high temperatures that would destroy high carbon and high speed steels. However they have to be handled carefully because of their brittleness.

REFRACTORINESS

This is the ability of a material to withstand high temperatures without softening and deforming under normal service conditions. Some refractories such as high-alumina brick and fireclays tend to soften gradually and may collapse at temperatures well below their fusion (melting) temperatures. Refractories made from clays containing a high proportion of silica to alumina are most widely used for furnace linings.

Test your knowledge 6.19
Select a suitable ferrous metal, and state its carbon content, for each of the following objects. Select a suitable non-ferrous metal for each of the following objects. Give reasons for your choice.

a. The body casting of a water pump
b. screws for clamping the electric cables in the terminals of a domestic electric light switch
c. a bearing bush
d. a deep drawn, cup-shaped component for use on land
e. a ship's fitting made by hot stamping.

ELECTRICAL PROPERTIES

As well as being used for weather resistant high-voltage insulators for overhead cables and sub-station equipment, ceramics are now being used for low-loss high-frequency insulators. For example they are being used for the dielectric in silvered ceramic capacitors for high-frequency applications.

In all the previous examples the ceramic material is *polycrystalline*. That is, the material is made up of a large number of very tiny crystals.

FIGURE 6.16
Thermosetting plastics and thermoplastics are used widely in the production of electrical components such as these micro-switches

For solid state electronic devices single crystals of silicon are grown under very carefully controlled conditions. The single crystal can range from 50 mm diameter to 150 mm diameter with a length ranging from 500 mm to 2,500 mm. These crystals are without impurities. They are then cut up into thin wafers and made into such devices as thermistors, diodes, transistors and integrated circuits. This is done by doping the pure silicon wafers with small, controlled amounts of carefully selected impurities.

Some types of impurity give the silicon *n-type* characteristics. That is they make the silicon electrically negative by increasing the number of electrons present. Other types of impurity give the silicon *p-type* characteristics. That is they make the silicon electrically positive by reducing the number of electrons present.

Thermosetting Plastics

Themosetting plastics are also known as *thermosets*. These materials are available in powder or granular form and consist of a synthetic resin mixed with a 'filler'. The filler reduces the cost and modifies the properties of the material. A colouring agent and a lubricant are also added. The lubricant helps the plasticised moulding material to flow into the fine detail of the mould.

The moulding material is subjected to heat and pressure in the moulds during the moulding process. The hot moulds not only plasticise the moulding material so that it flows into all the detail of moulds, the heat also causes a chemical change in the material.

This chemical change is called *polymerisation* or, more simply, 'curing'. Once cured, the moulding is hard and rigid. It can never again be softened by heating. If made hot enough it will just burn. Some thermosets and typical applications are summarised in Table 6.3.

Thermoplastics

Unlike the thermosets we have just considered, thermoplastics soften every time they are heated. In fact, any material trimmed from the mouldings can be ground up and recycled. They tend to be less rigid but tougher and more 'rubbery' than the thermosetting materials. Some thermoplastics and typical applications are summarised in Table 6.4.

Reinforced Plastics

The strength of plastics can be increased by reinforcing them with fibrous materials. Such materials include:

- *Laminated plastics (Tufnol)*. Fibrous material such as paper, woven cloth, woven glass fibre, etc. is impregnated with a thermosetting resin. The sheets of impregnated material are laid up in powerful hydraulic presses and they are heated and squeezed until they become solid and rigid sheets, rods, tubes, etc. This material has a high strength and good electrical properties. It can be machined with ordinary metal working tools and machines. Tufnol is used for making insulators, gears and bearing bushes.

Another view
Metals, both ferrous and non-ferrous, are conventional engineering materials and you will already be very familiar with their use. Other classes of material (such as ceramics, plastics and composites) are also widely used in engineering. You will find plenty of examples of their use in electrical and electronic engineering, chemical engineering, optical engineering, and so on.

Another view
Plastic materials are not particularly new. In fact, the first semi-synthetic plastics appeared in the 1860s, and plastics made out of natural polymers have been used for centuries. Many other plastic materials were developed in the twentieth century, and some were in mass production well before the Second World War.

Another view

Plastics are composed of natural, modified natural, or completely synthetic polymers—long-chain molecules that determine the properties of the resulting materials. Various additives are added to the base polymer to produce other desirable qualities such as colour, bulk or improved longevity. There are two main types of plastics: the relatively soft and pliable thermoplastics that can soften and flow when reheated, and the normally harder and more brittle thermosetting plastics that do not.

TABLE 6.3 Thermosetting plastics

Type	Applications
Phenolic resins and powders	The original 'Bakelite' type of plastic materials, hard, strong and rigid. Moulded easily and heat 'cured' in the mould. Unfortunately, they darken during processing and are only available in the darker and stronger colours. Phenolic resins are used as the 'adhesive' in making plywoods and laminated plastic materials (Tufnol)
Amino (containing nitrogen) resins and powders	The basic resin is colourless and can be coloured as required. Can be strengthened by paper-pulp fillers and are suitable for thin sections. Used widely in domestic electrical switchgear
Polyester resins	Polyester chains can be cross-linked by adding monomer such as styrene, when the polyester ceases to behave as a thermoplastic and becomes a thermoset. Curing takes place by internal heating due to chemical reaction and not by heating the mould. Used largely as the bond in the production of glass fibre mouldings
Epoxy resins	The strongest of the plastic materials used widely as adhesives, can be 'cold cast' to form electrical insulators and used also for potting and encapsulating electrical components.

Test your knowledge 6.20

Select a suitable plastic material for the following applications. Give reasons for your choice.

a. The moulded cockpit cover for a light aircraft

b. a moulded bearing for a computer printer

c. a rope for rock climbing

d. the body of a domestic light switch

e. encapsulating (potting) a small power transformer.

- *Glass reinforced plastics (GRP).* Woven glass fibre and chopped strand mat can be bonded together by polyester or by epoxy resins to form mouldings. These may range from simple objects such as crash helmets to complex hulls for ocean-going racing yachts. The thermosetting plastics used are cured by chemical action at room temperature and a press is not required. The glass fibre is laid up over plaster or wooden moulds and coated with the resin which is well worked into the reinforcing material. Several layers or 'plies' may be built up according to the strength required. When cured the moulding is removed from the mould. The mould can be used again. Note that the mould is coated with a release agent before moulding commences.

We shall return to this topic a little later when we describe composite materials as materials in their own right.

GENERAL PROPERTIES OF PLASTICS

Although the properties of plastic materials can vary widely, they all have some general properties in common.

Test your knowledge 6.21

Name and state the properties of:

a. TWO thermosetting plastics

b. TWO thermoplastics.

STRENGTH/WEIGHT RATIO

Plastic materials vary considerably in strength and some of the stronger (such as nylon) compare favourably with the metals. All plastics have a lower density than metals and, therefore, chosen with care and proportioned correctly their strength/weight ratio compares favourably with the light alloys.

CORROSION RESISTANCE

Plastic materials are inert to most inorganic chemicals and some are inert to all solvents. Thus they can be used in environments that are hostile to the most corrosion resistant metals and many naturally occurring non-metals.

TABLE 6.4 Thermoplastic materials

Type	Material	Characteristics
Acrylics	Polymethyl-methacrylate	Materials of the 'Perspex' or 'Plexiglass' types. Excellent light transmission and optical properties, tough, non-splintering and can be easily heat-bent and shaped. Excellent high-frequency electrical insulators.
Cellulose plastics	Nitro-cellulose	Materials of the 'celluloid' type. Tough, waterproof, and available as preformed sections, sheets and films. Difficult to mould because of their high flammability. In powder form nitro-cellulose is explosive.
	Cellulose acetate	Far less flammable than nitro-cellulose and the basis of photographic 'safety' film. Frequently used for moulded handles for tools and electrical insulators.
Fluorine plastics (Teflon)	Polytetrafluoro-ethylene (PTFE)	A very expensive plastic material, more heat resistant than any other plastic. Also has the lowest coefficient of friction. PTFE is used for heat-resistant and anti-friction coatings. Can be moulded (with difficulty) to produce components with a waxy feel and appearance.
Nylon	Polyamide	Used as a fibre or as a wax-like moulding material. Tough, with a low coefficient of friction. Cheaper than PTFE but loses its strength rapidly when raised above ambient temperature. Absorbs moisture readily, making it dimensionally unstable and a poor electrical insulator.
Polyesters (Terylene)	Polyethylene-teraphthalate	Available as a film or in fibre form. Ropes made from polyesters are light and strong and have more 'give' than nylon ropes. The film makes an excellent electrical insulator.
Vinyl plastics	Polythene	A simple material, relatively weak but easy to mould, and a good electrical insulator. Used also as a waterproof membrane in the building industry.
	Polypropylene	A more complicated material than polythene. Can be moulded easily and is similar to nylon in many respects. Its strength lies between polythene and nylon. Cheaper than nylon and does not absorb water.
	Polystyrene	Cheap and can be easily moulded. Good strength but tends to be rigid and brittle. Good electrical insulation properties but tends to craze and yellow with age.
	Polyvinylchloride (PVC)	Tough, rubbery, practically non-flammable, cheap and easily manipulated. Good electrical properties and used widely as an insulator for flexible and semi-flexible cables.

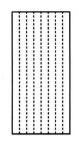

(a) Aligned continuous

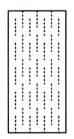

(b) Aligned discontinuous

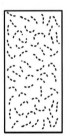

(c) Random discontinuous

FIGURE 6.17
Different fibre alignments in a composite material

ELECTRICAL RESISTANCE

All plastic materials are good electrical insulators, particularly at high frequencies. However their usefulness is limited by their softness and low heat resistance compared with ceramics. Flexible plastics such as PVC are useful for the insulation and sheathing of electric cables.

6.7 COMPOSITES

Composite materials are quite different from any of the materials that we have met so far. As their name suggests, composite materials are combinations of materials that, however, unlike their individual constituents, retain their separate identities and do not dissolve or merge together.

In structural applications composite materials have the following characteristics:

1. They generally consist of two or more physically distinct and mechanically separable materials.
2. They are made by mixing the separate materials in such a way as to achieve controlled and uniform dispersion of the constituents.
3. The mechanical properties of the composite material are superior to (and in many cases quite different from) the properties of the constituent materials.

A good example of a composite material is glass-reinforced plastic (GRP). This material combines glass fibres with epoxy resin. The latter material is relatively weak and brittle and, although the glass fibres are strong and stiff, they can only be effectively loaded in tension as single fibres. However, when combined into a composite material, the resin and fibre provide us with a strong, stiff material with excellent toughness characteristics.

The following materials are commonly used as fibres:

● alumina
● boron
● carbon
● glass
● polyethylene
● polyamide.

The following materials can be used as matrix materials:

● alumina
● aluminium
● epoxy
● polyester
● polypropylene.

FIGURE 6.18
Composites and tough plastics are used in the manufacture of the bodywork for this off-road vehicle

The fibres may be arranged within the matrix material in various ways depending on the properties required and the intended application (see Fig. 6.17). Note that, whenever the fibres are aligned in a specific direction, the properties of the material also become directional. Conversely, if the fibres are arranged in random orientation, the resulting material will have the same properties in all directions. This is an important point because we may sometimes require that a component has maximum strength in a particular direction. In this case the correct alignment of the fibres becomes extremely important (see Figs. 6.20 and 6.21).

Where the material is required to have equal properties in all directions, fibre mats can be constructed as shown in Figure 6.19. Table 6.5 shows the properties of some reinforced polyester materials.

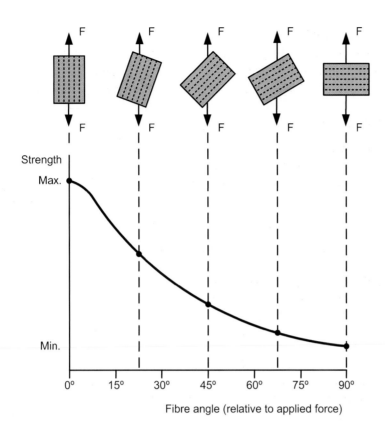

FIGURE 6.20
Variation of strength with angle of applied force

(a) Chopped strand mat

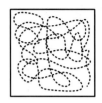

(b) Continuous filament mat

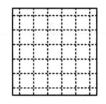

(c) Bi-directional woven mat

FIGURE 6.19
Construction of fibre mats

FIGURE 6.21
Composite materials are widely used in the aerospace industry where strong but lightweight aerodynamic profiles are required

Test your knowledge 6.22
Explain, with the aid of a diagram, how a composite material can be given directional mechanical properties.

Test your knowledge 6.23
List THREE characteristics of composite materials used in structural applications.

215

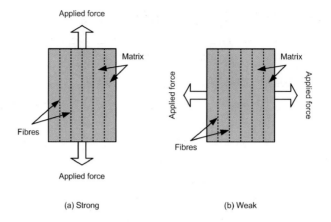

FIGURE 6.22
Effect of applying a force to a composite material with aligned fibres

(a) Strong   (b) Weak

TABLE 6.5 Properties of reinforced polyester

Type	% of glass fibre by weight	Tensile modulus (GPa)	Tensile strength (MPa)
Polyester (with no fibres added)	0	2 to 4	20 to 70
Polyester matrix with discontinuous, randomly aligned glass fibres	10 to 45	5 to 15	40 to 175
Polyester matrix with plain weave glass fibre cloth	45 to 65	10 to 20	250 to 350
Polyester matrix with long glass fibres	50 to 80	20 to 50	400 to 1250

ACTIVITY 6.4

Brief descriptions of five materials are as follows:

Material A   Supplied in sheets and in various colours; a good electrical insulator with excellent thermal forming properties and good impact strength; ideal for use in the manufacture of cases and enclosures

Material B   A rather expensive material; has very low friction and adhesion properties; suitable for use over a very wide temperature range; can be extruded or supplied as a coating; has excellent electrical properties

Material C   A thermoplastic material with good sliding properties; easily bonded, welded and machined; resistant to oils and grease; suitable for use in the manufacture of gears, pulleys and machine parts

Material D   A rather weak material supplied in opaque clear sheets or as a film; has very high electrical resistance and is ideal for use as an insulator

Material E   Supplied in clear sheets; a very rigid material; ideal for use as a cover for light fittings; windows and door panels.

a. Identify materials A to E by matching them to the following list: nylon, ABS, polycarbonate, PTFE, polythene.
b. Which of the materials would be suitable for the manufacture of (i) a roller bearing, (ii) a damp-proof membrane, and (iii) a 'non-stick' coating.

6.8 CODING AND IDENTIFYING MATERIALS

Materials are identified using various methods of coding and marking. For example, the British Standard for steel (BS 970) is based on a six-digit code.

The first three digits of the code indicate the type of steel. For example:

000–199	Carbon and carbon-manganese steels. The first three figures indicate 100 times the mean manganese content.
200–240	Free cutting steels. The second and third digits represent 100 times the minimum or mean sulphur content.
250	Silicon-manganese spring steels.
300–399	Stainless, heat resisting and valve steels.
500–999	Alloy steels with different groups of numbers within the range indicating different alloys, e.g. 500 to 519 indicate that nickel is the main alloying element, 520 to 539 when chromium is the main alloying element, and so on.

The fourth digit of the identifying code is a letter indicating specific requirements for the steel, for example:

A	The steel will be supplied to close limits of chemical composition (no mechanical or hardenability properties specified).
H	A combination of hardenability and chemical analysis.
M	A combination of mechanical properties and chemical analysis.
S	The specification relates to a stainless steel.

The fifth and sixth digit represent 100 times the mean carbon content of the steel (note that this does not apply to stainless steel specifications).

EXAMPLES

- 080A42 denotes a carbon-manganese steel, supplied to a specification based on chemical composition, containing 0.80% manganese, and with 0.42% carbon. This type of steel is extremely strong and is used for the manufacture of nuts, bolts, gears, pinions, shafts and rollers.
- 230M07 denotes a free cutting mild steel with 0.30% sulphur content supplied to a mechanical specification. This type of steel is used for the production of turned components where machineability and surface finish are important. Applications include automotive and general engineering

The American Iron and Steel Institute (AISI) standard is widely accepted in the United States as well as in many other countries. A four digit code is used for steels in which the first two digits indicate the type of steel and the remaining two digits indicate the percentage of carbon content. Further (and more comprehensive standards) have been developed by ASTM (American Society for Testing and Metals) and SAE (Society of Automotive Engineers), that provide for the classification of a wide range of metals and metal alloys. These standards are referred to as the UNS coding system.

Other standards used for steels in the UK and in Europe include BS EN 10079:1992 (Definition of steel products), BS EN 10027-1:1992 (Designation systems for steel: steel names, principal symbols), and BS EN 10027-2:1992 (Designation systems for steels: steel numbers).

Other systems are used for other metals and alloys. For example, the coding system for wrought aluminium alloys is specified by the Aluminium Association and it is based on four digits with the first indicating the principal alloying element, the second modifications to impurity limits, and the remainder indicating the percentage of aluminium content of the identification of specific alloys.

Test your knowledge 6.24
A steel is coded 070M20. What type of steel is this?

Key point
Various coding schemes conforming to national and international standards are used to identify metals and metal alloys. Examples of these coding schemes are BS 970 (used in the UK), and the EN and UNS schemes used in Europe and the USA (as well as many other countries). By referring to the metal coding we can determine not the type of metal and its constituents but also what the material can be used for.

Another view
There are approximate equivalents between the BS and AISI coding standards. For example, BS 970 080M30 is approximately equivalent to AISI 1030, whilst BS 970 080M40 is roughly equivalent to 1040, and so on.

ACTIVITY 6.5

Use the matSdata database to locate information on stainless steel. Use this information to answer the following questions:

1. Describe the appearance of the material.
2. What are the constituents of the material?
3. State TWO notable properties of the material.
4. What is the density of the material?
5. What is the melting point of the material?
6. What is the appearance of the material?
7. How does the tensile strength of stainless steel compare with other forms of steel?
8. State TWO applications for stainless steel.

Finally, it is important to note that colour codes are sometimes used to indicate metal and alloy types and metals suppliers usually provide information on how these are used and how they relate to national and international standards data.

With most forms of metal (see Fig. 6.14 on page 209) coding (using letters and numbers) is applied at intervals along the length of a tube, bar, or sheet. In conjunction with the definitive character marking, a colour coding may also be applied.

REVIEW QUESTIONS

1. Explain, in simple terms, each of the following properties of materials:
   a. strength
   b. toughness
   c. elasticity
   d. hardness
   e. rigidity.
2. Distinguish between the terms *ductility* and *malleability* when applied to engineering materials.
3. Sketch graphs to show how (a) tensile strength and (b) ductility varies with carbon content for a plain carbon steel. Label your graph.
4. Complete the chart shown in Figure 6.23.
5. Explain what is meant by the term *composite material*. Give THREE examples of common composite materials.
6. Name, and briefly describe the properties of, THREE main types of *polymer material*.
7. Name each of the forms of supply of material shown in Figure 6.24.
8. Name, and briefly describe the properties of, THREE *ceramic materials*.
9. Classify each of the materials listed below as either metals, polymers, ceramics or composite materials:
   a. aluminium
   b. clay
   c. polyvinyl chloride (PVC)
   d. rubber
   e. glass
   f. tungsten
   g. brick
   h. wood.
10. Give examples of suitable non-ferrous metals (with reasons for your choice) for each of the following applications:
    a. an instrument case used to house a portable electronic test set
    b. a water pump to be used with a marine engine

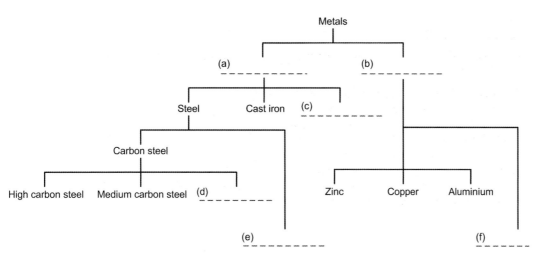

FIGURE 6.23
See Question 4

FIGURE 6.24
See Question 7

FIGURE 6.25
See Question 13

FIGURE 6.26
See Question 16

FIGURE 6.27
See Question 16

    c. a bus-bar to carry electric current in a steelworks

    d. a screw terminal used in an electric light fitting.

11. Distinguish between the terms *thermosetting plastic* and *thermoplastic* when applied to engineering materials.

12. Give examples of suitable plastic materials (with reasons for your choice) for each of the following applications:

    a. a car tow rope

    b. an engine cover for use on a light aircraft

    c. an electrical junction box

    d. the body of a 'ride-on' toy.

13. The control knob of the rotary switch shown in Figure 6.25 is made from a phenolic material. Explain what class of material this is and why it is used for this application.

14. State TWO examples of *ferromagnetic materials*.

15. Describe the essential properties of materials that will be used in each of the following electrical/electronic applications:

    a. the magnetic core of a transformer

    b. the field coil windings in a generator

    c. the insulating material between the plates of a capacitor

    d. the printed circuit board on which components are mounted.

16. State the materials used for the manufacture of (a) the sprocket shown in Figure 6.26 and (b) the hoze clip shown in Figure 6.27. Give reasons for selecting these materials.

219

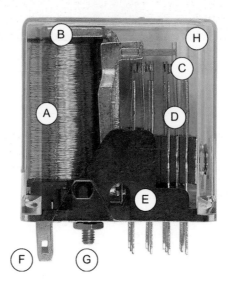

FIGURE 6.28
See Question 17

Component reference	Component name and/or function	Material(s) used in the manufacture of the component	Reason(s) for using this material
A	Coil		
B	Core and armature		
C	Contacts		
D	Spring		
E	Base		
F	Connecting tags		
G	Securing bolt and nut		
H	Enclosure		

TABLE 6.6 See Question 6.17

17. Figure 6.28 shows a miniature enclosed relay. Complete Table 6.6 showing the materials used for each of the main component parts of the relay and give reasons for using these materials.
18. Explain how steel is coded using BS 970. Give TWO example of typical codes used for different types of steel.

Computer Aided Drawing Techniques

SUMMARY

Computer Aided Drawing (CAD) is now used extensively throughout the engineering industry as a means of communicating drawing data to required standards. Two and three dimensional representations of components can be drawn and modified allowing the sharing of data from designer to customer. CAD data can be shared with CNC machines and Computer Aided Manufacturing (CAM) software, which may then assist in improving productivity, flexibility and quality of the final product. This unit will provide you with an introduction to CAD and to enable you to produce engineering drawings to given industry standards. It will also provide you with the necessary foundation to study CAD at a higher level. When you study this unit, you will be expected to produce engineering and assembly drawings as well as schematic circuit diagrams. You will also be expected to understand and apply the basic procedures of starting-up and closing down a CAD system and the storage, retrieval, modification and printing/plotting of data. The unit builds on some of the work that you carried out in Unit 1: *Working Safely and Effectively in Engineering,* and Unit 2: *Interpreting and Using Engineering Information.*

7.1 INTRODUCING COMPUTER AIDED DRAWING, DESIGN AND MANUFACTURE

Technology has had a huge impact on engineering and the engineering industry. Not only has it revolutionised the way that we design and manufacture products but it has also given us an exciting array of new materials, processes and components. We have already looked briefly at the production and use of engineering drawings in Unit 2. This unit further develops this theme and provides you with an opportunity to get to grips with the technology that's used to produce engineering drawings and designs. The prime mover in all of this is the widespread availability of the computer. We shall begin by introducing you to some of the terminology that we use.

Computer Aided Engineering (CAE)

You will probably already know a little about computer aided design (CAD) however this is just one aspect of a larger field known as *computer aided engineering* (CAE). Computer aided engineering is about automating *all* of the stages that go into providing an engineered

221

BTEC First Engineering: Mandatory and Selected Optional Units for BTEC Firsts in Engineering. DOI: 10.1016/B978-1-85617-685-9.00007-X

product or service. When applied effectively, CAE ties all of the functions within an engineering company together. Within a true CAE environment, information (i.e. data) is passed from one computer aided process to another. This may involve computer simulation, computer aided drawing (drafting), and computer aided manufacture (CAM).

The term, CAD/CAM, is used to describe the integration of computer aided design, drafting and manufacture. Another term, CIM (computer integration manufacturing), is often applied to an environment in which computers are used as a common link that binds together the various different stages of manufacturing a product, from initial design and drawing to final product testing.

Whilst all of these abbreviations can be confusing (particularly as some of them are often used interchangeably) it is worth remembering that 'computer' appears in all of them. What we are really talking about is the application of computers within engineering. Nowadays, the boundaries between the strict disciplines of CAD and CAM are becoming increasingly blurred and fully integrated CAE systems are becoming commonplace in engineering companies.

We have already said that CAD is often used to produce engineering drawings. Several different types of drawing are used in engineering. An example of a CAD drawing is shown in Figure 7.1.

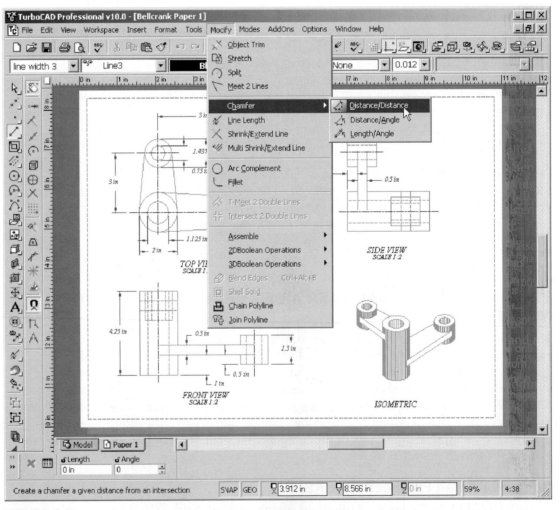

FIGURE 7.1
Drop-down menus and extensive toolbars are available in this powerful CAD package which incorporates both 2D and 3D features

Computer Aided Manufacture (CAM)

Computer aided manufacture (CAM) encompasses a number of more specialised applications of computers in engineering including computer integrated manufacturing (CIM), manufacturing system modelling and simulation, systems integration, artificial intelligence applications in manufacturing control, CAD/CAM, robotics, and metrology.

Computer aided engineering analysis can be conducted to investigate and predict mechanical, thermal and fatigue stress, fluid flow and heat transfer, and vibration/noise characteristics of design concepts to optimise final product performance. In addition, metal and plastic flow, solid modelling, and variation simulation analysis are performed to examine the feasibility of manufacturing a particular part.

In a modern engineering company, all of the machine tools within a particular manufacturing company may be directly linked to the CAE network through the use of centrally located floor managers which monitor machining operations and provides sufficient memory for complete machining runs.

Manufacturing industries rely heavily on computer controlled manufacturing systems. Some of the most advanced automated systems are employed by those industries that process petrol, gas, iron and steel. The manufacture of cars and trucks frequently involves computer-controlled robot devices. Industrial robots are used in a huge range of applications that involve assembly or manipulation of components.

The introduction of CAD/CAM has significantly increased productivity and reduced the time required to develop new products. When using a CAD/CAM system, an engineer develops the design of a component directly on the display screen of a computer. Information about the component and how it is to be manufactured is then passed from computer to computer within the CAD/CAM system. After the design has been tested and approved, the CAD/CAM system prepares sequences of instructions for computer numerically controlled (CNC) machine tools and places orders for the required materials and any additional parts (such as nuts, bolts or adhesives). The CAD/CAM system allows an engineer (or, more likely, a team of engineers) to perform all the activities of engineering design by interacting with a computer system (invariably networked) before actually manufacturing the component in question using one or more CNC machines linked to the CAE system.

> **Test your knowledge 7.1**
> What do each of the following abbreviations stand for?
>
> a. CAD
> b. CAM
> c. CAD/CAM
> d. CIM
> e. CNC
> f. CAE.

FIGURE 7.2
A large CNC milling machine

> **Test your knowledge 7.2**
> Explain the advantages of using CAD/CAM in the design and manufacture of an engineered product.

FIGURE 7.3
Some typical components manufactured using CNC processes and steel bar

FIGURE 7.4
Some typical parts manufactured using CNC processes on cast alloy components

7.2 2D CAD

Two-dimensional (2D) CAD packages are widely used by engineers for creating conventional 'flat' drawings. Typical of these 2D packages are 'industry standard' packages like Autosketch, AutoCAD, TurboCAD, DesignCAD and FastCAD.

Older 2D CAD programs accept text commands or a combination of buttons followed by coordinates or other parameters entered as text. More modern packages make more use of the graphical user interface (GUI) available in modern computers but they may still require precise dimensions to be entered from the keyboard during a drawing session. For example, in the A9CAD 2D CAD package the following commands draw a circle inscribed into a square with a side dimension of five units (see Fig. 7.5):

Command: RECTANGLE (by clicking on the rectangle tool button)

First corner: 0,0

Other corner: 10,10

This draws a square with a side of length 10 units starting at the origin, 0,0.

Command: CIRCLE (by clicking on the circle tool button)

Circle center point: 5,5

Circle radius: 5

This draws a circle with a radius of 5 units centred on 5,5.

Most modern CAD programs are reasonably intuitive and will allow you to quickly create drawings using standard templates and symbol libraries. A variety of drawing tools will be provided which will allow you to assign different properties to drawing entities (such as a dashed line or a hatched rectangle) or to snap, glue or group entities together. Intelligent connectors (which rebuild automatically when you reposition objects and avoid crossing other objects) allow you to easily create charts and schematics.

The features available from a modern 2D CAD package include the following:

- Various modes for creating lines, arcs, circles, ellipses, parallel line, angle bisectors, etc.
- Support (import and export) for standard CAD file formats (e.g. DXF).
- Standard and user-defined symbol libraries.
- Text in different fonts and sizes.
- Automatic dimensioning of distances, angles, diameters, and tolerances.
- Solid fills and hatching.

Another view
Your school or college will provide you with access to a modern CAD package so that you can gain some 'hands-on' experience with CAD and also so that you can provide evidence to show that you have met the assessment requirements for this unit. You may find that it takes some time to become proficient with using this software—so don't be too disappointed if your first efforts don't look too professional! With time (and a little effort) you should be able to produce engineering drawings to an acceptable standard.

- Support for layers and blocks (which can be manipulated as a separate entity).
- Selection and modification tools (move, rotate, mirror, trim, stretch, etc.).
- Snapping to grid and to objects (endpoints, centres, intersections, etc.).
- Multiple undo and redo levels.
- Support for various units including metric, imperial, degrees, radians, etc.
- Import and export of bitmaps and other images in various format (e.g. BMP, JPEG, PNG, etc.).
- OLE (Object Linking and Embedding) compatibility (this means that a CAD drawing can be inserted and edited within any other OLE-compatible Windows application).

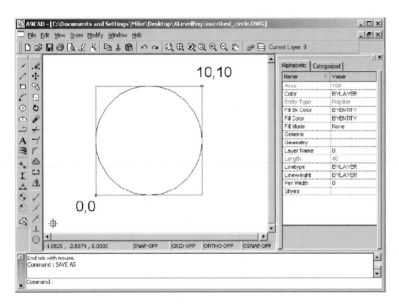

FIGURE 7.5
A circle of radius 5 units inscribed within a square having sides of 10 units drawn using the commands shown in the text

Another view
If you would like to practise some CAD skills at an early stage in this unit (and to help you get acquainted with the commands and techniques used in CAD packages) you can download a free copy of a basic CAD package, A9CAD, from the web address given in Appendix 5. A9CAD is a simple general-purpose two-dimensional CAD program that supports industry standard DWG and DXF formats. The package uses basic commands that are broadly compatible with those used in several of the most powerful and popular CAD packages.

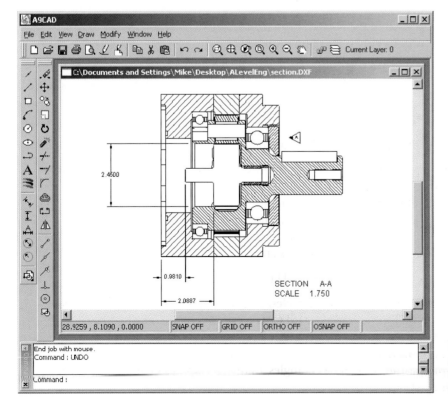

FIGURE 7.6
A more complex 2D drawing showing the use of hatching and dimensioning

7.3 3D CAD

A 3D CAD package produces a model of a component part or product which can be viewed from any desired angle. 3D CAD packages may operate in *wire-frame mode* (in which you will see a skeletal outline of the component or product, as shown in Fig. 7.7) or in *render mode* (in which you will see a more realistic shaded image, see Figs. 7.11 and 7.12). CAD packages will usually provide you with both *orthographic views* (see Fig. 7.8) and *isometric views* (as shown in Fig. 7.9). These views will often be displayed in multiple windows so that it is possible to view a model from several different points at the same time. In addition, more advanced packages provide a *camera view* that can be used to provide additional perspective and which will allow you to examine a component part or a product from any desired angle.

3D CAD packages use a three-axis coordinate system and a plane of reference (the *workplane*) as shown in Figure 7.10. In addition, some packages also allow you to define your own user coordinate system which has a different workplane that travels with an object and which is usually defined first in 2D mode.

The workplane is the plane in which a 2D object is initially created (i.e. as a 2D drawing). In 2D mode, you will normally do all your work in the same workplane but in 3D mode it is frequently necessary to change the workplane in order to perform all the required commands.

One of the most important tools for controlling the view of a 3D model is that which allows you to render an object so that it looks both solid and realistic. In normal *rendering* modes, all 3D objects are displayed as shaded (with or without hidden lines). Higher level rendering will enable you to view materials and textures, thus providing a more realistic image of what a component or your model will actually look like.

Test your knowledge 7.3
List SIX features of a modern 2D CAD package.

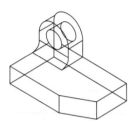

FIGURE 7.7
A 3D wire-frame view of a component

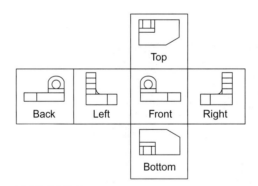

FIGURE 7.8
Orthographic views (see later) of the component shown in Figure 7.7

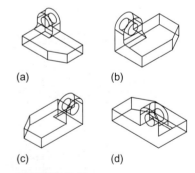

(a) (b)

(c) (d)

FIGURE 7.9
Isometric views (see later) of the object shown in Figure 7.7

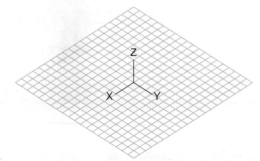

FIGURE 7.10
A 3D workplane
showing reference axes

In order to create a realistic rendered view lighting effects must be added and the positions of the light sources are usually made adjustable. In high-end packages, it may also be possible to further enhance the rendering of the image by assigning materials and luminance qualities to objects. This adds further realism to the finally generated image.

Test your knowledge 7.4
Sketch a 3D workplane and label the axes.

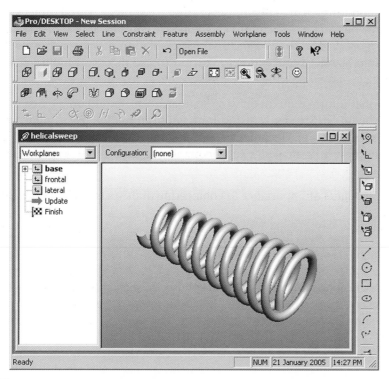

FIGURE 7.11
A 3D rendered model of a helical spring

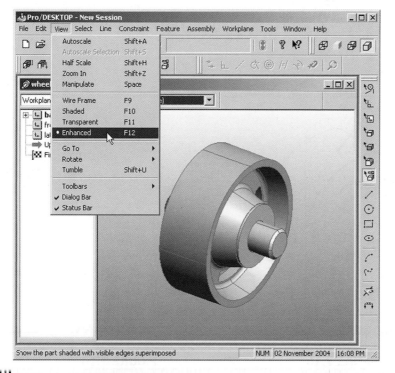

FIGURE 7.12
A rendered 3D representation of a flywheel

7.4 DRAWING STANDARDS AND CONVENTIONS

To avoid confusion, engineering drawings make use of nationally and internationally recognised symbols, conventions and abbreviations which are explained in the appropriate British Standards. Standard conventions are used in order to avoid having to draw out, in detail, common features in frequent use. Figure 7.13 shows a typical dimensioned engineering drawing. Some conventions can help save you a great deal of time and effort. For example, Figure 7.14(a) shows a pictorial representation of a screw thread whilst Figure 7.14(b) shows the standard convention for a screw thread that's much quicker and easier to draw!

If you need further information on drawing standards you should, in the first instance, take a look at the British Standards Institution's publication PP 8888, A guide for schools and colleges to BS 8888. This document provides some useful information that will help you improve your drawings and a copy should be available in your school or college.

Other British Standards of importance to engineering drafts-persons and designers are:

- BS 308 Engineering Drawing Practice.
- BS 4500 ISO Limits and Fits (these are used by mechanical and production engineers.
- BS 3939 Graphical symbols for electrical power, telecommunications and electronics diagrams.
- BS 2197 Specifications for graphical symbols used in diagrams for fluid power systems and components.
- BS 8888 Technical product specification.

Lines

The lines used in a drawing should be uniformly black, dense and bold. On any one drawing they should all be in pencil or in black ink. Pencil is quicker to use but ink prints out more clearly. Lines should be thick or thin as recommended below. Thick lines will normally be twice as thick as thin lines. Figure 7.15 shows the types of lines recommended in BS 308 for use in engineering drawing and what they should be used for.

> **Key point**
> Relevant British Standards (and their associated International Standards) are used to ensure that drawings can be read without confusion or misinterpretation. The standards used for engineering drawing relate to the use of symbols, conventions and abbreviations.

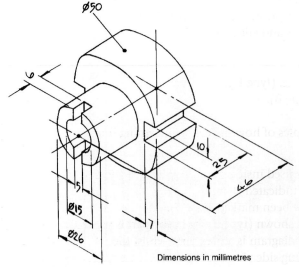

Dimensions in millimetres

FIGURE 7.13
A typical dimensioned engineering drawing

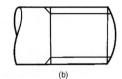

FIGURE 7.14
A screw thread (a) and corresponding drawing convention (b)

Line type	Example	Description	Application
A	————————	Continuous thick	Visible outlines and edges
B	————————	Continuous thin	Dimension, projection and leader lines, hatching, outlines of revolved sections, short centre lines, imaginary intersections
C	∿∿∿	Continuous thin irregular	Limits of partial or interrupted views and sections
D	⋀⋁⋀⋁	Continuous thin with zigzags	Limits of partial or interrupted views and sections
E	– – – – – – –	Dashed thin	Hidden outlines and edges
F	– · – · – · – · –	Chain thin	Centre lines, lines of symmetry, trajectories and loci, pitch lines and pitch circles
G	⌐· – · – ¬	Chain thin, thick at ends and changes of direction	Cutting planes
H	– ·· – ·· – ·· –	Chain thin double dashed	Outlines and edges of adjacent parts, outlines and edges of alternative and extreme positions of movable parts, initial outlines prior to forming, bend lines on developed blanks or patterns

FIGURE 7.15
Types of lines

Test your knowledge 7.5
What do each of the following British Standards relate to?

a. PP 8888
b. BS 3939.

Test your knowledge 7.6
Refer to the detail drawing shown in Figure 2.24 on page 55. Which British Standard is referred to?

Test your knowledge 7.7
Identify each of the line styles shown in Figure 7.16.

Sometimes the lines overlap in different views. When this happens the following order of priority should be observed:

- Visible outlines and edges (type A) take priority over all other lines.
- Next in importance are hidden outlines and edges (type E).
- Then cutting planes (type G).
- Next come centre lines (type F and B).
- Outlines and edges of adjacent parts, etc. (type H).
- Projection lines and shading lines (type B).

Figures 7.17 and 7.18 shows some examples of how the rules concerning lines should be applied:

- In Figure 7.17(a), hatching (type B) is used inside a continuous area (type A).
- In Figure 7.17(b), the limit of view is indicated (type C).
- In Figure 7.17(c), the centre lines have been marked (type F).
- In Figure 7.17(d), the hidden detail is shown (type E) when the part is viewed from the side (note that the left-hand diagram is a plan view whilst the right-hand diagram is the corresponding side view).
- In Figure 7.18, the component shown is able to move from its resting position A to position B. Its extreme position is shown using line type H.

1	————————
2	∿∿∿
3	————————
4	– · – · – · – · –
5	– – – – – – –

FIGURE 7.16
See Test your knowledge 7.7

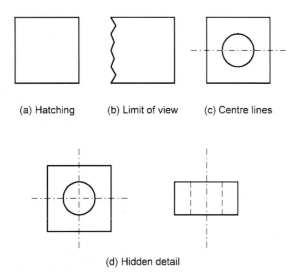

(a) Hatching   (b) Limit of view   (c) Centre lines

(d) Hidden detail

FIGURE 7.17
Use of different types of line

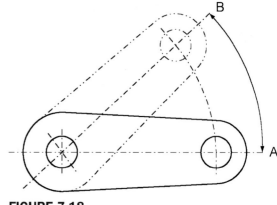

FIGURE 7.18
Use of different types of line

Leader lines are used when written information or dimensions need to be added to a drawing. Leader lines are thin lines (type B) and they end in an arrowhead or in a dot as shown in Figure 7.19(a). Arrowheads touch and stop on a line, whilst dots should always be used *within* an outline. The rules for arrowheads are as follows:

- When an arrowed leader line is applied to an arc it should be in line with the centre of the arc as shown in Figure 7.19(b).
- When an arrowed leader line is applied to a flat surface, it should be nearly normal to the lines representing that surface as shown in Figure 7.19(c).
- Long and intersecting leader lines should not be used, even if this means repeating dimensions as shown in Figure 7.19(d).
- Leader lines must not pass through the points where other lines intersect.
- Arrowheads should be triangular with their length some three times larger than the maximum width. They should be formed from straight lines and the arrowheads should be filled in. The arrowhead should be symmetrical about the leader line, dimension line or stem. It is recommended that arrowheads on dimension and leader lines should be some 3 mm to 5 mm long.
- Arrowheads showing direction of movement or direction of viewing should be typically between 7 mm to 10 mm long. The stem should be the same length as the arrowhead or slightly greater. It must never be shorter.

Letters and Numbers

The following conventions apply to the use of letters and numbers in formal engineering drawings:

STYLE

The style should be clear and free from embellishments. In general, capital letters should be used. Suitable styles are shown in Figure 7.20.

SIZE

The characters used for dimensions and notes on drawings should not be less than 3 mm tall. Title and drawing numbers should be at least twice as big.

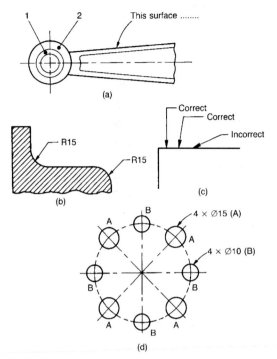

FIGURE 7.19
Examples of the use of leader lines

ABCDEFGHIJKLMNOPQRSTUVWXYZ
1234567890

ABCDEFGHIJKLMNOPQRSTUVWXYZ
1234567890

ABCDEFGHIJKLMNOPQRSTUVWXYZ
1234567890

ABCDEFGHIJKLMNOPQRSTUVWXYZ
1234567890

FIGURE 7.20
Suitable styles for letters and numbers in engineering drawings

Test your knowledge 7.8
Figure 7.21 shows some applications of leader lines with arrowheads and leader lines with dots. List the numbers and state whether the application is correct or incorrect. If incorrect explain (with sketches if required) how the application should be corrected.

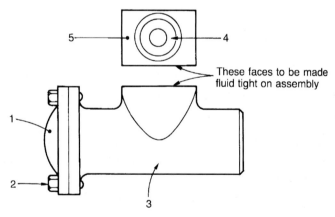

FIGURE 7.21
See Test your knowledge 7.8

Test your knowledge 7.9
Which of the following statements are TRUE when adding letters and numbers to an engineering drawing?

a. Only lower case letters should be used
b. Words should be emphasised by using underlining
c. The same size should be used for all of the letters and numbers
d. The style used for letters and numbers should be as simple as possible
e. Notes relating to a particular feature should be placed in the title block.

DIRECTION OF LETTERING

Notes and captions should be positioned so that they can be read in the same direction as the information in the title block. Dimensions have special rules and will be dealt with later.

LOCATION OF NOTES

General notes should all be grouped together and not scattered about the drawing. Notes relating to a specific feature should be placed adjacent to that feature.

EMPHASIS

Characters, words and/or notes should not be emphasised by underlining. Where emphasis is required the characters should be enlarged.

Finally, please remember that, in order to save space, symbols and abbreviations are frequently often used in formal engineering drawings. You should refer to BS 308 or PP 8888 if you need further information.

Drawing Conventions

We have already seen how standard conventions can help save time and effort when drawing a screw thread (take a look back at Figure 7.14). Conventions are a form of 'shorthand' that is used to speed up the drawing of common features that are in regular use. The full range of conventions and examples of their use can be found in appropriate standards so we will not waste space by listing them here. However by completing Activity 7.1 you will get to know some of the more common drawing conventions and this will help you to produce drawings more easily.

Dimensions

When a component is being dimensioned, the dimension lines and the projection lines should be thin full lines (type B). Where possible dimensions should be placed outside the outline of the object as shown in Figure 7.23(a). The rules are:

- Outline of object to be dimensioned in thick lines (type A).
- Dimension and projection lines should be half the thickness of the outline (type B).
- There should be a small gap between the projection line and the outline.
- The projection line should extend to just beyond the dimension line.
- Dimension lines end in an arrowhead that should touch the projection line to which it refers.
- All dimensions should be placed in such a way that they can be read from the bottom right-hand corner of the drawing.

The purpose of these rules is to allow the outline of the object to stand out prominently from all the other lines and to prevent confusion.

There are three ways in which a component can be dimensioned. These are:

- Chain dimensioning as shown in Figure 7.23(b).
- Absolute dimensioning (dimensioning from a datum) using parallel dimension lines as shown in Figure 7.23(c).
- Absolute dimensioning (dimensioning from a datum) using superimposed running dimensions as shown in Figure 7.23(d). Note the common origin (termination) symbol.

It is neither possible to manufacture an object to an exact size nor to measure an exact size. Therefore important dimensions have to be *toleranced*. That is, the dimension is given two sizes; an upper limit of size and a lower limit of size. Providing the component is made so that it lies between these limits it will function correctly. Information on Limits and Fits can be found in BS 4500.

ACTIVITY 7.1

Use reference material found in your school or college library (or from the Internet) to fill in the blank spaces shown in the chart of drawing conventions shown in Figure 7.22. BS PP 8888 (Drawing Standards for Schools and Colleges) will help you get started!

TITLE	SUBJECT	CONVENTION
External screw threads (details)		
Screw threads (assembly		
Compression springs		
Diamond knurling		
Square on shaft		
Holes on linear pitch		

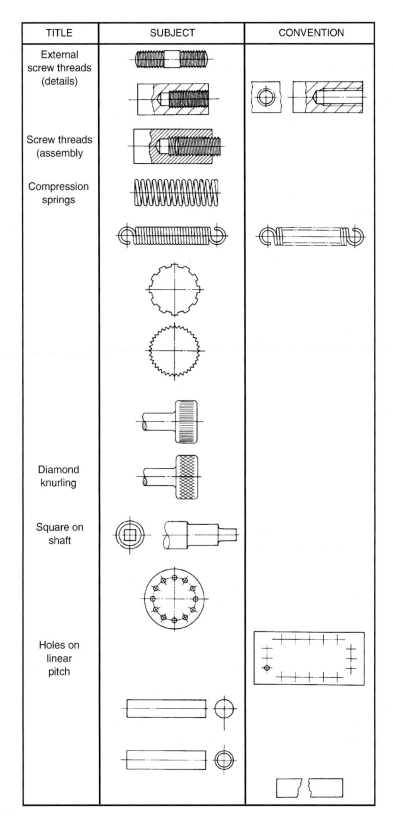

FIGURE 7.22
See Activity 7.1

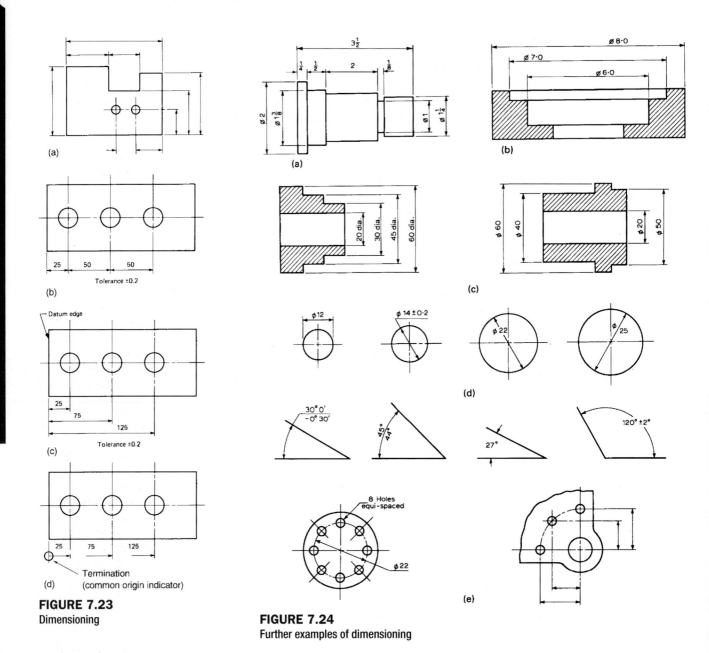

FIGURE 7.23
Dimensioning

FIGURE 7.24
Further examples of dimensioning

The method of dimensioning can also affect the accuracy of a component and produce some unexpected effects. Figure 7.23(b) shows the effect of chain dimensioning on a series of holes or other features. The designer specifies a common tolerance of ±0.2 mm. However, since this tolerance is applied to each and every dimension, the cumulative tolerance becomes ±0.6 mm by the time you reach the final, right-hand hole. Not what was intended! Therefore absolute dimensioning as shown in Figure 7.23(c) and Figure 7.23(d) is to be preferred in this example. With absolute dimensioning, the position of each hole lies within a tolerance of ±0.2 mm and there is no cumulative error.

Reference dimensions are usually specified without a tolerance. They are used for information only and not for production or inspection purposes. A reference dimension repeats a dimension or size already given or derived from other values shown on the drawing or related drawing. Reference dimensions are enclosed in brackets, for example (23.50). Further examples of dimensioning techniques are shown in Figure 7.24.

Sections and Sectional Views

Sections are used to show the hidden detail inside hollow objects more clearly than can be achieved using dashed thin (type E) lines. Figure 7.25 shows an example of a simple sectioned drawing. The *cutting plane* is the line A-A. In your imagination you remove everything to the left of the cutting plane, so that you only see what remains to the right of the cutting plane looking in the direction of the arrowheads. Another example is shown in Figure 7.26.

Figure 7.28 shows how to section an assembly. Note how solid shafts and the key are not sectioned. Also note that thin webs that lie on the section plane are not sectioned. When interpreting sectioned drawings, some care is required!

You will have noticed that the shading of sections and sectional views consists of sloping, thin (type B) lines. This is called hatching. The lines are equally spaced, slope at 45° and are not usually less than 4 mm apart. However when hatching very small areas the hatching can be reduced, but never less than 1 mm. The drawings in this book may look as though they do not obey these rules. Remember that they have been reduced from much bigger drawings to fit onto the pages.

Figure 7.27 shows the basic rules of hatching. The hatching of separated areas is shown in Figure 7.27(a). Separate sectioned areas of the same component should be hatched in the same direction and with the same spacing.

Figure 7.27(b) shows how to hatch assembled parts. Where the different parts meet on assembly drawings, the direction of hatching should be reversed. The hatching lines should also be staggered. The spacing may also be changed.

Figure 7.27(c) shows how to hatch large areas. This saves time and avoids clutter. The hatching is limited to that part of the area that touches adjacent hatched parts or just to the outline of a large part.

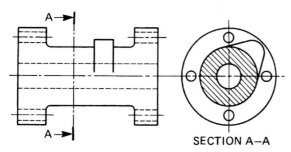

FIGURE 7.25
A simple sectioned drawing

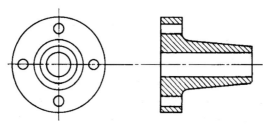

FIGURE 7.26
Another example of a sectioned drawing

Test your knowledge 7.10
Explain each of the following terms:

a. Reference dimension
b. Tolerance
c. Datum point
d. Datum feature.

Key point
In a *sectional view* you see the outline of the object at the cutting plane. You also see all the visible outlines seen beyond the cutting plane in the direction of viewing (therefore Figure 7.25 is a sectional view).

Key point
A *section* shows the outline of the object at the cutting plane. Visible outlines beyond the cutting plane in the direction of viewing are *not* shown. Therefore a section has no thickness.

Key point
A *cutting plane* shows where a section is made. Cutting planes consist of type G lines (a thin chain line that is thick at the ends and at changes of direction). The direction of viewing is indicated by arrows with large heads. The points of the arrowheads touch the thick portion of the cutting plane. The cutting plane is labelled by placing a capital letter close to the stems of the arrows.

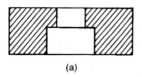

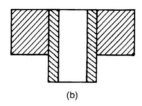

(a)

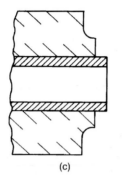

(b)

(c)

FIGURE 7.27
Rules of hatching

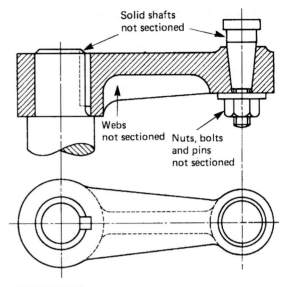

Solid shafts
not sectioned

Webs
not sectioned

Nuts, bolts
and pins
not sectioned

FIGURE 7.28
A sectioned assembly

7.5 PRODUCING AND STORING ENGINEERING DRAWINGS

Engineering drawings can be produced using a variety of different techniques. The choice of technique is dependent upon a number of factors such as:

SPEED

How much time can be allowed for producing the drawing. How soon the drawing can be commenced.

MEDIA

The choice will depend upon the equipment available (e.g., CAD or conventional drawing board and instruments) and the skill of the person producing the drawing.

COMPLEXITY

The amount of detail required and the anticipated amount and frequency of development modifications.

COST

Engineering drawings are not works of art and have no intrinsic value. They are only a means to an end and should be produced as cheaply as possible. Both initial and on-going costs must be considered.

PRESENTATION

This will depend upon who will see/use the drawings. Non-technical people can visualise pictorial representations better than orthographic drawings.

Key point

A *cutting plane* shows where a section is made. Cutting planes consist of type G lines (a thin chain line that is thick at the ends and at changes of direction). The direction of viewing is indicated by arrows with large heads. The points of the arrowheads touch the thick portion of the cutting plane. The cutting plane is labelled by placing a capital letter close to the stems of the arrows. The same letters are used to identify the corresponding section or sectional view.

Nowadays engineering drawings are increasingly produced using computer aided drawing techniques (CAD). Developments in software and personal computers have reduced the cost of CAD and made it more powerful. At the same time, it has become more 'user friendly'.

Computer aided drawing does not require the high physical skill required for manual drawing that often takes years of practice to achieve. CAD also has a number of other advantages over manual drawing. Let's consider some of these advantages:

ACCURACY

Points and dimensions can be fixed very accurately and do not depend on the draftsperson.

RADII

Radii can be made to blend with straight lines automatically.

REPETITIVE FEATURES

For example holes round a pitch circle do not have to be individually drawn but can be easily produced automatically by 'mirror imaging'. Again, some repeated, complex feature need only be drawn once and saved as a matrix. It can then be called up from the computer memory at each point in the drawing where it appears at the touch of a key.

EDITING

Every time you erase and alter a manually produced drawing the surface of the drawing is increasingly damaged. On a computer you can delete and redraw as often as you like with no ill effects.

STORAGE

No matter how large and complex the drawing, it can be copied, stored and transmitted digitally with no loss of detail.

PRINTS

Hard copy can be produced accurately and easily on laser printers, flat bed or drum plotters and to any scale. Colour prints can also be made.

Key point
The choice of technique used to make an engineering drawing depends on a number of factors including speed, the media to be used, the resources available, and the skill of the person making the drawing.

Key point
CAD offers many advantages over manual drawing techniques including accuracy, the ability to carry out repetitive tasks easily and quickly, and the ease by which drawings can be stored, edited, modified and shared with others.

Test your knowledge 7.11
List FOUR advantages of CAD compared with manual engineering drawings techniques.

ACTIVITY 7.2

It is important to know how to start up and close down a computer system used for CAD. Based on the CAD studio in your school or college, produce a checklist for new CAD users that:

a. lists the steps (including detailed instructions for each step) required to turn a PC on, log in (if required), start Windows, launch a CAD program and open a drawing file

b. list the steps (including detailed instructions for each step) required to save a drawing file, close a program, and shut the computer system down.

Your checklist instructions should take the form of an A4 poster that can be displayed alongside a PC and should use simple, straightforward language.

7.6 DRAWING PROJECTIONS

The different views used in engineering drawing are often referred to as *projections*. Several different projections are available according to particular needs and requirements. These range from the simple orthographic projection shown in Figure 7.30 to the more complicated perspective views that provide three-dimensional representations of an object or component.

Orthographic Projection

Orthographic projection is used to represent three-dimensional solid objects on the two-dimensional surface of a computer screen or sheet of drawing paper so that all the dimensions are true length and all the surfaces are true to shape. To achieve this when surfaces are inclined to the vertical or the horizontal we have to make use of additional *auxiliary views,* but more about these later. Let's keep things simple for the moment.

Figure 7.29 shows a brass component used in the waveguide assembly of a high-power radar. Figure 7.30 shows how this component is drawn (together with dimensions) using conventional orthographic drawing techniques. On its own this drawing does not convey enough information for anyone to actually manufacture or specify the component from an off-the-shelf supplier. At this point you might like to think about what's missing and how you could show the additional information that's needed to allow the component to be manufactured!

Oblique Drawing

The most obvious way of improving Figure 7.29 is to attempt to draw the component in *perspective* (i.e. in a 3D view). The easiest way of doing this is to use a simple pictorial technique called *oblique drawing.* Figure 7.31 shows a simple oblique drawing using both *cavalier oblique* and *cabinet oblique* projection.

It's important to note that regardless of whether cavalier oblique or cabinet oblique projection is used, the front view (called the *elevation*) is drawn true shape and size. Therefore this view should be chosen so as to include any circles or arcs so that these can be drawn with compasses when using manual drawing techniques or using the circle drawing tool when using CAD.

The lines forming the side views appear to travel away from you, so these are called *receders.* They are drawn at 45° to the horizontal using a 45° set-square if you are using manual drawing techniques. They may be drawn full-length if you are using *cavalier oblique* drawing or they may be drawn half-length if you are using *cabinet oblique* drawing. This latter method gives a more realistic representation.

Test your knowledge 7.12

Refer to Figures 7.29 and 7.30.

a. What are the external dimensions of the component?

b. What are the internal dimensions of the component?

c. What is the inner diameter of the circular groove in the face of the component?

d. What is the outer diameter of the circular groove in the face of the component?

e. What is the diameter of the component's four locating holes?

FIGURE 7.29
A brass component used in the waveguide assembly of a high-power radar

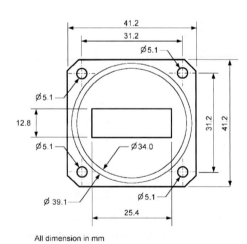

All dimension in mm

FIGURE 7.30
A conventional orthographic drawing of the component shown in Figure 7.29

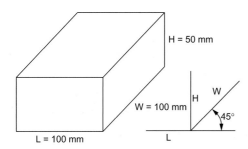

Cavalier oblique projection

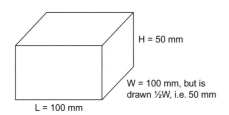

Cabinet oblique projection

FIGURE 7.31
A simple oblique drawing

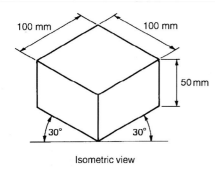

Isometric view

FIGURE 7.32
An isometric view

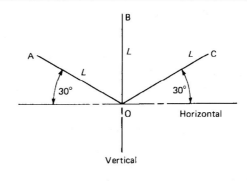

Vertical

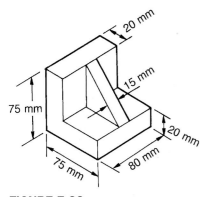

FIGURE 7.33
See Test your knowledge 7.13

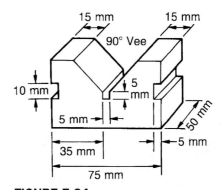

FIGURE 7.34
See Test your knowledge 7.13

Isometric Drawing

Isometric drawing provides us with another way of showing a 3D pictorial view of an object. Figure 7.32 shows an isometric view of the same box that we used for our examples of cavalier and cabinet oblique projection. You should take a little time to compare Figures 7.31 and 7.32 until you understand what the differences are!

When constructing the isometric view, the vertical lines should be drawn true length and the receders should be drawn to a special isometric scale. However this sort of accuracy is rarely required and, for all practical purposes, we can usually draw all the lines full size. As you can see, the receders are drawn at 30° to the horizontal for both the elevation and the end view.

First-angle Projection

Both oblique and isometric drawing projections provide a good way of showing the general features of a component. An alternative to using these techniques is providing a series of orthographic views. Often, just three views are required to show all the details required for manufacture or production of a part. Engineers use two orthographic drawing techniques called *first-angle projection* and *third-angle projection*. The former is called 'English projection' and the latter is called 'American projection'. We shall begin by looking at first-angle projection.

Figure 7.35(a) shows a simple component drawn in isometric projection. Figure 7.35(b) shows the same component as an orthographic drawing. This time we make no attempt to represent the component pictorially. Each view of each face is drawn out separately either full size or to the same scale. What is important is how we position the various views as this determines how we 'read' the drawing.

Take a careful look at Figure 7.35 and note the following features:

ELEVATION

This is the main view from which all the other views are positioned. You look directly at the side of the component and draw what you see.

PLAN

To draw this, you look directly down on the top of the component and draw what you see below the elevation (this is called a *plan view*).

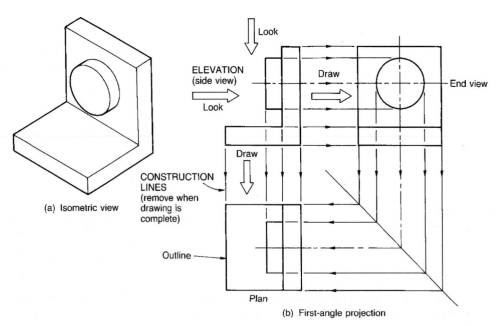

(a) Isometric view

(b) First-angle projection

FIGURE 7.35
An isometric view and its corresponding first-angle projection

END VIEW

This is sometimes called an *end elevation*. To draw this you look directly at the end of the component and draw what you see at the opposite end of the elevation. There may be two end views, one at each end of the elevation, or there may be only one end view if this is all that is required to completely depict the component. Figure 7.35 requires only one end view. When there is only one end view this can be placed at either end of the elevation depending upon which position gives the greater clarity and ease of interpretation. Whichever end is chosen the rules for drawing this view must be obeyed.

The *construction lines* shown in Figure 7.35 can be removed when the drawing is complete leaving the completed first-angle projection as shown in Figure 7.36.

Third-angle Projection

Third-angle projection is an alternative to first-angle projection. Figure 7.37 shows the same component as that used in Figure 7.35 but this time we have drawn it in third-angle projection.

Take a careful look at Figure 7.37 and note the following features:

ELEVATION

Again we have started with the elevation or side view of the component and, as you can see, there is no difference.

PLAN

Again we look down on top of the component to see what the plan view looks like. However, this time, we draw the plan view *above* the elevation. That is, in third-angle projection we draw all the views from where we look.

END VIEW

Note how the position of the end view is reversed compared with first-angle projection. This is because, like the plan view, we draw the end views at the same end from which we look at the component.

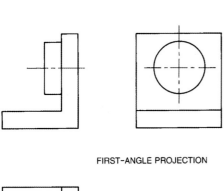

FIRST-ANGLE PROJECTION

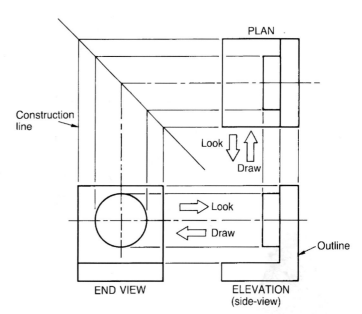

FIGURE 7.36
Completed first-angle projection

FIGURE 7.37
Third-angle projection

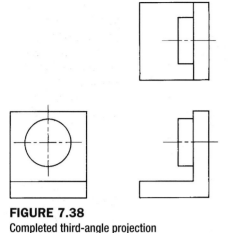

FIGURE 7.38
Completed third-angle projection

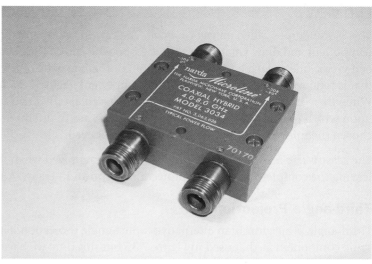

FIGURE 7.39
See Activity 7.3

Key point

Engineers use two orthographic drawing techniques called *first-angle projection* (English projection) and *third-angle projection* (American projection). Using these techniques, designers are able to draw three different views of a part that are referred to as the *elevation, end view* and *plan view*.

ACTIVITY 7.3

Figure 7.39 shows a component used in a microwave radio system. Sketch this component in:

a. cabinet oblique view
b. isometric view.

There is no need to draw the component accurately but your sketches MUST show the correct use of the drawing projection in each case. Label your sketches to show which projection has been used.

Once again, the *construction lines* shown in Figure 7.37 can be removed when the drawing is complete leaving the completed third-angle projection as shown in Figure 7.38.

7.7 GETTING STARTED WITH CAD

Hopefully you will now be keen to get started on some CAD drawings. In this final section we will show you how to produce a CAD drawing of the component shown in Figure 7.44 on page 246. We will be using A9CAD (see Appendix 6) as our drawing package but your school or college may provide you with alternative CAD software. The A9CAD user interface is shown in Figure 7.40 and the more commonly used commands are explained in Tables 7.1 and 7.2.

Start by launching A9CAD and then open a new drawing. Now take a close look at the isometric view of the component that you will be drawing. This is shown in Figure 7.44. The final result of your drawing should look something like the projection shown in Figure 7.45. Note that this comprises three separate views.

Starting from the bottom left corner (P1 in Fig. 7.46) you should draw a line in the direction shown. Because we need this to be accurate we shall avoid using the mouse for this and use text entry for commands and the coordinates for the points, P1 to P8. Click the left mouse

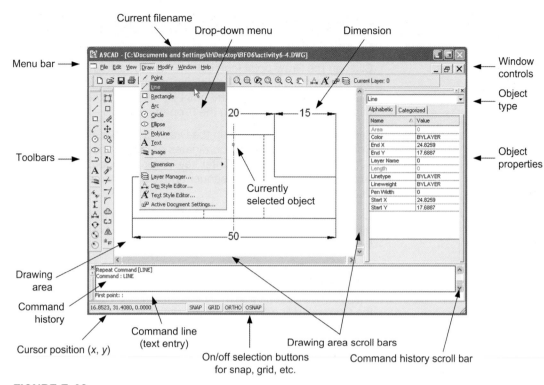

Current filename

Drop-down menu

Dimension

Menu bar →

Window controls

Object type

Toolbars →

Object properties

Currently selected object

Drawing area

Command history

Cursor position (x, y)

Command line (text entry)

On/off selection buttons for snap, grid, etc.

Drawing area scroll bars

Command history scroll bar

FIGURE 7.40
The A9CAD user interface

button in the Command line (just below the Command window) and enter the following:

Command: LINE
First point: 0,0   (this is P1)
Second point: 50,0   (this is P2)
Line next point: 50,10   (this is P3)
Line next point: 35,10   (this is P4)
Line next point: 35,20   (this is P5)
Line next point: 15,20   (this is P6)
Line next point: 15,10   (this is P7)
Line next point: 0,10   (this is P8)
Line next point: 0,0   (this is P1)

Finally, to complete the shape that you have just drawn, press the Esc key on your keyboard. The drawing should look like the one shown in Figure 7.47. To centre the shape in the drawing window you can use the Pan tool (click on the hand shape in the toolbar) and drag the shape into the centre of the window.

The coordinates that you entered to draw the last shape are known as *absolute coordinates*. They are all measured relative to the *zero reference point* (i.e. the point 0,0). Next we will add the two hidden lines but this time we will use *relative coordinates*.

Relative coordinates are measured relative to the last drawing point. In order to tell A9CAD that you are using relative coordinates rather than absolute coordinates, the @ symbol is used to precede the two coordinate values. The commands that you need to enter to construct the two hidden lines (see Fig. 7.48) are as follows:

Command: LINE
First point: 17.5,0   (this is P9)
Second point: @0,20   (this is P10)

243

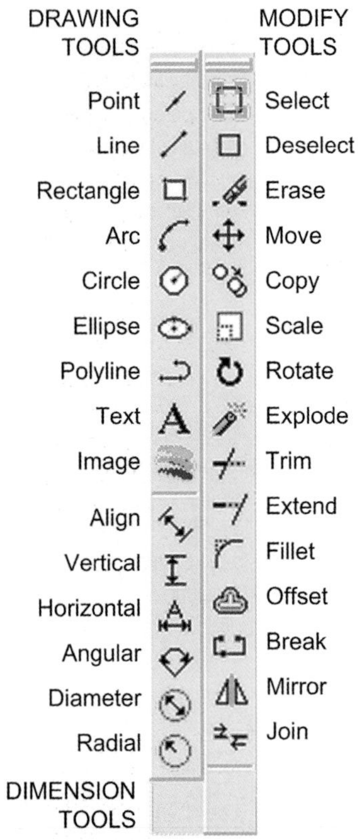

DRAWING TOOLS | MODIFY TOOLS

Point	/	☐	Select
Line	/	☐	Deselect
Rectangle	☐	✎	Erase
Arc	⌒	✛	Move
Circle	⊙	⁰ᵒ	Copy
Ellipse	⬭	▫	Scale
Polyline	⤵	↻	Rotate
Text	A	⟋	Explode
Image	▨	⊣	Trim
Align	↙	⊸	Extend
Vertical	I	⌐	Fillet
Horizontal	A	☁	Offset
Angular	◇	⊏⊐	Break
Diameter	⊘	⚎	Mirror
Radial	⊘	⇄	Join

DIMENSION TOOLS

FIGURE 7.41
The A9CAD drawing tools (top left), dimension tools (bottom left), and modify tools (right) are accessible from the toolbars

Another view
To change the drawing styles and settings of an active drawing, click on menu item Draw → Drawing Styles and Settings. A Drawing Styles and Settings dialog window will pop up. You can also change the properties of line, hatch, text and geometry.

TABLE 7.1 A9CAD drawing commands/tools

Command	Command use
Point	Click on the Draw point icon in the Drawing toolbar and then move the mouse to the desired location in the drawing and click. You can modify the point by selecting it and then changing the coordinates in the Property window.
Line	Click on the Draw line icon in the Drawing toolbar. From the Command line, enter the coordinates of the first point, e.g., 10,15, and then press the enter key. Enter the coordinates of the second point (and any subsequent points) the same way. Press the Esc key to finish.
Rectangle	Click on the Draw rectangle icon in the Drawing toolbar. From the Command line, enter the coordinates of the first corner and the other corner, respectively.
Arc	Click on the Draw arc icon in the Drawing toolbar. From the Command line, enter the coordinates of the arc centre point, e.g., 10,15, and then press the enter key. Follow the prompts in the Command line to enter the arc radius, start and end angles.
Circle	Click on the Draw circle icon in the Drawing toolbar. From the Command line, enter the coordinates of the circle centre point, e.g., 10,15, and then press the enter key. Follow the prompts in the Command line to enter the circle radius.
Ellipse	Click on the Draw ellipse icon in the Drawing toolbar. From the Command line, enter the coordinates of the axis1 end point and then press the enter key. Follow the prompts in the Command line to enter the distance to other axis.
Polyline	Click on the Draw polyline icon in the Drawing toolbar. From the Command line, enter the coordinates of the polyline start point and then press the enter key. Follow the prompts in the Command line to enter the coordinates of the other points. Right click the mouse (or press Esc) to end the job.
Image	Click on the Draw image icon in the Drawing toolbar. Select a BMP file to insert in your drawing. From the Command line, enter the coordinates of the start point for the image and the scale factor.
Text	Click on Draw text icon from Drawing toolbar. From the Command line, enter the coordinates of the start point and then press the key. Follow the prompts in the Command line to enter the rotation angle and text. The Text Style Editor will allow you to create a new text style or manage the existing text styles.

Press Esc to complete the first hidden line.

Command: LINE
First point: 32.5,0   (this is P11)
Second point: @0,20   (this is P12)

Press Esc to complete the second first line.

Your drawing should now look like the one shown in Figure 7.49. Note that the line will be solid (not dashed) but we will put this right next!

To change the style of the two hidden lines (from solid to dashed) we need to select each line in turn and then change their *linetype.* To select the left-hand hidden line you need to

TABLE 7.2 A9CAD modifying and editing commands/tools

Command	Command use
Erase	Select the object that you need to erase by clicking the mouse on the object and then click on the Erase icon in the Modify toolbar. The selected object will be removed.
Copy	Select the object that you need to copy and then click on the Copy icon in the Modify toolbar. From the Command line, enter first the coordinates of the 'Copy from point' and then the coordinates of the 'Copy to point'. The selected object will be copied to the specified location. Like most other commands, you can also perform this operation using the mouse.
Move	Select the object that you need to move and click on the Move icon in the Modify toolbar. From the Command line, enter first the coordinates of the 'Copy from point' and then the 'Copy to point'. The selected object will be moved to the specified location.
Scale	Select the object that you want to scale and then click on the Scale icon in the Modify toolbar. From the Command line, enter first the coordinates of the base point and then the scale factor. The selected object will be moved to the specified location and scaled according to the specified scale factor.
Rotate	Select the object you desire to rotate and then click on the Rotate icon in the Modify toolbar. From the Command line, enter first the coordinates of the base point and then the rotation angle (in radians). The selected object will be moved to the specified location and rotated to the specified angle.
Explode	Select the object (e.g. a polyline) that you need to explode and then click on the Explode icon in the Modify toolbar. The polyline will become several regular lines.
Trim	As you edit your drawing, you may find some lines or arcs need to be shortened. The Trim command will help you cut those edges. For example, to trim to the drawing shown in Figure 7.42, click on the Trim icon in the Modify toolbar then select line B before right-clicking the mouse. Next select line A and then click on Regenerate from the View drop-down menu.
Extend	Extend allows you to extend an object to the other object. For example, to extend the lines in Figure 7.43, click on the Extend icon in the Modify toolbar then select line X before right-clicking the mouse. Next select line Y and then click on Regenerate from the View drop-down menu.
Fillet	Click on the Fillet icon. Select the first and the second object in order. The edges between these two objects will be filleted.
Offset	Click on the Offset icon. Select the object on which you need to perform the offset. Enter the offset distance using the Command line before clicking on the offset side on the screen. A new object will be added to the drawing. Offset provides you with a means of easily drawing parallel lines.
Break	Click on the Break tool button. Select the object that you want to break. Select the first and second point by clicking on the object. The object will be broken.
Mirror	Select the object you desire to mirror. Hit Mirror button. Enter the coordinates of the first and second point. A newly mirrored object will be added to the drawing. It is also possible to perform this operation using the mouse.

Another view

To change A9CAD general settings, click on menu item File → General Settings. A General Settings dialog window will pop up in which you can make changes on the text field of Cursor Size, Grip Size, Pick Size or Background colour before clicking on the Apply button. If you don't wish to make any changes, you can keep the current settings by clicking on the Cancel button.

Another view

You may sometimes find that the A9CAD drawing window has not reacted to a command that you've entered. If this is the case, you should click on View and then Redraw or Regenerate in order to refresh the drawing window.

(a) Before

(b) After

FIGURE 7.42
The A9CAD Trim command

(a) Before

(b) After

FIGURE 7.43
The A9CAD Extend
command

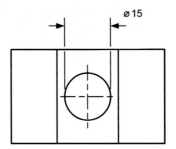

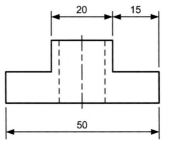

FIGURE 7.45
The three different views that
should appear in your CAD
drawing. You will be starting
with the bottom left view.

Not to scale
All dimensions in mm

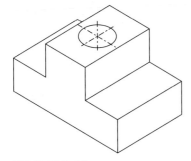

FIGURE 7.44
Isometric view of the component that
you will be drawing

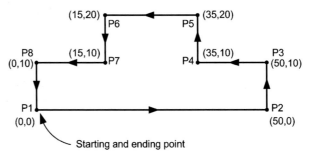

Starting and ending point

FIGURE 7.46
Sequence of points and lines required to draw the shape

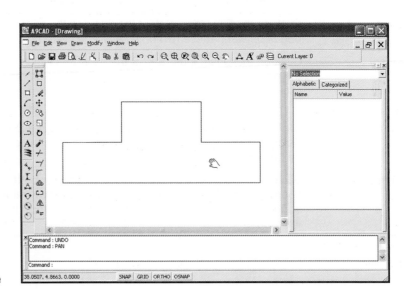

FIGURE 7.47
The completed shape

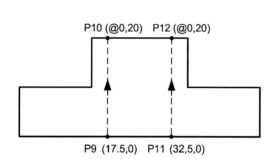

FIGURE 7.48
Sequence of points and lines required to draw the
two hidden lines

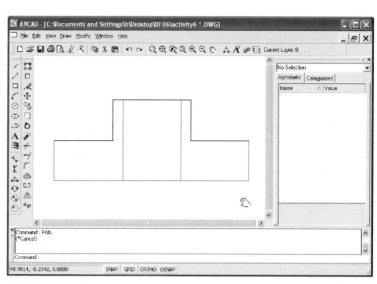

FIGURE 7.49
The completed hidden lines

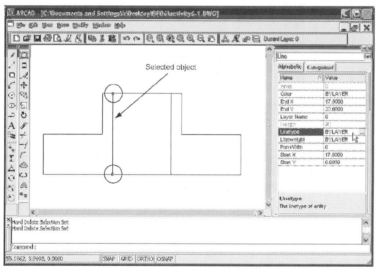

FIGURE 7.50
Selecting the left-hand hidden line

left-click the mouse on it. When you've done, two small blue squares will appear at each end
of the line (see Fig. 7.50). The two squares show you that the line has been selected. When
an object has been selected, its properties will appear in the Properties window. In this case
the selected object is a line (as shown in Fig. 7.50). We need to edit the properties in order to
change the type of line so left-click on Linetype and change the properties from BYLAYER to
Dashed as shown in Figure 7.51. Repeat this procedure for the right-hand hidden line.

Next we will add the centre line (see Fig. 7.52) using the following commands:

Command: LINE
First point: 25,−3   (this is P13)
Second point: @0,26   (this is P14)

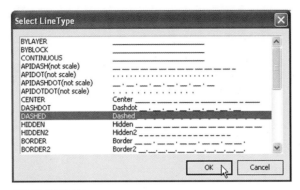

FIGURE 7.51
Changing the line type from BYLAYER to Dashed

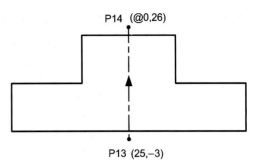

FIGURE 7.52
Sequence of points and lines required to draw the centre line

FIGURE 7.53
Dimension Style Editor window

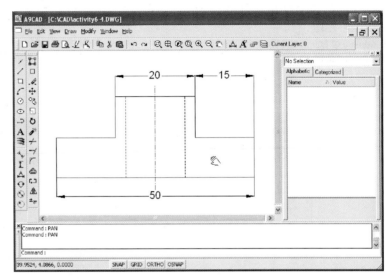

FIGURE 7.54
The completed view (including dimensions)

Then press Esc to complete the centre line.

Note how the centre line is extended so that it begins and ends just outside the original shape. You can see how the centre line extends outside the shape by looking at Figure 7.54.

The next stage is that of adding the dimension lines to the drawing. However, before you do this it's a good idea to set the dimensions styles by clicking on the Dimension Style Editor tool from the toolbar that runs along the top of the window. You need to make the following changes (as shown in Fig. 7.53):

Arrow size: 2
Decimal precision: 0
Text Height: 2

Then click on Apply and OK to make the changes and close the Dimension Style Editor Window. You can now add the dimensions (using the Horizontal Dimension tool button), clicking on the lines that need to be dimensioned and then dragging the dimension line clear of the drawing.

The completed view (including dimensions) is shown in Fig. 7.54. You now need to complete the drawing by adding the other two views (see Fig. 7.45).

Another view

To avoid losing your hard work due to a software error (program crash) or an interruption of the mains supply, it is good practice to Save your work regularly. You can do this by clicking on File → Save or File → Save As. Use File → Save As if you want to set or change the file name.

7.8 GETTING STARTED WITH AUTOCAD

Many schools and colleges will encourage you to use an industry standard CAD package, such as AutoCAD, for your drawings. This will help you to understand the power, complexity and sophisticated features available in modern CAD software. AutoCAD is available in several different versions including AutoCAD LT (a lower cost, less sophisticated package) and a student version that's designed for use in education.

The AutoCAD 2010 user interface is shown in Fig. 7.55. The drawing area (I) is in the centre of the AutoCAD window. Above the drawing area you will find the command ribbon (F) and

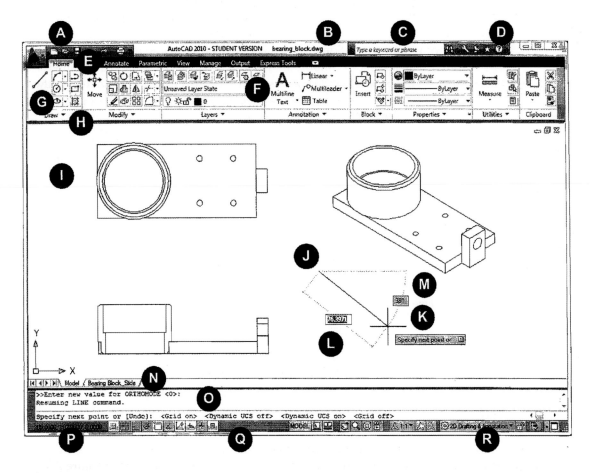

Key:

A	Filing commands (new, open and save)	J	Start of line
B	Drawing filename	K	End of line (at crosshair cursor)
C	Search box	L	Line length
D	Maximise and minimise	M	Line angle (relative to horizontal)
E	Quick access toolbar	N	View tabs (currently model view)
F	Command ribbon	O	Command window (text entry)
G	Line command button	P	Drawing coordinates
H	Drop down command menus	Q	Status bar
I	Drawing area	R	Workspace pop-up menu

FIGURE 7.55
The AutoCAD 2010 drawing interface

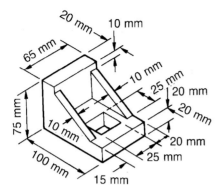

FIGURE 7.56
See Activity 7.5

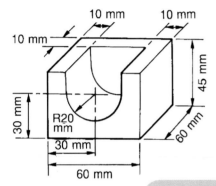

FIGURE 7.57
See Activity 7.6

quick access toolbar (E). This arrangement makes it quick and easy to select the particular drawing tools that you need. Note that the filename for your drawing is shown at the top of the window (B) and that the usual new, open and save file commands are available at the top left of the screen (A).

AutoCAD has an excellent help facility which can be accessed via the help button (D). There is also a search window (C) where you can enter a search word or expression. By default you will draw using 'model space'. When you later wish to lay out your drawing for display or printing you can select a layout or 'paper space' view. You can easily move between these views by clicking on the view tabs (N).

AutoCAD provides you with several different ways of executing most of its drawing commands. Sometimes it is more convenient to draw with the mouse but, where accuracy is important, you can enter precise values using either the command window (O) or one of the prompts that appear directly in the drawing area when executing drawing commands (L and M). After a little practice you will find executing a particular command becomes reasonably intuitive.

Finally, it is worth remembering that you can easily abandon a command by pressing the Escape (Esc) key on the keyboard or you can simply click on the undo button at the top of the screen. As before, it's important to save your work regularly and ensure that you have a backup of any files that you have created!

ACTIVITY 7.4

Complete the CAD drawing of the component shown in Figure 7.44 on page 246 by adding the two remaining views. Include a title block, then save and print the completed drawing.

ACTIVITY 7.5

Use a CAD drawing package to produce three fully dimensioned orthographic views of the component shown in isometric view in Figure 7.56. Include a title block, then save and print the completed drawing.

ACTIVITY 7.6

Use a CAD drawing package to produce three fully dimensioned orthographic views of the component shown in cabinet oblique view in Figure 7.57. Include a title block, then save and print the completed drawing.

ACTIVITY 7.7

Use a CAD drawing package to draw the block schematic diagram of a radio receiver shown in Figure 7.58. Include a title block then save and print the completed drawing.

ACTIVITY 7.8

Use a CAD drawing package to draw the electronic circuit diagram of the power supply shown in Figure 7.59. Include a title block then save and print the completed drawing.

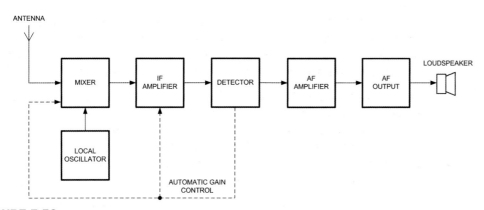

FIGURE 7.58
See Activity 7.7

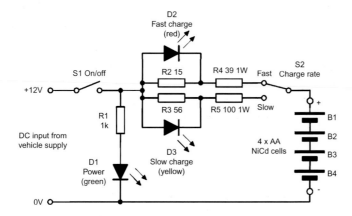

FIGURE 7.59
See Activity 7.8

REVIEW QUESTIONS

1. Give THREE examples of the use of *computer aided engineering* (CAE).
2. List THREE advantages of using CAD/CAM.
3. Distinguish between 2D and 3D CAD and describe a typical application of each.
4. List FIVE features of a modern 2D CAD package.
5. In relation to a 2D CAD system, explain the terms:
   a. absolute coordinates
   b. relative coordinates.
6. In relation to a 3D CAD system, explain the terms:
   a. wire frame drawing
   b. rendering.
7. Explain the difference between an *orthographic* and an *isometric* view.
8. Sketch a 3D workplane and label the axes.
9. Sketch the standard type of line used to represent:
   a. a visible outline
   b. hidden detail
   c. a centre line
   d. the limit of a partial view.
10. Name TWO British Standards that relate to engineering drawing and explain why they are necessary.

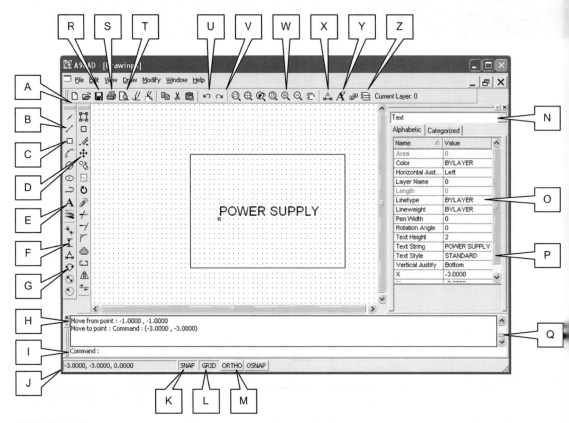

FIGURE 7.60
See Question 13

11. Explain what is meant by an *origin* or *datum point*.
12. List the steps required to start up and close down a typical CAD program.
13. Figure 7.60 shows a CAD program being used to draw a block schematic diagram. Identify the features marked A to Z and explain what each feature is used for.

Electronic Circuit Construction

SUMMARY

Whether we are aware of it or not, electronics is part of our everyday lives. It controls the energy that we use in our homes, manages the engines and instruments in our vehicles, navigates our ships and aircraft, allows us to communicate directly with anywhere in the world and provides us with information and entertainment. This unit provides you with an opportunity to build and test some simple electronic circuits and help you to understand how they work. It will also help you to identify common types of electronic component (such as resistors, capacitors, diodes and transistors) and how they work together. As part of the assessment of this unit, your tutor will require you to construct and test an electronic circuit. You will be expected to use basic electronic test equipment (such as multimeters, oscilloscopes and logic probes) to confirm that the circuit operates correctly and also to identify and rectify any faults that might be present. This chapter is designed to provide you with a starting point for this practical task. It is designed to introduce you to the components and constructional techniques that you will be using as well as how to use electronic test equipment in order to confirm that your circuit is working correctly.

8.1 SAFETY

When working on electronic circuits, personal safety (both yours and of those around you) should be paramount in everything that you do. Hazards can exist within many circuits—even those that, on the face of it, may appear to be totally safe. Inadvertent misconnection of a supply, incorrect earthing, reverse connection of a large-value electrolytic capacitor and incorrect component substitution can all result in serious hazards to personal safety as a consequence of fire, explosion or the generation of toxic fumes.

Potential hazards can be easily recognised and it is well worth making yourself familiar with them but perhaps the most important point to make is that electricity acts very quickly and you should always think carefully before working on circuits where mains or high voltages (i.e. those over 50 V, or so) might be present. Failure to observe this simple precaution can result in the very real risk of electric shock.

Voltages in many items of electronic equipment, including all items that derive their power from the a.c. mains supply, are at a level which can cause sufficient current flow in the body

BTEC First Engineering: Mandatory and Selected Optional Units for BTEC Firsts in Engineering. DOI: 10.1016/B978-1-85617-685-9.00008-1

TABLE 8.1 Different levels of electric shock

Current	Effect on the body
less than 1 mA	Not usually noticeable
1 mA to 2 mA	Threshold of perception (a slight tingle may be felt)
2 mA to 4 mA	Mild shock (effects of current flow are felt)
4 mA to 10 mA	Serious shock (shock is felt as pain)
10 mA to 20 mA	Nerve paralysis (unable to let go)
20 mA to 50 mA	Breathing may stop
more than 50 mA	Burns and possible heart failure

to disrupt normal operation of the heart. The threshold will be even lower for anyone with a defective heart. Bodily contact with mains or high-voltage circuits can thus be lethal. The most critical path for electric current within the body (i.e. the one that is most likely to stop the heart) is that which exists from one hand to the other. The hand-to-foot path is also dangerous but somewhat less dangerous than the hand-to-hand path.

So, before you start to work on an item of electronic equipment, it is essential not only to *switch off* but to *disconnect the equipment at the mains* by removing the mains plug. If you have to make measurements or carry out adjustments on a piece of working (or 'live') equipment, a useful precaution is that of using one hand only to perform the adjustment or to make the measurement. Your 'spare' hand should be placed safely away from contact with anything metal (including the chassis of the equipment which may, or may not, be earthed).

The severity of electric shock depends upon several factors including the magnitude of the current, whether it is alternating or direct current, and its precise path through the body. The magnitude of the current depends upon the voltage which is applied and the resistance of the body. The electrical energy developed in the body will depend upon the time for which the current flows. The duration of contact is also crucial in determining the eventual physiological effects of the shock. As a rough guide, and assuming that the voltage applied is from the 250 V 50 Hz a.c. mains supply, the effects listed in Table 8.1 are typical. It is, however, important to note that the figures are quoted as a guide—there have been cases of lethal shocks resulting from contact with much lower voltages and at relatively small values of current. The upshot of all this is simply that *any potential in excess of 50 V should be considered dangerous*. Lesser potentials may, under unusual circumstances, also be dangerous. As such, it is wise to get into the habit of treating all electrical and electronic circuits with great caution.

Fuses

A fuse is simply an electrical conductor that is designed to fail when the current passing through it exceeds a certain value. Most fuses comprise a thin wire conductor fixed inside a glass or ceramic tube and terminated at each end by caps which provide an effective electrical connection with the associated fuse-holder. Fuses are usually designed to protect a battery or an AC. mains supply against the effects of a catastrophic failure (such as a short-circuit across the supply connections). Fuses can also help to protect the equipment itself (or its power supply) from the effects of excessive current which might otherwise cause over-heating and permanent damage to electronic components and PCB tracks.

Fuses are designed to rupture and break a circuit when the rated current is exceeded. You should always follow the manufacturer's recommendations when selecting and fitting fuses. Some typical fuse ratings and applications are shown in Table 8.2.

Key point

Before you start to work on an item of electronic equipment, it is essential not only to *switch off* but to *disconnect the equipment at the mains* by removing the mains plug.

TABLE 8.2 Typical ratings for supply fuses		
Application	Power rating	Fuse rating
Soldering iron	25 W	1 A
Personal computer	350 W	3 A
Halogen lighting	500 W	5 A
Electric fire	2 kW	13 A

Another view
A fuse can normally pass its rated current for an indefinite period. However, when the rated current is exceeded, the fuse will blow in an increasingly shorter time interval.

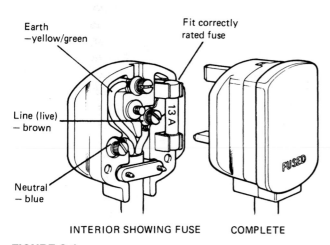

FIGURE 8.1
Fitting a fuse to a mains plug

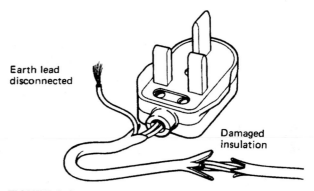

FIGURE 8.2
Mains plugs should be regularly examined for defective wiring and other faults

Various types of fuse and fusible device are fitted to modern electronic equipment. They include the following types:

- Fuses designed for fitting into mains plugs (i.e. the 13 A domestic plug universally found in the UK which may be fitted with 1 A, 3 A, 5 A and 13 A cartridge fuses).
- 'Quick-blow' fuses.
- 'Slow-blow' (or *time delay*) fuses.

Quick-blow fuses are usually designed to break within a time interval of about one second. Time-delay fuses, on the other hand, are actually designed to withstand currents that are well in excess of their rated current values for short periods of time (typically 5 to 10s). This type of fuse is used in equipment where there may be a large *surge* of current on 'switch-on'.

Residual Current Devices

We have already said that, if a person comes into contact with a conductor that is live and at the same time is earthed through the other hand or feet, a current will flow dependent on the resistance of the path through the person's body. To avoid this situation, it is possible to incorporate protection into an electrical supply known as a *residual current device* (RCD) or a *residual current circuit breaker* (RCCB). This protection device works by sensing the imbalance of current that occurs when current flows between the line conductor and earth. Typical currents at which an RCD will operate are 30 mA or 100 mA (the former giving a higher level of protection—see Table 8.1). Note that the RCD incorporates an in-built test facility and this should be checked from time to time in order to ensure that the RCD is working.

Another view
The basic principle behind the fuse is simply that, by making the fuse the *weakest* link in the current path, it will fail before anything else does. This is, unfortunately, sometimes NOT the case—simply because the action of rupturing (i.e. literally melting) even the thinnest of wire conductors takes a small time. Other components, such as transistors and integrated circuits, can fail much more rapidly!

ACTIVITY 8.1

Use your school or college library or sources on the Internet to locate information on how to deal with someone who has received an electric shock. Produce an A4 poster giving step-by-step instructions that can be displayed in an electronic workshop. Make sure that your poster includes details on how to switch off the mains supply.

Test your knowledge 8.1
Fuses of 1 A, 3 A, 5 A, and 13 A are available. Which of these should be used in each of the following applications?

a. an electric kettle
b. an audio system
c. a microwave oven
d. a desk lamp.

ACTIVITY 8.2

Refer to the section on Safety when soldering from Unit 1 (see pages 29 and 30 and answer the following questions:

1. What is the typical range of temperatures found at the bit of a soldering iron?
2. Why should fume extraction equipment be used when soldering?
3. What can cause damage to the eyes when soldering and how can this be avoided?
4. Why are low-voltage soldering irons usually preferred?
5. What is ESD and how can it be avoided?

Test your knowledge 8.2
Explain the purpose of an RCD. At what current does this typically operate?

Other Hazards

Apart from the risk of electric shock, various other hazards are present in an electronic workshop or laboratory. These include the use of hand tools (including soldering and de-soldering equipment) and chemicals used to develop and etch printed circuit boards. In order to reduce risk to yourself and others around you, it is important to be aware of these hazards and also to observe the Health and Safety precautions given to you by your tutor.

8.2 ELECTRONIC COMPONENTS

Electronic circuits are made from a number of individual electronic components that work together to provide the desired function. Some of the most common types are listed below:

- A *cell* is a source of direct current (d.c.) electrical energy. Primary cells have a nominal potential of 1.5 V each. They cannot be recharged and are disposable. Secondary cells are rechargeable. Lead–acid cells have a nominal potential of 2 V and nickel cadmium (NiCd) cells have a nominal potential of 1.2 V. Cells are often connected in series to form a battery.
- *Batteries* consist of a number of cells connected in series to increase the overall potential. A 12 V car battery consists of six lead–acid secondary cells of 2 V each.
- *Fuses* protect the circuit in which they are connected from excess current flow. This can result from a fault in the circuit, from a fault in an appliance connected to the circuit or from too many appliances being connected to the same circuit. The current flowing in the circuit tends to heat up the fuse wire. When the current reaches some pre-determined value the fuse wire melts and breaks the circuit so the current can no longer flow. Without a fuse the circuit wiring could overheat and cause a fire.
- *Resistors* are used to control the magnitude of the current flowing in a circuit. The resistance value of the resistor may be fixed or it may be variable. Variable resistors may be pre-set or they may be adjustable by the user. The electric current does work in flowing through the resistor and this heats up the resistor. The resistor must be chosen so that it can withstand this heating effect and sited so that it has adequate ventilation.
- *Capacitors*, like resistors, may be fixed in value or they may be preset or variable. Capacitors store electrical energy but, unlike secondary cells, they may be charged or discharged almost instantaneously. The stored charge is much smaller than the charge stored by a secondary cell. Large value capacitors are used to smooth the residual ripple from the

rectifier in a power pack. Medium value capacitors are used for coupling and decoupling the stages of audio frequency amplifiers. Small value capacitors are used for coupling and decoupling radio frequency signals and they are also used in tuned (resonant) circuits.

- *Inductors* act like electrical 'flywheels'. They limit the build up of current in a circuit and try to keep the circuit running by putting energy back into it when the supply is turned off. They are used as current limiting devices in fluorescent lamp units. They are used as chokes in telecommunications equipment. They are also used together with capacitors to make up resonant (tuned) circuits in telecommunications equipment.

- *Transformers* are used to step-up or step-down the voltage of an alternating current. Inductors and transformers cannot be used in direct current circuits. The input side of a transformer is referred to as the primary whilst the output side is referred to as the secondary. Note that, ignoring any losses, the same power appears at the secondary as is applied to the primary.

- *Ammeters* indicate the current flowing in a circuit. They are always wired in series with the circuit so that the current being measured can flow through the meter.

- *Voltmeters* indicate the potential difference (voltage) between two points in a circuit. To do this they are always wired in parallel across that part of the circuit where the potential is to be measured.

- *Switches* are used to control the flow of current in a circuit. They can only open or close the circuit. So the current either flows or it doesn't.

- *Diodes* act like non-return valves—they allow current to flow in one direction only (as indicated by the arrow on the diode symbol). Diodes are used to *rectify* alternating current (a.c.) and convert it into direct current (d.c).

- *Transistors* are used in amplifiers, oscillators and in switching circuits. Transistors work by amplifying current—the output current from the *collector* or *emitter* being a magnified version of the current applied to the *base*.

- *Integrated circuits* consist of all the components necessary to produce amplifiers, oscillators, logic circuits, memory devices and a host of other devices built on a single slice of silicon; each 'chip' being housed in a single compact package.

Key point
Electronic circuits are made from a number of individual electronic components (such as resistors and capacitors) that work together to provide the desired function.

8.3 CIRCUIT DIAGRAMS

Circuit diagrams use standard symbols and conventions to represent the components and wiring used in an electronic circuit. Visually, they bear very little relationship to the physical layout of a circuit but, instead, they provide us with a 'theoretical' view of the circuit. In this section we show you how to find your way round.

To be able to understand a circuit diagram you first need to be familiar with the symbols that are used to represent the components and devices. A selection of some of the most commonly used symbols are shown in Figure 8.3. It's important to be aware that there are a few (thankfully quite small) differences between the symbols used in circuit diagrams of American and European origin.

As a general rule, the input to a circuit should be shown on the left of a circuit diagram and the output shown on the right. The supply (usually the most positive voltage) is normally shown at the top of the diagram and the common, 0 V, or ground connection is normally shown at the bottom. This rule is not always obeyed, particularly for complex diagrams where many signals and supply voltages may be present. The rules for showing connections that join and those that cross over one another are shown in Figure 8.4.

Note also that, in order to simplify a circuit diagram (and avoid having too many lines connected to the same point) multiple connections to common, 0 V, or ground may be shown using the appropriate symbol (see Fig. 8.3). The same applies to supply connections that may be repeated (appropriately labelled) at various points in the diagram. Now let's look at some examples of schematic circuit diagrams using the component symbols that we met earlier.

Key point
Circuit diagrams use standard symbols and conventions to represent the components and wiring used in an electronic circuit. Visually, they bear very little relationship to the physical layout of a circuit but, instead, they provide us with a 'theoretical' view of the circuit.

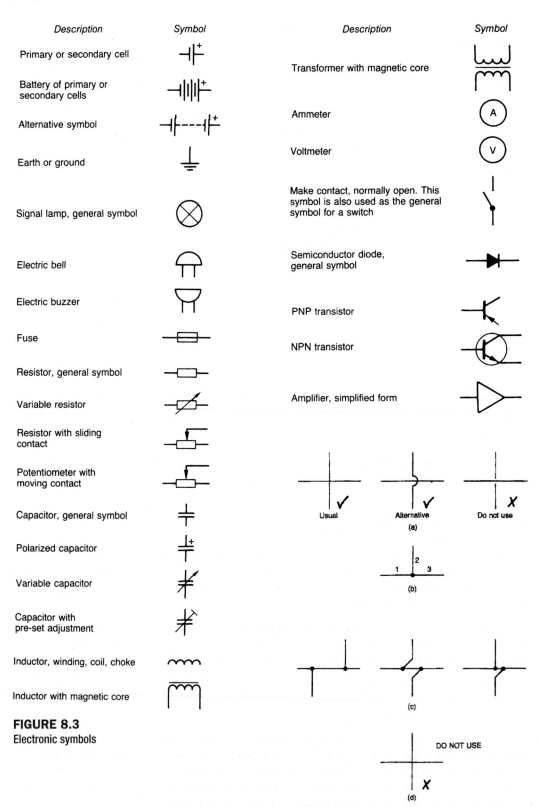

FIGURE 8.3
Electronic symbols

FIGURE 8.4
Junction and crossing connections

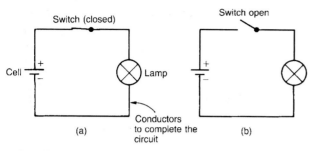

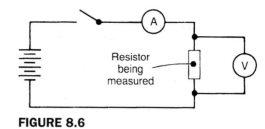

FIGURE 8.6
Circuit for determining resistance

FIGURE 8.5
A simple electric circuit

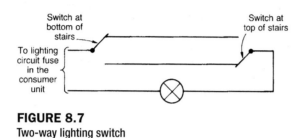

FIGURE 8.7
Two-way lighting switch

Figure 8.5 shows a very simple circuit that satisfies the above requirements. In Figure 8.5(a) the switch is 'closed' therefore the circuit as a whole is also a closed loop. This enables the electrons that make up the electric current to flow from the source of electrical energy through the lamp and back to the source of energy ready to circulate again. In Figure 8.5(b) the switch is 'open' and the circuit is no longer a closed loop. The circuit is broken and the electrons can no longer circulate. We normally draw our circuits with the switches in the 'open' position so that the circuit is not functioning and is 'safe'.

Figure 8.6 shows a simple battery operated circuit for determining the resistance of a fixed value resistor. The resistance value is obtained by substituting the values of current and potential into the Ohm's Law formula, $R = V/I$. The current in amperes is read from the ammeter and the potential in volts is read from the voltmeter. Note that the ammeter is wired in series with the resistor so that the current can flow through it. The voltmeter is wired in parallel with the resistor so that the potential can be read across it. This is always the way these instruments are connected.

Figure 8.7 shows a circuit for operating the light over the stairs in a house. The light can be operated either by the switch at the bottom of the stairs or by the switch at the top of the stairs. Can you work out how this is achieved? The switches are of a type called 'single-pole double-throw' (SPDT). The circuit is connected to the mains supply. It is protected by a mains circuit breaker (MCB) or fuse in the *consumer unit*. This unit contains the main switch and all the fuses for the house.

Figure 8.8 shows the circuit of a two-stage transistorised amplifier, complete with a mains power unit that provides the 9 V supply that the transistors require. Note that the power supply uses a *bridge rectifier* arrangement comprising four diodes, D_1 to D_4 and that the a.c. mains input is connected to the step-down transformer's primary winding (shown on the extreme right of the circuit diagram). Table 8.3 lists and names the components.

Figure 8.9 shows an amplifier which provides a voltage gain of 10 based on a single *operational amplifier* chip. Such an amplifier would have similar performance to that shown in Figure 8.8 but, as you can see, fewer components are required. Therefore it is cheaper and quicker to make. The integrated circuit contains many electronic components—Figure 8.10 will give you some idea of just how many components there are inside a 741 integrated circuit!

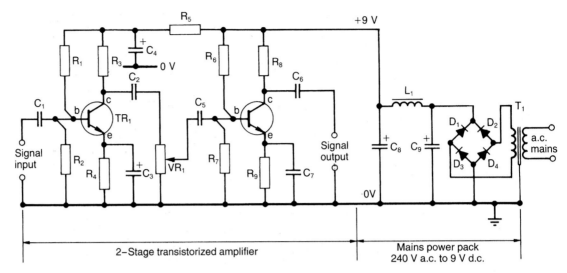

FIGURE 8.8
A two-stage transistor amplifier

TABLE 8.3 Component list for the two-stage transistor amplifier

Component	Description
R_1–R_9	Fixed resistors
VR_1	Variable resistor
C_1–C_9	Capacitors
D_1–D_4	Diodes
TR_1, TR_2	Transistors
T_1	Mains transformer
L_1	Inductor (*choke*)

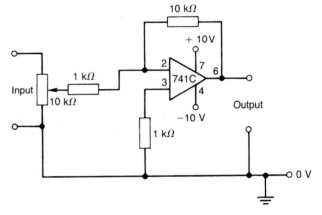

FIGURE 8.9
A single-chip amplifier circuit

ACTIVITY 8.3

Draw a simple circuit for powering four electric lamps from a battery. It must be possible to turn each lamp on or off independently. A master switch must also be provided so that the whole circuit can be turned on or off. A fuse must be included in the circuit in order to prevent the battery being overloaded. Label your diagram clearly.

ACTIVITY 8.4

Draw a schematic circuit diagram for a battery charger having the following features:

- the primary circuit of the transformer (i.e. the side that is connected to the a.c. mains supply) is to have an on/off switch, a fuse and an indicator lamp
- the secondary circuit of the transformer is to have a single rectifier diode connected in series with a variable resistor to control the charging current, a fuse and an ammeter to indicate the charging current.

Label your diagram (identifying each component) and include a full component list for the battery charger.

Test your knowledge 8.3

How many individual transistors are incorporated in a 741 integrated circuit? How many of these are NPN and how many are PNP?

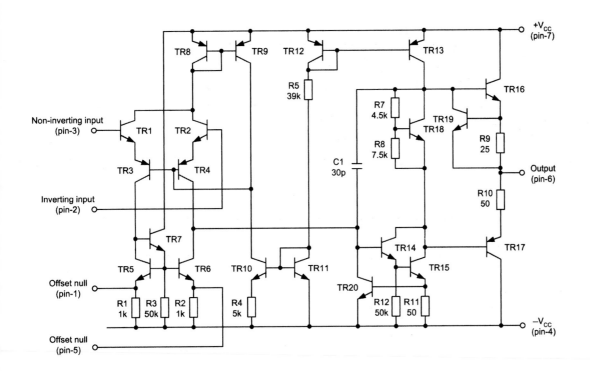

FIGURE 8.10
Internal circuit of a 741 integrated circuit operational amplifier

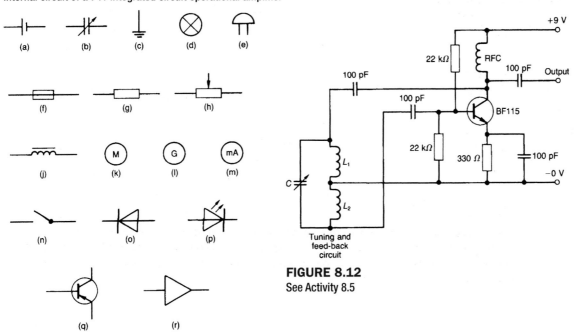

FIGURE 8.11
See Test your knowledge 8.4

FIGURE 8.12
See Activity 8.5

ACTIVITY 8.5

Figure 8.12 shows an electronic circuit.

a. Produce a component list that numbers and names each of the components (include values where given).

b. Suggest what the circuit might be used for (Hint: the circuit only has an output!).

Test your knowledge 8.4

Figure 8.11 shows some common electronic component symbols. Name them and briefly explain what they do.

8.4 COMPONENT IDENTIFICATION

This section provides you with information that will help you to identify common electronic components from their physical appearance and markings. To help you become familiar with a wider range of electronic components you may find it useful to refer to the catalogues and data sheets provided by electronic component manufacturers and distributors.

FIGURE 8.13
Various types of switch. From left to right: a mains rocker switch, a single-pole single-throw (SPST or changeover) miniature toggle switch, a double-pole double-throw (DPDT) slide switch, an SPDT push-button (wired for use as an SPST push-button), and a miniature PCB mounting DPDT push-button (with a latching action)

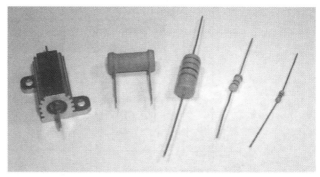

FIGURE 8.14
A selection of resistors including high-power metal clad (extreme left), ceramic wirewound, carbon and metal film types (extreme right) with values ranging from $15\,\Omega$ to $4.7\,k\Omega$

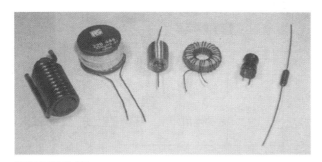

FIGURE 8.15
A selection of small inductors with ferrite cores and values ranging from $15\,\mu H$ to $1\,mH$

FIGURE 8.16
A selection of common types of carbon and wirewound variable resistors/potentiometers with values ranging from $100\,\Omega$ to $500\,k\Omega$

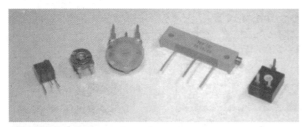

FIGURE 8.17
A selection of common types of standard and miniature pre-set resistors/potentiometers designed for printed circuit board (PCB) mounting. Values of the components shown range from $100\,\Omega$ to $5\,k\Omega$

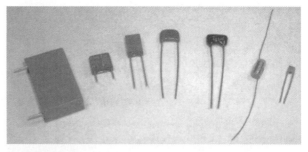

FIGURE 8.18
A selection of capacitors (including polyester, polystyrene, ceramic and mica types) with values ranging from $10\,pF$ to $470\,nF$

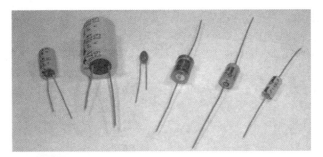

FIGURE 8.19
A selection of electrolytic (polarised) capacitors with values ranging from 1 μF to 470 μF

Another view
Transformers are often rated in terms of 'volt-amperes' (VA). As an example, a transformer that delivers 15 V at 2 A would have a rating of 30 VA.

FIGURE 8.20
An air-spaced variable capacitor. This component (used for tuning an AM radio) has two separate variable capacitors (each of 500 pF maximum) operated from a common control shaft

FIGURE 8.21
A selection of transformers with power ratings from 0.1 VA to 100 VA

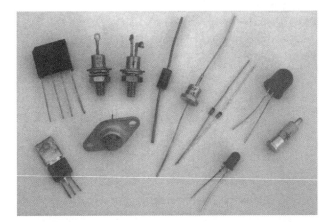

FIGURE 8.22
A selection of diodes including rectifiers, a bridge rectifier (top left), silicon-controlled rectifier (thyristors), signal and zener diodes, and two light emitting diodes (LED)

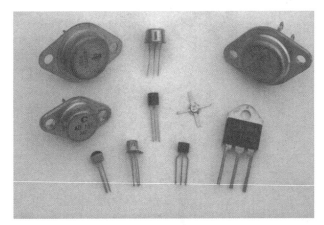

FIGURE 8.23
A selection of transistors include small-signal transistors (bottom centre and left), power transistors (top left, top right and bottom right) and high frequency (centre) types

Test your knowledge 8.5
Identify the components marked A, B, C and D in Figure 8.25.

FIGURE 8.24
Various integrated circuit packages including dual-in-line (DIL) and surface mounted types. Note that I've included a couple of resistors in this photograph so that you can get an idea of the relative size!

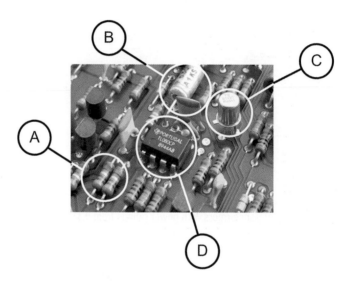

FIGURE 8.25
See Test your knowledge 8.5

8.5 COMPONENT COLOUR CODES AND MARKINGS

Carbon and metal oxide resistors are normally marked with colour codes which indicate their value and tolerance. Two methods of colour coding are in common use; one involves four coloured bands (see Fig. 8.26) while the other uses five colour bands (see Fig. 8.27).

EXAMPLE 8.1

A resistor is marked with the following coloured stripes: brown, black, red, silver. What is its value and tolerance?

See Figure 8.28.

EXAMPLE 8.2

A resistor is marked with the following coloured stripes: red, violet, orange, gold. What is its value and tolerance?

See Figure 8.29.

Some types of resistor have markings based on a system of coding defined in BS 1852. This system involves marking the position of the decimal point with a letter to indicate the multiplier

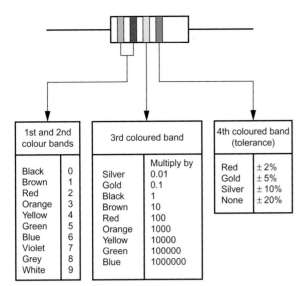

FIGURE 8.26
Four band resistor colour code

1st and 2nd colour bands	
Black	0
Brown	1
Red	2
Orange	3
Yellow	4
Green	5
Blue	6
Violet	7
Grey	8
White	9

3rd coloured band	Multiply by
Silver	0.01
Gold	0.1
Black	1
Brown	10
Red	100
Orange	1000
Yellow	10000
Green	100000
Blue	1000000

4th coloured band (tolerance)	
Red	± 2%
Gold	± 5%
Silver	± 10%
None	± 20%

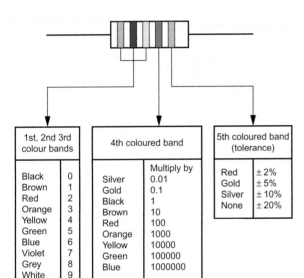

FIGURE 8.27
Five band resistor colour code

1st, 2nd 3rd colour bands	
Black	0
Brown	1
Red	2
Orange	3
Yellow	4
Green	5
Blue	6
Violet	7
Grey	8
White	9

4th coloured band	Multiply by
Silver	0.01
Gold	0.1
Black	1
Brown	10
Red	100
Orange	1000
Yellow	10000
Green	100000
Blue	1000000

5th coloured band (tolerance)	
Red	± 2%
Gold	± 5%
Silver	± 10%
None	± 20%

concerned as shown in Table 8.4. A further letter is then appended to indicate the tolerance as shown in Table 8.5.

The vast majority of capacitors employ written markings which indicate their values, working voltages, and tolerance (see Fig. 8.30). The most usual method of marking resin dipped polyester (and other) types of capacitor involves quoting the value (μF, nF or pF), the tolerance (often either ±10% or ±20%), and the working voltage (often using _ and ~ to indicate d.c. and a.c. respectively). Several manufacturers use two separate lines for their capacitor markings and these have the following meanings:

First line: capacitance (pF or μF) and tolerance (K = ±10%, M = ±20%)
Second line: rated d.c. voltage and code for the insulating dielectric material

A three-digit code is commonly used to mark monolithic ceramic capacitors. The first two digits of this code correspond to the first two digits of the value while the third digit is a multiplier which gives the number of zeros to be added to give the value in picofarads. Other capacitors may use a colour code similar to that used for marking resistor values (see Fig. 8.31).

Test your knowledge 8.6
A resistor is marked with the following coloured bands: green, blue, red, gold. What is its value and tolerance?

Test your knowledge 8.7
A resistor is marked with the following coloured bands: brown, green, black, red, red. What is its value and tolerance?

Test your knowledge 8.8
A resistor is marked with the following legend: 4R7K. What is its value and tolerance?

Test your knowledge 8.9
A ceramic capacitor is marked '103 K'. What is its value?

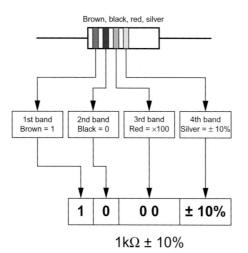

Brown, black, red, silver

1st band Brown = 1	2nd band Black = 0	3rd band Red = ×100	4th band Silver = ± 10%

1	0	0 0	± 10%

$1k\Omega \pm 10\%$

FIGURE 8.28
See Example 8.1

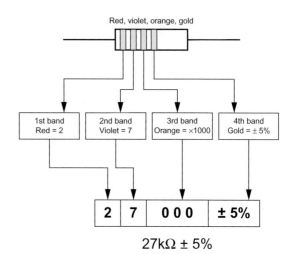

Red, violet, orange, gold

1st band Red = 2	2nd band Violet = 7	3rd band Orange = ×1000	4th band Gold = ± 5%

2	7	0 0 0	± 5%

$27k\Omega \pm 5\%$

FIGURE 8.29
See Example 8.2

TABLE 8.4 BS1852
Resistor multiplier markings

Letter	Multiplier
R	1
K	1,000
M	1,000,000

TABLE 8.5 BS1852
Resistor tolerance markings

Letter	Multiplier
F	±1%
G	±2%
J	±5%
K	±10%
M	±20%

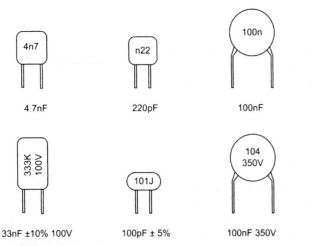

4n7	n22	100n
4.7nF	220pF	100nF

333K 100V	101J	104 350V
33nF ±10% 100V	100pF ± 5%	100nF 350V

FIGURE 8.30
Examples of typical capacitor markings

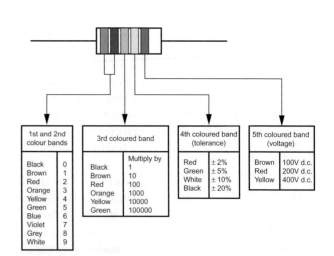

1st and 2nd colour bands		3rd coloured band		4th coloured band (tolerance)		5th coloured band (voltage)	
Black	0		Multiply by	Red	± 2%	Brown	100V d.c.
Brown	1	Black	1	Green	± 5%	Red	200V d.c.
Red	2	Brown	10	White	± 10%	Yellow	400V d.c.
Orange	3	Red	100	Black	± 20%		
Yellow	4	Orange	1000				
Green	5	Yellow	10000				
Blue	6	Green	100000				
Violet	7						
Grey	8						
White	9						

FIGURE 8.31
Capacitor colour code

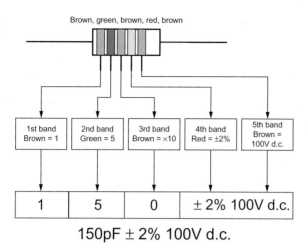

1st band
Brown = 1

2nd band
Green = 5

3rd band
Brown = ×10

4th band
Red = ±2%

5th band
Brown =
100V d.c.

| 1 | 5 | 0 | ± 2% 100V d.c. |

150pF ± 2% 100V d.c.

FIGURE 8.32
See Example 8.3

EXAMPLE 8.3

A tubular capacitor is marked with the following coloured stripes: brown, green, brown, red, brown. What is its value, tolerance, and working voltage?

See Figure 8.32.

As with capacitors, the vast majority of inductors use written markings to indicate values, working current, and tolerance. Some small inductors are marked with coloured stripes to indicate their value and tolerance (in which case the standard colour values are used and inductance is normally expressed in microhenries).

8.6 CABLES AND HARDWARE

Figure 8.33 shows some typical cables and hardware for electronic equipment.

a. This shows examples of matrix board, strip board and a printed circuit board. The matrix board (1) is a panel of laminated plastic perforated with a grid of holes. Pins can be fixed in the holes at convenient places for the attachment of such components as resistors and capacitors. They are merely attachment points and do not form part of the circuit. The strip board (2) is like a matrix board but is copper faced in strips on one side. The holes are the same pitch as the pins of integrated circuits which can be soldered into position. The components are placed on the insulated side of the board, with the copper strips underneath. The wire leads pass through the holes in the board and are soldered on the underside. The copper tracks have to be cut wherever a break in the circuit is required. Printed circuit boards (3) are custom made for a particular circuit and are designed to give the most efficient layout for the circuit. The components are installed in the same way as for the strip board. Assembly will be considered in greater detail later in this unit.

b. This is a typical flexible mains lead with PVC sheathing and colour coded PVC insulation. In selecting such a cable, the only factors you have to consider are its current handling capacity and its colour, providing the insulation is rated for mains use.

c. This is a signal cable suitable for audio frequency analogue signals and for data processing signals. The conductors only need to have a limited current handling capacity, so many conductors can be carried in one cable. The conductors are surrounded by an

Another view
Signals in digital logic circuits are either on or off. These two states are referred to as logic 1 (or *high*) and logic 0 (or *low*). In circuits containing logic gates, the two logic states are represented by voltages (typically +5V for logic 1 and 0V for logic 0).

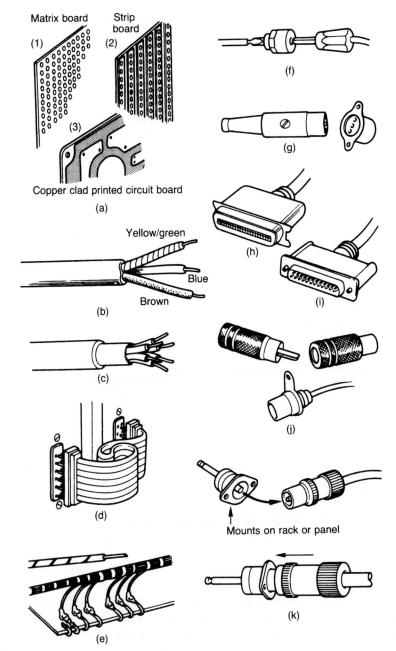

FIGURE 8.33
Typical cables and hardware

earthed metal braid to prevent pick-up of external interference and corruption of the signal.

d. This is a ribbon cable widely used in the manufacture of computers and the interconnection of computers and printers. Unlike the signal cable described earlier it is not screened against interference. However, it is cheaper and more easily terminated.

e. Single cored PVC insulated wire is useful for making up wiring harnesses and for flying leads on PVC boards.

f. This shows a 'banana' plug and socket. These are used with single cored, flexible conductors for low-voltage power supply connections.

g. This shows a DIN type plug and socket. These are used in conjunction with multi-cored screened signal cables. They are available for 3-way to 8-way connections inclusive and are designed so that the plug can only be inserted into the socket in one position.

h. This shows a 36-way Centronics plug as widely used for making connections to the parallel port of a computer printer.

i. This shows a 25-way D-type plug as widely used for making connections to the parallel output port of a computer.

j. This shows a selection of phono type plugs and sockets. These are widely used for making signal lead connections to audio amplifiers.

k. This shows a typical coaxial plug and socket. These are used at radio frequencies for connecting aerial leads to radio and television receivers.

8.7 LOGIC CIRCUITS

Logic gates are circuits designed to produce the basic logic functions, AND, OR, etc. that are used in digital circuits and computers.

The British Standard (BS) and American Standard (MIU/ANSI) symbols for some basic logic gates are shown in Figure 8.34. Note that, whilst inverters and buffers each have only one input, exclusive-OR gates have two inputs and the other basic gates (AND, OR, NAND and NOR) are commonly available with up to eight inputs.

BUFFERS

Buffers do not affect the logical state of a digital signal (i.e., a logic 1 input results in a logic 1 output whereas a logic 0 input results in a logic 0 output). Buffers are normally used to provide extra current drive at the output but can also be used to regularise the logic levels present at an interface.

INVERTERS

Inverters are used to complement the logical state (i.e., a logic 1 input results in a logic 0 output and vice versa). Inverters also provide extra current drive and, like buffers, are used in interfacing applications where they provide a means of regularising logic levels present at the input or output of a digital system.

AND GATES

AND gates will only produce a logic I output when all inputs are simultaneously at logic l. Any other input combination results in a logic 0 output.

OR GATES

OR gates will produce a logic l output whenever any one, or more, inputs are at logic l. Putting this another way, an OR gate will only produce a logic 0 output whenever all of its inputs are simultaneously at logic 0.

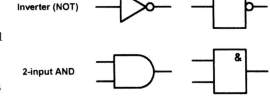
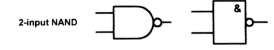

LOGIC FUNCTION	MIL/ANSI	BS
Buffer		
Inverter (NOT)		
2-input AND		&
2-input NAND		&
2-input OR		≥1
2-input NOR		≥1
2-input XOR		=1
2-input XNOR		=1

FIGURE 8.34
Logic gate symbols

269

Test your knowledge 8.11
Identify each of the logic gate symbols shown in Figure 8.35.

NAND GATES

NAND (or NOT-AND) gates will only produce a logic 0 output when all inputs are simultaneously at logic 1. Any other input combination will produce a logic 1 output. A NAND gate, therefore, is nothing more than an AND gate with its output inverted. The circle shown at the output is used to denote this inversion.

NOR GATES

NOR gates will only produce a logic 1 output when all inputs are simultaneously at logic 0. Any other input combination will produce a logic 0 output. A NOR gate, therefore, is simply an OR gate with its output inverted. A circle is again used to indicate inversion.

EXCLUSIVE-OR GATES

Exclusive-OR gates will produce a logic 1 output whenever either one of the inputs is at logic 1 and the other is at logic 0. Exclusive-OR gates produce a logic 0 output whenever both inputs have the same logical state (i.e., when both are at logic 0 or when both outputs are at logic 1). Any other input combination results in a logic 0 output.

Logic gates are often supplied in packages where several identical logic gates are made available. Figure 8.36 shows two common logic devices. One of these (the 4001UBE) is based on complementary metal oxide semiconductor (CMOS) technology whilst the other (74LS00) uses transistor transistor logic (TTL). Both devices are packaged in a 14-pin dual-in-line (DIL) package with the pin connections (viewed from the top) shown in Figure 8.36.

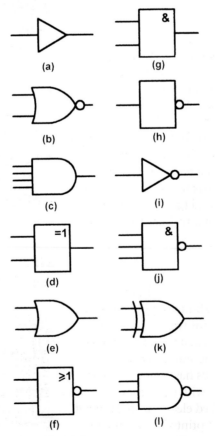

FIGURE 8.35
See Test your knowledge 8.11

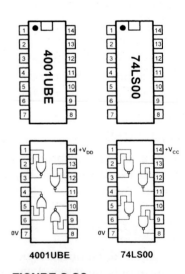

4001UBE   **74LS00**

FIGURE 8.36
Packages and pin connections for two common logic devices

8.8 CIRCUIT CONSTRUCTION

There are several different ways of building an electronic circuit. The method that you choose depends on a number of factors, including the resources available to you and whether you are building a 'one-off' prototype or a large number of identical circuits. The methods that are available to you include:

POINT-TO-POINT WIRING

With the advent of miniature components, printed circuit boards and integrated circuits, point-to-point wiring construction is a construction technique that is nowadays considered obsolete. The example shown in Figure 8.37 is the underside of a valve amplifier chassis dating back to the early 1960s. Unless you are dealing with a very small number of components or have a particular desire to use *tag strips* and *group boards*, point-to-point wiring is not a particularly attractive construction method these days!

FIGURE 8.37
Point-to-point wiring

BREADBOARD CONSTRUCTION

Breadboard construction is often used for assembling and testing simple circuit prior to production of a more permanent circuit using a stripboard or printed circuit board. The advantage of this technique is that changes can be quickly and easily made to a circuit and all of the components can be re-used. The obvious disadvantages of breadboard construction are that it is unsuitable for permanent use and also unsuitable for complex circuits (i.e. circuits with more than half a dozen, or so, active devices or integrated circuits). Figure 8.38 shows a simple logic circuit and two LEDs assembled and ready for testing.

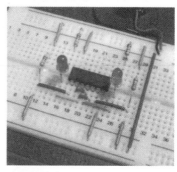

FIGURE 8.38
Breadboard construction

MATRIX BOARD CONSTRUCTION

Figure 8.39 shows matrix board construction. This low-cost technique avoids the need for a printed circuit but is generally only suitable for one-off prototypes. A matrix board consists of an insulated board into which a matrix of holes are drilled with copper tracks arranged as strips on the reverse side of the board. Component leads are inserted through the holes and soldered into place. Strips (or tracks) are linked together with short length of tinned copper wire (inserted through holes in the board and soldered into place on the underside of the board). Tracks can be broken at various points as appropriate. The advantage of this technique is that it avoids the need for a printed circuit board (which may be relatively expensive and may take some time to design). Disadvantages of matrix board construction are that it is usually only suitable for one-off production and the end result is invariably less compact than a printed circuit board. The matrix board shown in Figure 8.39 is a simple oscilloscope calibrator based on a PIC chip.

FIGURE 8.39
Matrix board construction

PRINTED CIRCUIT BOARDS

Printed circuit board construction (see Fig. 8.40) technique is ideal for high volume manufacture of electronic circuits where speed and repeatability of production are important. Depending on the complexity of a circuit, various types of printed circuit board are possible. The most basic form of printed circuit (and one which is suitable for home construction) has copper tracks on one side and components mounted on the other. More complex printed circuit boards have tracks on both sides (they are referred to as 'double-sided') whilst boards with up to four layers are used for some of the most sophisticated and densely packed electronic equipment (for example, computer motherboards). The double-sided printed circuit board shown in Figure 8.40 is part of an instrument landing system (ILS) fitted to a large passenger aircraft. A further example of printed circuit construction is shown in

FIGURE 8.40
Printed circuit construction

FIGURE 8.41
Printed circuit construction at UHF

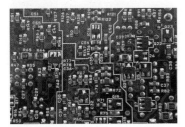

FIGURE 8.42
PCB using surface mounted devices

Figure 8.41. This shows 'zig-zag' inductive components fabricated using the copper track surface together with leadless 'chip' capacitors and a leadless transistor which is soldered directly to the surface of the copper track.

SURFACE MOUNTING

Figure 8.42 shows an example of surface mounting construction. This technique is suitable for sub-miniature leadless components. These are designed for automated soldering directly to pads on the surface of a printed circuit board. This technique makes it possible to pack in the largest number of components into the smallest space but, since the components require specialised handling and soldering equipment, it is not suitable for home construction nor is it suitable for hand-built prototypes. The example shown in Figure 8.42 is part of the signal processing circuitry used in a large computer monitor.

Using Matrix Boards and Stripboards

Matrix boards and stripboards are ideal for simple prototype and one-off electronic circuit construction. The distinction between matrix boards and stripboards is simply that the former has no copper tracks and the user has to make extensive use of press-fit terminal pins which are used for component connection. Extensive inter-wiring is then necessary to link terminal pins together. This may be carried out using sleeved tinned copper wire (of appropriate gauge) or short lengths of PVC-insulated 'hook-up' or *equipment wire*.

Like their matrix board counterparts, stripboards are also pierced with a matrix of holes which, again, are almost invariably placed on a 0.1 in pitch. The important difference, however, is that stripboards have copper strips bounded to one surface which link together rows of holes along the complete length of the board. The result, therefore, is something of a compromise between a 'naked' matrix board and a true printed circuit. Compared with the matrix board, the stripboard has the advantage that relatively few wire links arc required and that components can be mounted and soldered directly to the copper strips without the need for terminal pins.

The following steps are required when laying out a circuit for stripboard construction:

1. Carefully examine a copy of the circuit diagram and check that you have all of the components needed to build the circuit (you will need to know the size of these components and the pin orientation for each of them).
2. Mark all components to be mounted 'off-board' and identify using appropriate letters and/or numbers, e.g. SK, pin-2) all points at which an 'off-board' connection is to be made.
3. Identify any multiple connections required between integrated circuits or between integrated circuits and connectors. Arrange such components in physical proximity and with such orientation that will effectively minimise the number of links required.
4. Identify components that require special attention, such as those that might require heatsinks (see Fig. 8.43) or those that might need to be aligned with external hardware (such as control shafts, connectors, etc.).
5. Keep inputs and outputs at opposite ends of the stripboard. This not only helps maintain a logical circuit layout (progressing from input to output) but, in high gain circuits, it may also be instrumental in preventing instability due to unwanted *feedback*.
6. Use standard sizes of stripboard wherever possible. Where boards have to be cut to size, it is usually more efficient to align the strips along the major axis of the board.
7. Consider the means of mounting the stripboard. If it is to be secured using bolts and threaded spacers (or equivalent) it will be necessary to allow adequate clearance around the mounting holes.

Test your knowledge 8.12
Describe THREE different methods of building an electronic circuit.

8. Produce a rough layout for the stripboard first using paper ruled with squares, the corners of the squares representing the holes in the stripboard. This process can be carried out 'actual size' using 0.1 in. graph paper or suitably enlarged by means of an appropriate choice of paper. For preference, it is wise to choose paper with a feint blue or green grid as this will subsequently disappear after photocopying leaving you with a 'clean' layout.

9. Identify all conductors that will be handling high currents (i.e. those in excess of 1 A) and use adjacent strips connected in parallel at various points along the length of the board.

10. Identify the strips that will be used to convey the supply rails: as far as possible these should be continuous from one end of the board to the other. It is often convenient to use adjacent strips for supply and 0 V (or 'ground') since decoupling capacitors can easily be distributed at strategic points. Ideally, such capacitors should be positioned in close proximity to the positive supply input pin to all integrated circuits which are likely to demand sudden surges of current (e.g. 555 timers, comparators, integrated circuit power amplifiers).

11. In high-frequency circuits, link all unused strips to 0 V at regular points. This promotes stability by ensuring that the 0 V rail is effective as a common rail.

12. Minimise, as far as possible, the number of links required. These should be made on the upper (component) side of the stripboard. Only in exceptional cases should links be made on the underside (foil side) of the board.

13. Experiment with positioning of integrated circuits (it is good practice, though not essential, to align them all in the same direction). In some cases, logic gates may be exchanged from package to package (or within the same package) in order to minimise strip usage and links. (If you have to resort to this dodge, do not forget to amend the circuit diagram!)

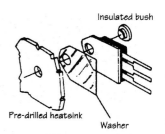

FIGURE 8.43
A typical mounting arrangement for a transistor or integrated circuit requiring a heatsink

When the stripboard layout is complete, it is important to carefully check it against the circuit diagram. Not only can this save considerable frustration at a later stage but it can be instrumental in preventing some costly mistakes. In particular, you should follow the positive supply and 0 V (or 'ground') strips and check that all chips and other devices have supplies. A useful technique is that of using coloured pencils to trace the circuit and stripboard layout; associating each line in the circuit diagram with a physical interconnection on the stripboard. Colours can be used as follows:

Positive supply rails	Red
Negative supply rails	Black
Common 0 V rail	Green
Signal	Yellow/Pink/White or Grey
Off-board connections	Orange/Violet
Mains wiring	Brown and blue

Assembly should be a fairly straightforward process. The sequence used for stripboard assembly will normally involve mounting sockets for integrated circuits first followed by transistors, diodes, resistors, capacitors and other passive components. Finally, terminal pins and links should be fitted before making the track breaks. On completion, the board should be carefully checked, paying particular attention to all polarised components (e.g. diodes, transistors and electrolytic capacitors). Figure 8.44 shows a typical stripboard layout together with matching component overlay.

Using Printed Circuits

Printed circuit boards (PCB) comprise copper tracks bonded to an epoxy glass or synthetic resin bounded paper (SRBP) board. The result is a neat and professional looking circuit

273

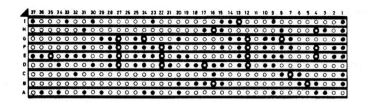

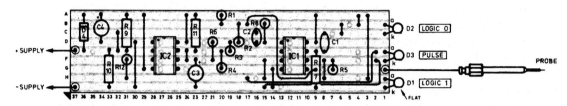

FIGURE 8.44
Stripboard layout and component overlay for a simple logic probe circuit

that is ideal for prototype as well as production quantities. Printed circuits can easily be duplicated or modified from original master artwork and the production techniques are quite simple. Your school or college will probably have all of the necessary equipment and materials available for you to use.

The following steps should be followed when laying out a printed circuit board:

1. Carefully examine a copy of the circuit diagram and check that you have all of the components needed to build the circuit (you will need to know the size of these components and the pin orientation for each of them).
2. Mark all components that are to be mounted 'off-board' and, using appropriate letters and/or numbers (e.g. SK1, pin-2), identify all points at which an 'off-board' connection is to be made.
3. Identify any multiple connections required between integrated circuits or between integrated circuits and edge connectors. Arrange such components in physical proximity and with such orientation that will effectively minimise the number of links required.
4. Identify components that require special attention (such as those that require heatsinks or have special screening requirements). Ensure that such components are positioned sensibly bearing in mind their particular needs.
5. Keep inputs and outputs at opposite ends of the PCB wherever possible. This not only helps maintain a logical circuit layout (progressing from input) but, in high gain circuits, it may also be instrumental in preventing instability.
6. Use the minimum board area consistent with a layout which is not cramped. In practice, and to prevent wastage, you should aim to utilise as high a proportion of the PCB surface area as possible. In the initial stages, however, it is wise to allow some room for manoeuvre as you may want to make subsequent modifications to the design.
7. Unless the design makes extensive use of PCB edge connectors, try to ensure that a common 0V foil is run all round the periphery of the PCB. This has a number of advantages not the least of which is the fact that it will then be relatively simple to find a route to the 0V rail from almost anywhere on the board.
8. Consider the means of mounting the PCB and, if it is to be secured using bolts and threaded spacers (or equivalent) you should ascertain the number and location of the holes required. You may also wish to ensure that the holes coincide with the 0V foil, alternatively where the 0V rail is not to be taken to chassis ground, it will be

necessary to ensure that the PCB mounting holes occur in an area of the PCB that is clear of foil.

9. Commence the PCB design in rough first using paper ruled with squares. This process can be carried out 'actual size' using 0.1 in. graph paper or suitably enlarged by means of an appropriate choice of paper. For preference, it is wise to choose paper with a feint blue or green grid as this will subsequently disappear after photocopying leaving you with a 'clean' layout.

10. Using the square grid as a guide, try to arrange all components so that they are mounted on the standard 0.1 in. matrix. This may complicate things a little but is important if you should subsequently wish to convert the design using computer-aided design (CAD) techniques (see later).

11. Arrange straight runs of track so that they align with one dimension of the board or another. Avoid haphazard track layout.

12. Identify all conductors that will be handling significant current (i.e. in excess of about 500 mA) and ensure that tracks have adequate widths. Table 8.6 will provide you with a rough guide. Note that, as a general rule, the width of the 0 V track should be at least *twice* that used for any other track.

13. Identify all conductors that will be handling high voltages (i.e. those in excess of 150 V d.c. or 100 V r.m.s. a.c.) and ensure that these are adequately spaced from other tracks. Table 8.7 will provide you with a rough guide.

14. Identify the point at which the principal supply rail is to be connected. Employ extra wide track widths (for both the 0 V and supply rail) in this area and check that decoupling capacitors are placed as close as possible to the point of supply connection. Check that other decoupling capacitors are distributed at strategic points around the board. These should be positioned in close proximity to the positive supply input pin to all integrated circuits that are likely to demand sudden transient currents (e.g. 555 timers, comparators, integrated circuit power amplifiers). Ensure that there is adequate connection to the 0 V rail for each decoupling capacitor that you use. Additional decoupling may also be required for high-frequency devices and you should consult semiconductor manufacturers' recommendations for specific guidance.

15. Fill unused areas of PCB with 'land' (areas of foil which should be linked to 0 V). This helps ensure that the 0 V rail is effective as a *common* rail, minimises use of the etchant, helps to conduct heat away from heat producing components, and furthermore, is essential in promoting stability in high-frequency applications.

16. Lay out the 0 V and positive supply rails first. Then turn your attention to linking to the pads used for connecting the off-board components. It is also important to minimise, as far as possible, the number of links required and avoid using any links in the 0 V rail.

17. Experiment with positioning of integrated circuits (it is good practice, though not essential, to align them all in the same direction). In some cases, logic gates may be exchanged from package to package (or within the same package) in order to minimize track runs and links. (If you have to resort to this dodge, do not forget to amend the circuit diagram.)

TABLE 8.6 Current and PCB track width

Current	Minimum track width
Less than 500 mA	0.6 mm
0.5 A to 1.5 A	1.6 mm
1.5 A to 3 A	3 mm
3 A to 6 A	6 mm
6 A to 12 A	10 mm

TABLE 8.7 Voltage and PCB track spacing

Voltage	Minimum track spacing
Less than 50 V	0.6 mm
50 V to 150 V	1 mm
150 V to 300 V	1.6 mm
300 V to 600 V	3 mm
600 V to 1 kV	5 mm

18. Be aware of the pin spacing used by components and try to keep this consistent throughout. With the exception of the larger wirewound resistors (which should be mounted on ceramic stand-off pillars) axial lead components should be mounted flat against the PCB (with their leads bent at right angles). Axial lead components should not be mounted vertically.

19. Do not forget that tracks may be conveniently routed beneath other components. Supply rails in particular can be routed between opposite rows of pads of DIL integrated circuits; this permits very effective supply distribution and decoupling.

20. Minimise track runs as far as possible and maintain constant spacing between parallel runs of track. Corners should be radiused and acute internal and external angles should be avoided. In exceptional circumstances, it may be necessary to run a track between adjacent pads of a DIL integrated circuit. In such cases, the track should not be a common 0 V path, neither should it be a supply rail.

Figure 8.45 shows examples of good and bad practice associated with PCB layout while Figure 8.46 shows an example of a PCB layout and matching component overlay that embodies most of the techniques and principles discussed.

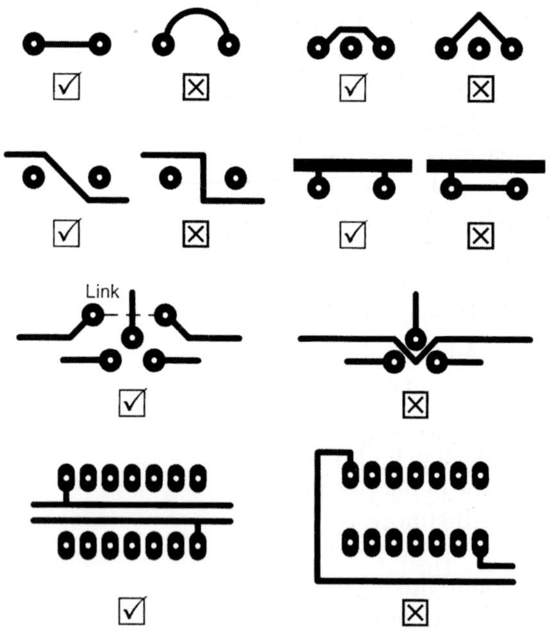

FIGURE 8.45
Some examples of good and bad practice in PCB layout

As with stripboard layouts, it is well worth devoting some time to checking the final draft PCB layout before starting on the master artwork. This can be instrumental in saving much agony and heartache at a later stage. The same procedure should be adopted as described on page 272 (i.e. simultaneously tracing the circuit diagram and PCB layout).

The next stage depends upon the actual PCB production. Four methods are commonly used for prototype and small-scale production. These may be summarised as follows:

a. Drawing the track layout directly on the copper surface of the board using a special pen filled with etch resist ink. The track layout should, of course, conform as closely as possible with the draft layout.

b. Laying down etch resist transfers of tracks and pads on to the copper surface of the PCB following the same layout as the draft but appropriately scaled.

c. Producing a transparency (using artwork transfers of tracks and pads) conforming to the draft layout and then applying photographic techniques.

d. Using a PCB layout CAD package (see page 278).

Methods (a) and (b) have the obvious limitation that they are a strictly one-off process. Method (a) is also extremely crude and only applicable to very simple boards. Method (c) is by far the most superior and allows one to re-use or modify the master artwork transparency and produce as many further boards as are required. The disadvantage of the method is that it is slightly more expensive in terms of materials (specially coated copper board is required) and requires some form of ultra-violet exposure unit. This device normally comprises a light-tight enclosure into the base of which one or more ultra-violet tubes are fitted. Smaller units are available which permit exposure of boards measuring 250 mm × 150 mm while the larger units are suitable for boards of up to 500 mm × 350 mm. The more expensive exposure units are fitted with timers which can be set to determine the actual exposure time. Low-cost units do not have such a facility and the operator has to refer to a clock or wristwatch in order to determine the exposure time.

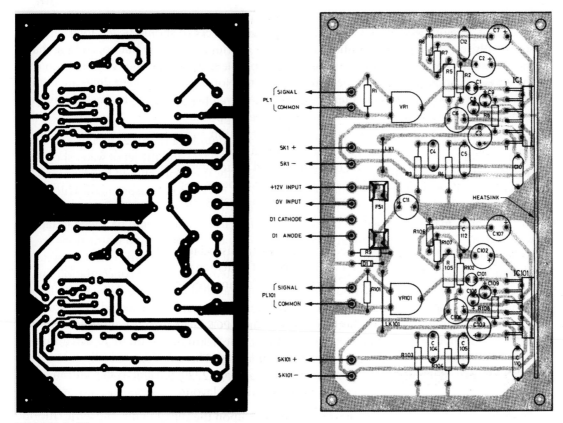

FIGURE 8.46
Example of a manually produced PCB layout for an automotive stereo amplifier. The copper foil layout is shown on the left and the matching component overlay is shown on the right

In use, the 1:1 master artwork (in the form of opaque transfers and tape on translucent polyester drafting film) is placed on the glass screen immediately in front of the ultra-violet tubes (taking care to ensure that it is placed so that the component side is uppermost). The opaque plastic film is then removed from the photo-resist board (previously cut roughly to size) and the board is then placed on top of the film (coated side down). The lid of the exposure unit is then closed and the timer set (usually for around four minutes but see individual manufacturer's recommendations. The inside of the lid is lined with foam which exerts an even pressure over the board such that it is held firmly in place during the exposure process.

It should be noted that, as with all photographic materials, sensitised copper board has a finite shelf-life. Furthermore, boards should ideally be stored in a cool place at a temperature of between 2°C and 10°C. Shelf-life at 20°C will only be around twelve months and thus boards should be used reasonably promptly after purchase.

Following exposure for the correct time, the board should be removed and immersed in a solution of sodium hydroxide that acts as a developer. The solution should be freshly made and the normal concentration required is obtained by mixing approximately 500 ml of tap water (at 20°C) with one tablespoon of sodium hydroxide crystals. A photographic developing tray (or similar shallow plastic container) should be used to hold the developer. Note that care should be taken when handling the developer solution and the use of disposable gloves should be used. This process should be carried out immediately after exposure and care should be taken not to allow the board to be further exposed under room lights.

The board should be gently agitated while immersed in the developer and the ensuing process of development should be carefully watched. The board should be left for a sufficiently long period for the entire surface to be developed correctly but not so long that the tracks lift. Development times will depend upon the temperature and concentration of the developer and on the age of the sensitized board. Normal development times are in the region of 30 to 90 seconds and after this period the developed image of the track layout (an etch-resist positive) should be seen.

After developing the board it should be carefully washed under a running tap. It is advisable not to rub or touch the board (to avoid scratching the surface) and the jet of water should be sufficient to remove all traces of the developer. Finally, the board should be placed in the etchant which is a ferric chloride solution ($FeCl_3$). For obvious reasons, ferric chloride is normally provided in crystalline form (though at least one major supplier is prepared to supply it on a 'mail order basis' in concentrated liquid form) and should be added to tap water (at $20°C$) following the instructions provided by the supplier. If no instructions are given, the normal quantities involved are 750 ml of water to 500 g of ferric chloride crystals. Etching times will also be very much dependent upon temperature and concentration but, for a fresh solution warmed to around $40°C$ the time taken should typically be ten to fifteen minutes. During this time the board should be regularly agitated and checked to ascertain the state of etching. The board should be removed as soon as all areas not protected by resist have been cleared of copper; failure to observe this precaution will result in 'undercutting' of the resist and consequent thinning of tracks and pads. Where thermostatically controlled tanks are used, times of five minutes or less can be achieved when using fresh solution.

Great care should be exercised when working with ferric chloride. Disposable gloves should be worn and care must be taken to avoid spills and splashes. After cooling, the ferric chloride solution may be stored (using a sealed plastic container) for future use. In general, 750 ml of solution can be used to etch around six to ten boards of average size; the etching process taking longer as the solution nears the end of its working life. Finally, the exhausted solution must be disposed of with great care (it should not be poured into an ordinary mains drainage system). Your tutor will advise on what should be done with it.

Having completed the etching process, the next stage involves thoroughly washing, cleaning, and drying the printed circuit board. After this, the board will be ready for drilling. Drilling will normally involve the services of a 0.6 mm or 1 mm twist drill bit for standard component leads and integrated circuit pins. Larger drill bits may be required for the leads fitted to some larger components (e.g. power diodes) and mounting holes. Drilling is greatly simplified if a special PCB drill and matching stand can be enlisted. Alternatively, provided it has a bench stand, a standard electric drill can be used. Problems sometimes arise when a standard drill or hand drill is unable to adequately grip a miniature twist drill bit. In such cases you should make use of a miniature pin chuck or a drill fitted with an enlarged shank (usually of 2.4 mm diameter).

PCB CAD Packages

The task of laying out a PCB and producing master artwork (both for the copper foil side of a board and for the component overlay) is greatly simplified by the use of a dedicated PCB CAD package. Some examples of the use of these packages are shown in Figures 8.47 and 8.48.

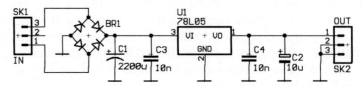

FIGURE 8.47
Circuit of a simple 5 V regulated power supply based on a 78L05 integrated circuit voltage regulator

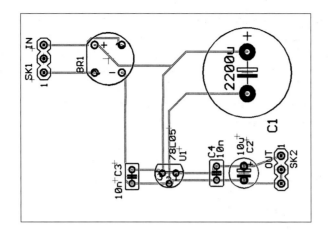

FIGURE 8.48
PCB track layout for the circuit shown in Figure 8.47 produced by an auto-routing CAD package (shown prior to optimisation)

8.9 TEST EQUIPMENT AND MEASUREMENTS

For practical measurements on electronic circuits it is often convenient to combine the functions of a voltmeter, ammeter and ohmmeter into a single instrument (known as a multi-range meter or simply a *multimeter*). In a conventional multimeter as many as eight or nine measuring functions may be provided with up to six or eight ranges for each measuring function.

Besides the normal voltage, current and resistance functions, some meters also include facilities for checking transistors and measuring capacitance. Most multi-range meters normally operate from internal batteries and thus they are independent of the mains supply. This leads to a high degree of portability which can be all-important when measurements are to be made away from a laboratory or workshop.

Analogue Multimeters

Analogue instruments employ conventional moving coil meters (see Figure 8.49) and the display takes the form of a pointer moving across a calibrated scale. This arrangement is not so convenient to use as that employed in digital instruments because the position of the pointer is rarely exact and may require interpolation. Analogue instruments do, however, offer some advantages not the least of which lies in the fact that it is very easy to make adjustments to a circuit whilst observing the relative direction of the pointer; a movement in one direction representing an increase and in the other a decrease. Despite this, the principal disadvantage of many analogue meters is the rather cramped, and sometimes confusing, scale calibration. To determine the exact reading requires first an estimation of the pointer's position and then the application of some mental arithmetic based on the range switch setting.

Digital Multimeters

Unlike their analogue counterparts, digital multimeters are usually extremely easy to read and have displays that are clear, unambiguous, and capable of providing a very high resolution. It is thus possible to distinguish between readings that are very close. This is just not possible with an analogue instrument. Typical analogue and digital meters are shown in Figure 8.50.

Digital multi-range meters offer a number of significant advantages when compared with their more humble analogue counterparts. The display fitted to

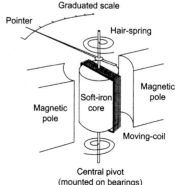

FIGURE 8.49
A moving coil meter movement

FIGURE 8.50
Typical analogue (left) and digital (right) multimeters

a digital multi-range meter usually consists of a 3½-digit seven-segment display—the ½ simply indicates that the first digit is either blank (zero) or 1. Consequently, the maximum indication on the 2 V range will be 1.999 V and this shows that the instrument is capable of offering a resolution of 1 mV on the 2 V range (i.e. the smallest increment in voltage increment that can be measured is 1 mV). The resolution obtained from a comparable analogue meter would be of the order of 50 mV, or so, and thus the digital instrument provides a resolution that is many times greater than its analogue counterpart.

Figure 8.51 shows the controls and display provided by a simple analogue multi-range meter. The mode switch and range selector allow you to select from a total of twenty ranges and eight measurement functions. These functions are:

- DC voltage (DC, V)
- DC current (DC, A)
- AC voltage (AC, V)
- AC current (AC, A)
- Resistance (OHM)
- Capacitance (CAP)
- Continuity test (buzzer)
- Transistor current gain (h_{FE}).

DC VOLTAGE MEASUREMENT

Figure 8.52 shows how to make DC voltage measurements using a digital multi-range meter. The red and black test leads are connected to the 'V-OHM' and 'COM' sockets respectively. In Figure 8.52, the mode switch and range selector is set to DCV, 200 V, and the display indicates a reading of 124.5 V.

DC CURRENT MEASUREMENTS

Figure 8.53 shows how to make a DC current measurement. Here, the red and black test leads are connected to the 'mA' and 'COM' sockets respectively. The mode switch and range selectors are set to DC, 200 mA, and the display indicates a reading of 85.9 mA.

DC HIGH-CURRENT MEASUREMENT

In common with simple analogue multi-range meters, the meter uses a shunt which is directly connected to a separate '10 A' terminal. Figure 8.54 shows the connections, mode

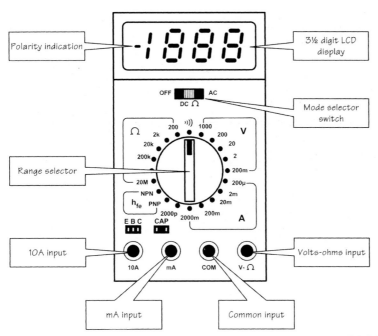

FIGURE 8.51
Digital multimeter display and controls

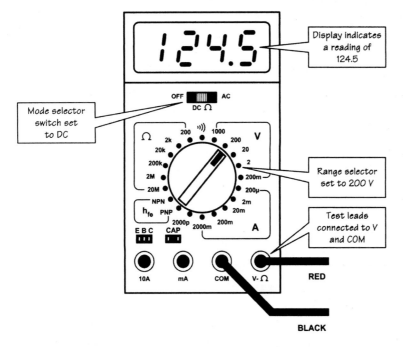

FIGURE 8.52
Digital multimeter set to the DC, 200V range

switch and range selector settings to permit high-current DC measurement. The mode switch and range selectors are set to DC, 2000 mA (2 A) and the red and black test leads are connected to '10 A' and 'COM' respectively. The display indicates a reading of 2.99 A.

AC VOLTAGE MEASUREMENT

Figure 8.55 shows how to make AC voltage measurements. Once again, the red and black test leads are connected to the 'V-OHM' and 'COM' sockets respectively. In Figure 8.55, the mode switch and range selectors are set to AC, 10 V, and the display indicates a reading of 1.736 V.

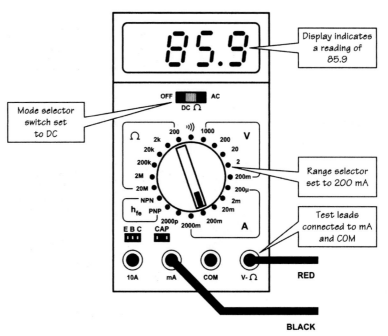

FIGURE 8.53
Digital multimeter set to the DC, 200 mA range

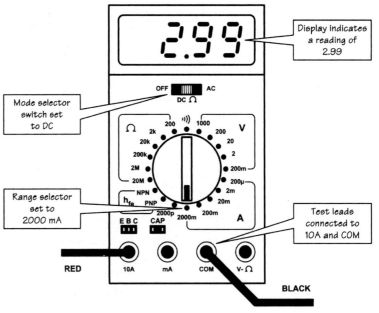

FIGURE 8.54
Digital multimeter set to the DC, 10 A range

RESISTANCE MEASUREMENT

Figure 8.56 shows how to make resistance measurements. As before, the red and black test leads are connected to 'V-OHM' and 'COM' respectively. In Figure 8.56, the mode switch and range selectors are set to OHM, 200 Ω, and the meter indicates a reading of 55.8 Ω. Note that is not necessary to 'zero' the meter by shorting the test probes together before taking any measurements (as would be the case with an analogue meter).

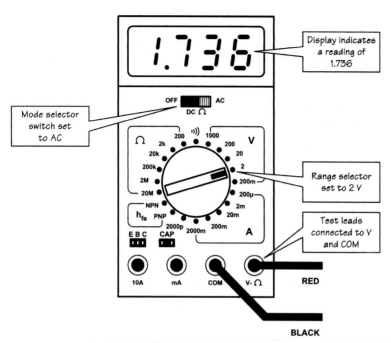

FIGURE 8.55
Digital multimeter set to the AC, 2V range

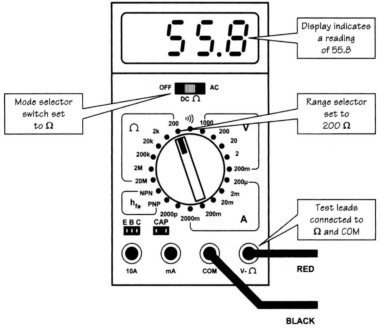

FIGURE 8.56
Digital multimeter set to the OHM, 200 Ω range

CAPACITANCE MEASUREMENT

Many modern digital multi-range meters incorporate a capacitance measuring although this may be limited to just one or two ranges. Figure 8.57 shows how to carry out a capacitance measurement. The capacitor on test is inserted into the two-way connector marked 'CAP' whilst the mode switch and range selector controls are set to DC, 2000 pF. The display indication shown in Figure 8.57 corresponds to a capacitance of 329 pF.

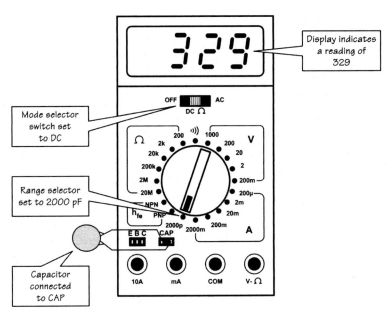

FIGURE 8.57
Digital multimeter set to the 2000 pF capacitance range

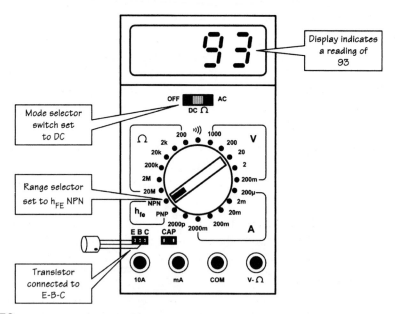

FIGURE 8.58
Digital multimeter set to the transistor current gain, NPN range

TRANSISTOR CURRENT GAIN (h_{FE}) MEASUREMENT

Many modern digital multi-range meters also provide some (laid; basic) facilities for checking transistors. Figure 8.58 shows how to measure the current gain (h_{FE}) of an NPN transistor. The transistor is inserted into the three-way connector marked 'EBC', taking care to ensure that the emitter lead is connected to 'E', the base lead to 'B' and the collector lead to 'C'. The mode switch and range selector controls are set to DC, NPN respectively. The display indication in Figure 8.58 shows that the device has a current gain of 93. This means that, for the device in question, the ratio of collector current (I_C) to base current (I_B) is 93.

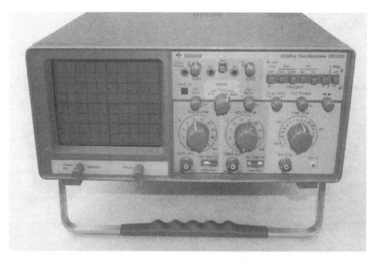

FIGURE 8.59
A typical bench oscilloscope

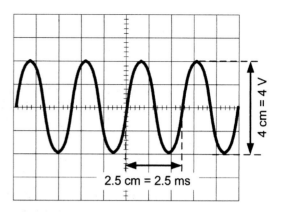

Timebase: 1 ms/cm
Vertical attenuator: 1 V/cm
FIGURE 8.60
Using an oscilloscope graticule

Oscilloscopes

An oscilloscope (see Fig. 8.59) is an extremely comprehensive and versatile item of test equipment which can be used in a variety of measuring applications, the most important of which is the display of time related voltage waveforms.

The oscilloscope display is provided by a *cathode ray tube* (CRT) that has a typical screen area of 8 cm × 10 cm. The CRT is fitted with a *graticule* that may either be integral with the tube face or a separate translucent sheet. The graticule is usually ruled with a 1 cm grid to which further bold lines may be added to mark the major axes on the central viewing area. Accurate voltage and time measurements may be made with reference to the graticule, applying a scale factor derived from the appropriate range switch.

A word of caution is appropriate at this stage, however. Before taking meaningful measurements from the CRT screen it is absolutely essential to ensure that the front panel variable controls are set in the *calibrate* (CAL) position. Results will almost certainly be inaccurate if this is not the case!

The use of the graticule is illustrated by the' following example. An oscilloscope screen is depicted in Figure 8.60. This diagram is reproduced actual size and the fine graticule markings are shown every 2 mm along the central vertical and horizontal axes.

The oscilloscope is operated with all relevant controls in the 'CAL' position. The timebase (horizontal deflection) is switched to the 1 ms/cm range and the vertical attenuator (vertical deflection) is switched to the 1 V/cm range. The overall height of the trace is 4 cm and thus the peak-peak voltage is 4 × 1 V = 4 V. Similarly, the time for one complete cycle (period) is 2.5 × 1 ms = 2.5 ms. One further important piece of information is the shape of the waveform that, in this case, is sinusoidal.

The front panel layout for a typical general purpose two-channel oscilloscope is shown in Figures 8.6l and 8.62. The controls and adjustments are summarised in Table 8.8 and Figures 8.61 to 8.64.

Logic Probes

The simplest, and by far the most convenient, method of checking simple logic circuits involves the use of a logic probe. This invaluable tool comprises a compact hand-held probe fitted with LEDs that indicate the logical state of its probe tip.

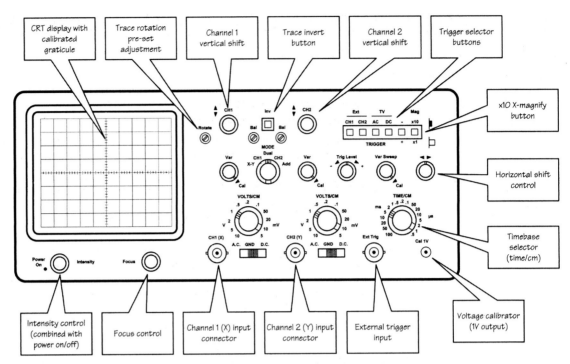

FIGURE 8.61

Front panel controls and displays on a typical dual-channel oscilloscope

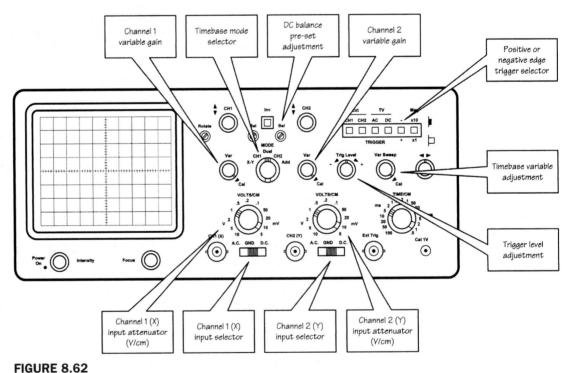

FIGURE 8.62

Front panel controls and displays on a typical dual-channel oscilloscope

Unlike multimeters, logic probes can distinguish between lines which are actively pulsing and those which are in a permanently tri-state condition. In the case of a line which is being pulsed, the logic 0 and logic 1 indicators will both be illuminated (though not necessarily with the same brightness) whereas, in the case of a tri-state line neither indicator should be illuminated.

TABLE 8.8 Oscilloscope controls and adjustments

Control	Adjustment
Cathode ray tube display	
Focus	Provides a correctly focused display on the CRT screen.
Intensity	Adjusts the brightness of the display.
Astigmatism	Provides a uniformly defined display over the entire screen area and in both x and y-directions. The control is normally used in conjunction with the focus and intensity controls.
Trace rotation	Permits accurate alignment of the display with respect to the graticule.
Scale illumination	Controls the brightness of the graticule lines.
Horizontal deflection system	
Timebase (time/cm)	Adjusts the timebase range and sets the horizontal time scale. Usually this control takes the form of a multi-position rotary switch and an additional continuously variable control is often provided. The 'CAL' position is usually at one, or other, extreme setting of this control.
Stability	Adjusts the timebase so that a stable displayed waveform is obtained.
Trigger level	Selects the particular level on the triggering signal at which the timebase sweep commences.
Trigger slope	This usually takes the form of a switch that determines whether triggering occurs on the positive or negative going edge of the triggering signal.
Trigger source	This switch allows selection of one of several waveforms for use as the timebase trigger. The options usually include an internal signal derived from the vertical amplifier, a 50 Hz signal derived from the supply mains, and a signal which may be applied to an External Trigger input.
Horizontal position	Positions the display along the horizontal axis of the CRT.
Vertical deflection system	
Vertical attenuator (V/cm)	Adjusts the magnitude of the signal attenuator (V/cm) displayed (V/cm) and sets the vertical voltage scale. This control is invariably a multi-position rotary switch; however, an additional variable gain control is sometimes also provided. Often this control is concentric with the main control and the 'CAL' position is usually at one, or other, extreme setting of the control.
Vertical position	Positions the display along the vertical axis of the CRT.
a.c.-d.c.-ground	Normally an oscilloscope employs d.c. coupling throughout the vertical amplifier; hence a shift along the vertical axis will occur whenever a direct voltage is present at the input. When investigating waveforms in a circuit one often encounters a.c. superimposed on d.c. levels; the latter may be removed by inserting a capacitor in series with the signal. With the a.c-d.c.-ground switch in the d.c. position a capacitor is inserted in the input lead, whereas in the DC position the capacitor is shorted. If ground is selected, the vertical input is taken to common (0 V) and the oscilloscope input is left floating. This last facility is useful in allowing the accurate positioning of the vertical position control along the central axis. The switch may then be set to d.c. and the magnitude of any d.c. level present at the input may be easily measured by examining the shift along the vertical axis.
Chopped-alternate	This control, which is only used in dual beam oscilloscopes, provides selection of the beam splitting mode. In the chopped position, the trace displays a small portion of one vertical channel waveform followed by an equally small portion of the other. The traces are, in effect, sampled at a relatively fast rate, the result being two apparently continuous displays. In the alternate position, a complete horizontal sweep is devoted to each channel alternately.

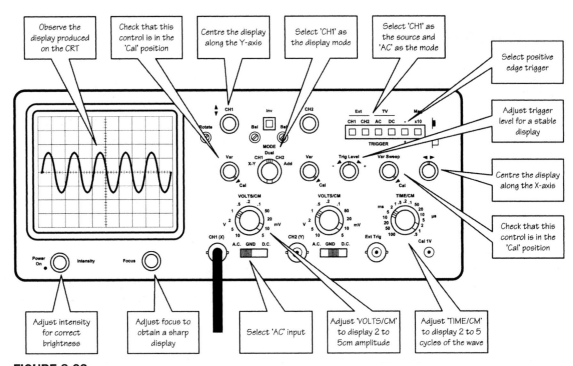

FIGURE 8.63
Procedure for adjusting the controls to display a sinusoidal waveform (single-channel mode)

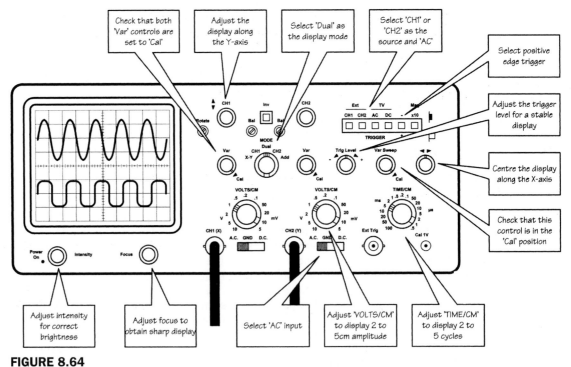

FIGURE 8.64
Procedure for adjusting the controls to display two waveforms (dual-channel mode)

Logic probes usually also provide a means of displaying pulses having a very short duration which may otherwise go undetected. This is accomplished by the inclusion of a pulse stretching circuit (i.e. a monostable). This elongates short duration pulses so that a visible indication is produced on a separate 'pulse' LED.

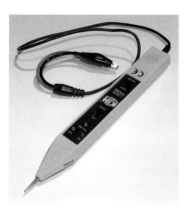

FIGURE 8.65

A versatile logic probe with logic level and pulse indicating facilities. Note the crocodile leads for connecting to the TTL or CMOS supply and ground connections

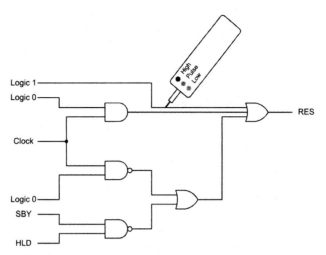

FIGURE 8.66

Using a logic probe to trace logic levels in a simple logic circuit (note that the supply and ground connections are not shown)

Logic probes invariably derive their power supply from the circuit under test and are connected by means of a short length of twin flex fitted with insulated crocodile clips. While almost any convenient connecting point may be used, the positive supply and ground terminals make ideal connecting points which can be easily identified.

A typical logic probe is shown in Figure 8.65. This circuit uses a *comparator* to sense the logic 0 (or *low*) and logic 1 (or *high*) levels and a timer that acts as a *monostable pulse stretcher* to indicate the presence of a pulse input rather than a continuous logic 0 or logic 1 condition. Figure 8.66 shows how a logic probe can be used to check a simple logic circuit. In use, the probe is simply moved from point to point and the logic level is displayed using the LED indicators fitted to the logic probe. The indicated logic levels can then be compared with the expected levels.

REVIEW QUESTIONS

1. Explain the purpose and principle of operation of a residual current device (RCD).
2. Identify the electronic component shown in Figure 8.67. Explain what this component is used for.
3. List THREE hazards associated with soldering.
4. State TWO advantages of using 'stripboard' construction methods for building a prototype electronic circuit.
5. Explain what is means by the term 'integrated circuit' and give THREE examples of where such devices are used.
6. Determine the peak-peak voltage and periodic time of the waveforms shown in Figure 8.68.
7. Sketch the circuit symbols for:
   a. a fixed resistor
   b. a variable capacitor
   c. a diode
   d. an NPN transistor
   e. an AND gate (with two inputs)
   f. a NOR gate (with four inputs)
   g. a transformer.
8. What indication is provided by the digital multimeter shown in Figure 8.69?

FIGURE 8.67
See Question 2

289

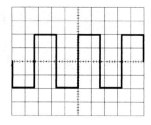

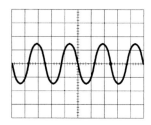

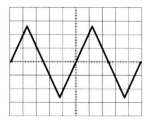

(a) Timebase: 1 ms/cm
Y-attenuator: 1 V/cm

(b) Timebase: 50 ns/cm
Y-attenuator: 50 mV/cm

(c) Timebase: 100 ms/cm
Y-attenuator: 50 V/cm

FIGURE 8.68
See Question 6

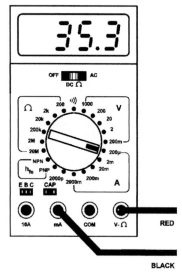

FIGURE 8.69
See Question 8

9. Explain, with the aid of a diagram, how a logic probe can be used to check the operation of an OR gate.
10. A resistor is marked with the following coloured bands: orange, white, red, gold. What is the value and tolerance of the resistor?
11. A capacitor is marked '104 250 V'. What is the value and working voltage for this capacitor?
12. Produce a component list for the electronic circuit shown in Figure 8.70. What type of transistors are used in this circuit?
13. Design a printed circuit board layout for the circuit shown in Figure 8.70. Sketch a track view and component overlay for the PCB assuming standard size components.
14. Briefly describe each of the stages necessary to manufacture the PCB in Question 13.

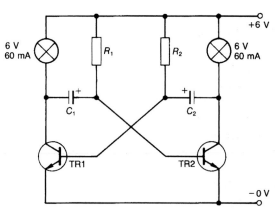

FIGURE 8.70
See Question 12

Abbreviations

2D	two-dimensional
3D	three-dimensional
AC	alternating current
ACW	anticlockwise
ADC	analogue-to-digital converter
AF	audio frequency
ALU	arithmetic and logic unit
AM	amplitude modulation
ASCII	American standard code for information interchange
BCD	binary coded decimal
BJT	bipolar junction transistor
BM	bending moment
BS	British Standard
CAD	computer aided design
CAE	computer aided engineering
CAM	computer aided manufacture
CG	centre of gravity
CIMS	computer integrated manufacturing system
CL	centre-line
CMOS	complementary metal oxide semiconductor
CNC	computer numerical control
COSHH	Control of Substances Hazardous to Health
CPM	critical path method
CPU	central processing unit
CRT	cathode ray tube
CSA	cross-sectional area
CW	clockwise
DAC	digital-to-analogue converter
DC	direct current
DFM	digital frequency meter
DIL	dual in-line
DIM	dimension
DMM	digital multi-meter
DPDT	double-pole double-throw
DPST	double-pole single-throw
EMF	electromotive force
ESD	electrostatic sensitive device
FEA	finite element analysis
FEM	finite element model
FET	field effect transistor

291

BTEC First Engineering: Mandatory and Selected Optional Units for BTEC Firsts in Engineering. DOI: 10.1016/B978-1-85617-685-9.00014-7

FM	frequency modulation
FMS	flexible manufacturing system
GA	general arrangement (drawing)
HASAWA	Health and Safety at Work Act
HEX	hexadecimal
HSE	Health and Safety Executive
HTML	hypertext mark-up language
HTTP	hypertext transfer protocol
I/O	input/output
IP	internet protocol
ISP	Internet service provider
JIT	just-in-time
KE	kinetic energy
LCD	liquid crystal display
LDR	light dependent resistor
LED	light emitting diode
LHS	left-hand side
LSB	least-significant bit
LVDT	linear variable differential transformer
MCB	mains circuit breaker
MDF	medium density fibreboard
MOS	metal oxide semiconductor
MSB	most-significant bit
MSDS	material safety data sheet
MTBF	mean time between failure
MTTF	mean time to failure
PAT	portable appliance tester
PC	personal computer
PCB	printed circuit board
PDS	product design specification
PDF	portable document format
PE	potential energy
PEEK	polyetheretherketone
PERT	programme evaluation and review technique
PLA	programmed logic array
PLC	programmable logic controller
PLD	programmed logic device
PPE	personal protective equipment
PROM	programmable read-only memory
PSI	pounds (force) per square inch
PTFE	polytetraflouroethylene
PVC	polyvinylchloride
PWM	pulse width modulated
QA	quality assurance
QC	quality control
R&D	research and development
RAM	random access memory
RCD	residual current device
RF	radio frequency
RHS	right-hand side
RIDDOR	Reporting of Injuries Diseases and Dangerous Occurrences
RMS	root mean square

ROM	read-only memory
RPM	revolutions per minute
SHM	simple harmonic motion
S/N	signal-to-noise
SI	System International
SNR	signal-to-noise ratio
SPDT	single-pole double-throw
SPST	single-pole single-throw
TQ	total quality
TQM	total quality management
TTL	transistor-transistor logic
UDL	uniformly distributed load
UHF	ultra high frequency
USB	Universal Serial Bus
URL	uniform resource locator
VA	volt-amperes
VDU	visual display unit
WD	work done

Using the Casio fx-83ES Calculator

The Casio fx-83ES calculator is currently widely available and is highly recommended for use by engineering students on BTEC First and BTEC National courses. The calculator has over 200 functions and it incorporates a 'natural display' that supports input and output using mathematical notation, such as fractions, roots, etc.

INITIALISING THE CALCULATOR

The keystrokes shown in Figure 1 can be used to return the calculator's settings and modes to their initial (default) values. Note that this operation will also clear any data from the calculator's memory. The fx-83's Clear menu is shown in Figure 3.

SETTING THE CALCULATOR MODE

The fx-83ES provides three different calculation modes; COMP is used for general calculations, STAT is used for statistical calculations, and TABLE is used to generate a table of values based on a given expression. For the BTEC First award, you will only need to use the COMP mode. The Mode menu is shown in Figure 4 and the keystrokes required to set the calculator to the COMP mode are shown in Figure 5.

FIGURE 1
Key entry required to initialise the calculator

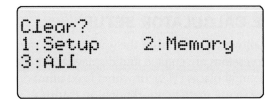

FIGURE 3
The fx-83's Clear menu will allow you to reset and initialise the calculator

FIGURE 2
The Casio fx-83ES
scientific calculator

295

BTEC First Engineering: Mandatory and Selected Optional Units for BTEC Firsts in Engineering. DOI: 10.1016/B978-1-85617-685-9.00015-9

Another view
When you first power up the fx-83ES calculator it will be in 'MathIO' mode. This may be awkward for use with many basic engineering calculations. To change the mode to the more conventional 'LineIO' mode, you need to press the SHIFT button followed by MODE and 2.

TABLE 1 The display symbols used on the fx-83

Display	Meaning
S	The keypad has been shifted by pressing the SHIFT key. The keypad will unshift and this indicator will disappear when you press a key
A	The alpha input mode has been entered by pressing the ALPHA key. The alpha input mode will be cancelled and this indicator will disappear when you press a key
M	This indicates that a value has been stored in the calculator's memory
STO	The calculator is waiting for the input of a variable name. The calculator will then assign a value to this variable. The indication appears after you press SHIFT RCL STO
RCL	The calculator is waiting for the input of a variable name. The calculator will then recall the value of this variable. The indication appears after you press RCL
STAT	The calculator is in the statistical mode
D	The default unit for angles is degrees
R	The default unit for angles is radians
G	The default unit for angles is grads
FIX	The calculator has been set to display a fixed number of decimal places
SCI	The calculator has been set to display a number of significant digits
Math	The calculator has been set to the Maths input/output mode
▼▲	Calculation history is available and can be replayed (or there is more data above or below the display)
Disp	The display currently shows an intermediate result of a multi-statement calculation

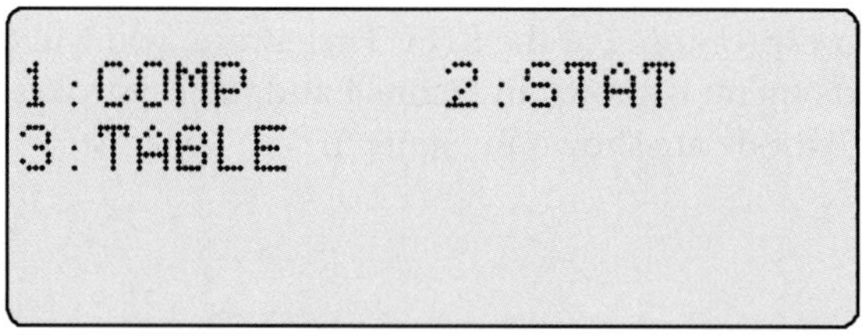

```
1:COMP      2:STAT
3:TABLE
```

FIGURE 4
The fx-83's Mode menu will allow you to change the calculator's mode

FIGURE 5
Key entry for setting the calculator to COMP mode

CONFIGURING THE CALCULATOR SETUP

The fx-83ES Setup menu allows you to control the way in which calculations are performed as well as the way that expressions are entered and displayed. The menu will let you work with a fixed number of decimal places (FIX), or with a fixed number of significant digits (SCI). You can also choose to use mathematical notation (MthIO) or conventional (line-based) notation (LineIO). The Setup menu is shown in Figure 6 whilst the keystrokes required to set the calculator to LineIO mode are shown in Figure 7.

SHIFT
SETUP
2
(LineIO)

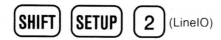

FIGURE 6
Key entry for setting the calculator to COMP mode

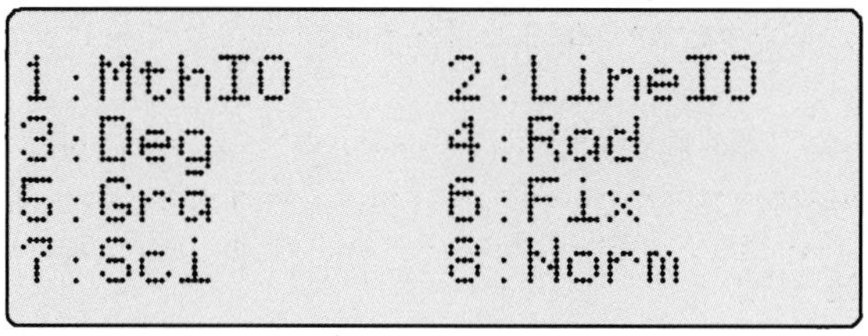
1:MthIO
2:LineIO
3:Deg
4:Rad
5:Gra
6:Fix
7:Sci
8:Norm

FIGURE 7
The fx-83's Mode menu will allow you to change the calculator's mode

List of BTEC Level 2 units

Unit 1:   Working Safely and Effectively in Engineering
Unit 2:   Interpreting and Using Engineering Information
Unit 3:   Mathematics for Engineering Technicians
Unit 4:   Applied Electrical and Mechanical Science
Unit 5:   Engineering Maintenance Procedures
Unit 6:   Preparing and Controlling Engineering Manufacturing Operations
Unit 7:   Electronic Devices and Communication Applications
Unit 8:   Selecting Engineering Materials
Unit 9:   Engineering Assembly Methods and Techniques
Unit 10:   Computer Aided Drawing Techniques
Unit 11:   Operation and Maintenance of Mechanical Systems and Components
Unit 12:   Operation and Maintenance of Electrical Systems and Components
Unit 13:   Operation and Maintenance of Electronic Systems and Components
Unit 14:   Selecting and Using Secondary Machining Techniques to Remove Material
Unit 15:   Part Programming CNC Machines
Unit 16:   Application of Welding Processes
Unit 17:   Fabrication Techniques and Sheet Metal Work
Unit 18:   Engineering Marking Out
Unit 19:   Electronic Circuit Construction
Unit 20:   Using Specialist Secondary Machining Techniques
Unit 21:   Production Planning for Engineering
Unit 22:   Application of Quality Control and Measurement in Engineering
Unit 23:   Casting and Moulding Engineering Components
Unit 24:   Operation and Maintenance of Fluid Power Systems and Components
Unit 25:   Applying Continuous Improvement and Problem-solving Techniques
Unit 26:   Workplace Organisation and Standard Operating Procedures
Unit 27:   PC Hardware and Software Installation and Configuration
Unit 28:   Mobile Communications Technology
Unit 29:   Mathematics for Engineering Technicians

BTEC First Engineering: Mandatory and Selected Optional Units for BTEC Firsts in Engineering. DOI: 10.1016/B978-1-85617-685-9.00016-0

Conversion Table: Inches to mm

Fractional inches			Decimal inches	Millimetres
		1/64	0.0156	0.396
	1/32		0.0313	0.793
		3/64	0.0469	1.190
1/16			0.0625	1.587
		5/64	0.0781	1.984
	3/32		0.0938	2.381
		7/64	0.1094	2.778
1/8			0.1250	3.175
		9/64	0.1406	3.571
	5/32		0.1563	3.968
		11/64	0.1719	4.365
3/16			0.1875	4.762
		13/64	0.2031	5.159
	7/32		0.2188	5.556
		15/64	0.2344	5.953
1/4			0.2500	6.350
		17/64	0.2656	6.746
	9/32		0.2813	7.143
		19/64	0.2969	7.540
5/16			0.3125	7.937
		21/64	0.3281	8.334
	11/32		0.3438	8.731
		23/64	0.3594	9.128
3/8			0.3750	9.525
		25.64	0.3906	9.921
	13/32		0.4063	10.318
		27/64	0.4219	10.715
7/16			0.4375	11.112
		29/64	0.4531	11.509
	15/32		0.4688	11.906
		31/64	0.4844	12.303

(Continued)

301

BTEC First Engineering: Mandatory and Selected Optional Units for BTEC Firsts in Engineering. DOI: 10.1016/B978-1-85617-685-9.00017-2

Fractional inches			Decimal inches	Millimetres
1/2			0.5000	12.700
		33/64	0.5156	13.096
	17/32		0.5313	13.493
		35/64	0.5469	13.890
9/16			0.5625	14.287
		37/64	0.5781	14.684
	19/32		0.5938	15.081
		39/64	0.6094	15.478
5/8			0.6250	15.875
		41/64	0.6406	16.271
	21/32		0.6563	16.668
		43/64	0.6719	17.065
11/16			0.6875	17.462
		45/64	0.7031	17.859
	22/32		0.7188	18.256
		47/64	0.7344	18.653
3/4			0.7500	19.050
		49/64	0.7656	19.446
	25/32		0.7813	19.843
		51/64	0.7969	20.240
13/16			0.8125	20.637
		53/64	0.8281	21.034
	27/32		0.8438	21.431
		55/64	0.8594	21.828
7/8			0.8750	22.225
		57/64	0.8906	22.621
	29/32		0.9063	23.018
		59/64	0.9219	23.415
15/16			0.9375	23.812
		61/64	0.9531	24.209
	31/32		0.9688	24.606
		63/64	0.9844	25.003
1			1.0000	25.400

Data on Selected Engineering Materials

Material	Density (Mg/m^3)	Young's modulus (Gpa)	Strength (Mpa)	Ductility (%)	Toughness K_{IC} (Mpa m$^{1/2}$)	Specific modulus (GPa)/ (Mg/m^3)	Specific strength (MPa)/ (Mg/m^3)
Ceramics							
Alumina	3.87	382	332	0	4.9	99	86
Magnesia	3.60	207	230	0	1.2	58	64
Silicon nitride		166	210	0	4.0		
Zirconia	5.92	170	900	0	8.6	29	152
β-Sialon	3.25	300	945	0	7.7	92	291
Metals							
Aluminium	2.70	69	77	47	~30	26	29
Aluminium alloy	2.83	72	325	18	~25–30	25	115
Brass	8.50	100	550	70	–	12	65
Nickel alloy	8.18	204	1200	26	~50–80	25	147
Steel mild	7.86	210	460	35	~50	27	59
Titanium alloy	4.56	112	792	20	~55–90	24	174
Polymers							
Epoxy	1.12	4	50	4	1.5	4	36
Nylon 6.6	1.14	2	70	60	3–4	18	61
Polyetheretherketone	1.30	4	70		1.0	3	54
Polymethylmethacrylate	1.19	3	50	3	1.5	3	42
Polystyrene	1.05	3	50	2	1.0	3	48
Polyvinylchloride (rigid)	1.70	3	60	15	4.0	2	35

BTEC First Engineering: Mandatory and Selected Optional Units for BTEC Firsts in Engineering. DOI: 10.1016/B978-1-85617-685-9.00018-4

Useful Web Addresses

PUBLISHER'S WEBSITE

Answers to Review Questions for Units 3 and 4:
http://textbooks.elsevier.com

AUTHOR'S WEBSITE

Additional resources to support this book (including downloadable CAD packages, materials database and spreadsheet tools):
http://www.key2study.com

COMPUTER AIDED DESIGN (UNIT 10)

Autodesk (AutoCAD):
http://www.autodesk.co.uk

IMSI Design (DesignCAD)
http://www.imsidesign.com

PTC (Pro/Engineer and Pro/Desktop):
http://www.ptc.com

ProgeSOFT (progeCAD):
http://www.progesoft.com

SolidWorks (SolidWorks 3D CAD):
http://www.solidworks.co.uk

TurboCAD (2D and 3D CAD):
http://www.turbocad.co.uk

ENGINEERING RESOURCES

British Standards Institute:
http://www.bsigroup.com

Engineering Council:
http://www.engc.org.uk

Engineering Industries Association:
http://www.eia.co.uk

Institution of Engineering and Technology:
http://www.theiet.org

Institute of Engineering Designers:
http://www.ied.org.uk

305

BTEC First Engineering: Mandatory and Selected Optional Units for BTEC Firsts in Engineering. DOI: 10.1016/B978-1-85617-685-9.00019-6

Institute of Mechanical Engineers:
http://www.imeche.org.uk

Database of material properties, including polymers, metals, ceramics, semiconductors, etc. (free access to basic features):
http://www.matweb.com

Worldwide composites search engine:
http://www.wwcomposites.com

ELECTRONIC COMPONENT SUPPLIERS (UNITS 7 AND 19)

Farnell Electronic Components:
http://www.farnell.co.uk

Greenweld:
http://www.greenweld.co.uk

Maplin Electronics:
http://www.maplin.co.uk

Quasar Electronics:
http://www.quasarelectronics.com

Rapid Electronics:
http://www.rapidonline.com

RS Components:
http://www.rswww.com

Answers to Selected Test Your Knowledge Questions

CHAPTER 3

3.1

a. -11
b. 12
c. -45
d. 77
e. -5

3.2

a. only

3.3

1. Hertz
2. Coulomb
3. Newton
4. m/s^2
5. I
6. 40 W
7. $10\,m/s^2$
8. 150 Coulombs

3.4

1. 0.05 V
2. 1,250 mm
3. 9,740 MHz
4. 0.44 kV
5. 75 μm
6. 15,620 kHz
7. 0.57 mA
8. 220 nF
9. 0.47 MΩ
10. 0.254 mm

3.5

1. 13.5 metric tonnes
2. 74,568 mph

3.

3. 5,455.32 litres
4. 193.05 kPa

3.6

(b), (c), (d), (f), (g) and (i)

3.7

a. 10.0489
b. 2.2222
c. 6.488095
d. 751.071
e. 1.14417
f. 0.775
g. 35.426
h. 9.256
i. 7.29
j. 1.4681

3.8

1. 9.8
2. 1.18
3. 26.67
4. 8,200
5. 41.58

3.9

1. 27 laps
2. 27.33 laps

3.10

45 psi

3.11

1. $\rho = \dfrac{m}{V}$

2. 0.1292 kg/m³

BTEC First Engineering: Mandatory and Selected Optional Units for BTEC Firsts in Engineering. DOI: 10.1016/B978-1-85617-685-9.00020-2

3.12

All five expressions are true!

3.13

a. 0.25
b. 100,000
c. 6

3.14

a. $1.66 \times 10^5\,\text{N}$
b. $5.15 \times 10^2\,\text{m/s}$
c. $3.77 \times 10^{-7}\,\text{F}$
d. $5.2 \times 10^{-7}\,\text{A}$
e. $4.75 \times 10^{-2}\,\text{W}$
f. $2.2 \times 10^4\,\Omega$

3.15

a. $2.65 \times 10^3\,\text{N}$
b. $525 \times 10^{-3}\,\text{V}$
c. $22 \times 10^{-3}\,\Omega$
d. $65 \times 10^3\,\text{m}$
e. $825.5 \times 10^3\,\text{Hz}$
f. $6.5 \times 10^{-3}\,\text{A}$

3.16

a. 2.65 kN
b. 525 mV
c. 22 mΩ
d. 65 km
e. 825.5 kHz
f. 6.5 mA

3.17

1. a. 1.995E4
  b. 7.5E −3
2. a. 1.59125×10^2
  b. 1.915×10^{-6}

3.18

a. 790×10^3
b. -30.9×10^{-3}
c. 405
d. $26.4705882 \times 10^{-3}$

3.19

a. 13
b. 7
c. 1
d. −4.2

3.21

1. $C = \dfrac{1}{2\pi f X}$

2. $u = \dfrac{s}{t} - \dfrac{at}{2}$

3.22

1. $1.11 \times 10^4\,\text{Pa}$
2. $3.68\,\text{m/s}^2$

3.23

1. 17
2. 0.8 kg

3.24

$v = 2\,\text{m/s}$
$a = 0.6\,\text{m/s}^2$

3.25

b. (i) 225 m/s
  (ii) 362.5 m/s
  (iii) 3.2 s

3.26

a. 3.8 m
b. 5.7 s

3.27

a. $8\,\text{m}^2$
b. $5\,\text{m}^2$
c. $7\,\text{m}^2$
d. $2.625\,\text{m}^2$
e. $1,525\,\text{mm}^2$

3.28

$A = 45°, B = 60°, C = 75°$

3.29

a. $3\,\text{m}^2$
b. $2.5\,\text{m}^2$
c. $2,175\,\text{mm}^2$

3.30

1. $283.53\,\text{mm}^2$
2. 34,557.5 km

3.31

$5.089\,\text{m}^3$

3.32

$1.672\,m^3$

3.33

$84,960\,mm^3$

3.34

$2.095\,m^3$

3.35

a. 0.707
b. 0.966
c. 0.364
d. 1.423
e. 0.389

3.36

1. **a.** 0.96
   b. 3.054
2. **a.** 11.46°
   b. 143.25°

3.37

260 mm

3.38

27° and 153°, ±2.2 m

CHAPTER 4

4.1

3 A

4.2

18 V

4.3

37.5 Ω

4.4

20 Ω

4.5

0.44 Ω

4.6

88 W

4.7

0.675 W (or 675 mW)

4.8

2.88 MJ

4.9

$1.08 \times 10^{-6}\,Wb$ (or $1.08\,\mu Wb$)

4.10

$50 \times 10^{-6}\,Wb$ (or $50\,\mu Wb$)

4.11

12 A

4.12

1.875 N

4.13

0.03 T

4.14

1.225 V

4.19

16.5 V

4.20

7.032 N

4.21

Horizontal = 17.82 N
Vertical = 9.08 N

4.22

5.796 kN

4.23

290 N

4.24

13 N, 67.4°

4.25

62.43 kPa

4.26

5 N

4.27

1.875 kN

4.28

4.3 m/s^2

4.29

194.58 mph (or 87 m/s)

4.30

0.8 m/s^2

4.31

27.5 m/s

CHAPTER 5

5.2

a. Ampere
b. Volt
c. Ohm
d. Hertz
e. bits per second.

5.3

a. millivolt
b. microamp
c. kilohertz
d. megabit per second.

5.4

1. 0.15 V
2. 0.25 Mbps
3. 75 mA
4. 0.025 mA
5. 75 kHz.

5.5

1. 25 mV
2. 0.25 Mbps
3. 0.025 mA
4. 560 kΩ
5. 0.455 MHz.

5.6

a. Sine wave
b. 1.5 V
c. 3 V
d. 250 Hz
e. 4 ms.

5.7

a. Pulse waveform
b. 5 mA

c. 1 µs
d. 1 MHz.

5.8

12 V

5.9

5

5.11

E12

5.12

a. Light dependent resistor (LDR)
b. Preset (adjustable) capacitor
c. Preset (adjustable) resistor
d. Battery
e. Electrolytic capacitor.

5.13

a. Variable resistor
b. Variable capacitor
c. Inductor (fixed)
d. Capacitor (fixed)
e. Transformer (iron or steel cored).

5.14

a. Diode
b. PNP bipolar junction transistor (BJT)
c. Light emitting diode (LED)
d. Light sensitive diode (photodiode)
e. N-channel junction gate field effect transistor (JFET).

5.15

a. Electrolytic capacitor
b. Transistor (BJT or FET)
c. Multipole connector
d. Resistor (fixed)
e. Inductor (ferrite cored).

5.16

a. Single-pole single-throw (SPST) switch
b. Normally open (NO) push-button switch
c. One-pole five-way switch
d. Double-pole single-throw (DPST) switch.

5.17

a. Thermocouple
b. Light dependent resistor (LDR)

c. Rotating vane flow rate sensor
d. Strain gauge
e. Float switch
f. Moving coil microphone.

5.18

a. Electric motor and gearbox
b. Moving coil loudspeaker
c. Buzzer or piezoelectric transducer
d. Light emitting diode (LED).

5.19

1. Dual operational amplifier
2. Pin-4 (negative) and pin-8 (positive)
3. Pin-1 and pin-7
4. '+' is the non-inverting input and '−' is the inverting input.

5.20

a. Two-input AND gate
b. Two-input NOR gate
c. Two-input OR gate
d. Two-input NAND gate.

5.22

1. 270 s (or 4 minutes and 30 seconds), time to fully charge approx. 22 minutes
2. 851 Ω.

5.24

1. 0.8 J (or 800 mJ)
2. 31.6 V.

5.27

1. 111011 and 1101100
2. 1B and A7.

5.28

HI JOHN